Vibration and Oscillation of Hydraulic Machinery

Hydraulic Machinery Book Series

HYDRAULIC MACHINERY BOOK SERIES

- **Hydraulic Machinery Systems**
  Editors: Prof D K Liu, Prof V Karelin
- **Hydraulic Design of Hydraulic Machinery**
  Editor: Prof H Radha Krishna
- **Mechanical Design and Manufacturing of Hydraulic Machinery**
  Editor: Prof Mei Z Y
- **Transient Phenomena of Hydraulic Machinery**
  Editors: Prof S Pejovic, Dr A P Boldy
- **Cavitation of Hydraulic Machinery**
  Editors: Prof Li S C, Prof H Murai
- **Erosion and Corrosion of Hydraulic Machinery**
  Editors: Prof Duan C G, Prof V Karelin
- **Vibration and Oscillation of Hydraulic Machinery**
  Editor: Prof H Ohashi
- **Testing of Hydraulic Machinery**
  Editor: Prof P Henry
- **Control of Hydraulic Machinery**
  Editor: Prof H Brekke

HYDRAULIC MACHINERY BOOK SERIES

International Editorial Committee
Chairman, **Duan C G** , Secretary, **A P Boldy**

Vibration and Oscillation of Hydraulic Machinery

H. OHASHI, Editor

LONDON AND NEW YORK

First published 1991 by Ashgate Publishing

2 Park Square, Milton Park, Abingdon, Oxon OX14 4RN
711 Third Avenue, New York, NY 10017, USA

Routledge is an imprint of the Taylor & Francis Group, an informa business

First issued in paperback 2016

A CIP catalogue record for this book is available from the British Library and the US Library of Congress.

ISBN 13: 978-1-85628-185-0 (hbk)
ISBN 13: 978-1-138-26742-8 (pbk)

CONTENTS

Preface

The publishing of this book series on Hydraulic Machinery, organised and edited by the International Editorial Committee for Book Series on Hydraulic Machinery (IECBSHM), marks the results of successful cooperation between the Committee, the authors and the editors of each volume.

The Editorial Committee consists of 35 scholars from 20 countries. More than 100 academics and engineers from 23 countries have participated in the compilation of the book series. The volumes reflect the latest developments, gained from many countries, in concepts, techniques, and experiences related to specific areas of hydraulic machinery. This is a great joint exercise by so many experts on a world wide basis that will inevitably bring impetus to technical achievement and progress within the hydraulic machinery industry and promote understanding and cooperation among scholars and professional societies throughout the world.

The authors have devoted considerable time and energy to complete the manuscripts and even more time was occupied by the editors in revising and correlating the individual volumes. Dr A P Boldy, Secretary IECBSHM, contributed greatly to the overall editing and preparation of the final manuscripts. Mr J G R Hindley, of Avebury Technical, offered many suggestions and remained helpful in every phase of the editorial and publishing work.

Great assistance in the preparation of the manuscripts from numerous persons in different parts of the world have been received, without which the publication of the book series would not have been possible. I would like to express appreciation both from myself and on behalf of the Editorial Committee to these people, the list of names being too lengthy to include here.

Duan Chang Guo
Chairman IECBSHM

Foreword of the Editor

The present book *Vibration and Oscillation of Hydraulic Machinery* is a volume in the Book Series on Hydraulic Machinery organized by the International Editorial Committee. It deals with the vibration and oscillation problems which are hazardous to the safety and reliable operation of hydraulic machinery. It restricts its scope, however, to the problems which are caused by a mechanical or hydraulic excitation induced by the rotating machine itself or by a self-exciting mechanisms inherent in a specific flow pattern in the machine. It excludes, therefore, so-called transient phenomena from the scope, where the interaction between the piping and the machine plays the predominant role. These phenomena are treated in the volume on hydraulic transients of this Book Series.

Hydraulic machinery, or pumps and hydraulic turbines to be concrete, is the machine which converts shaft power to hydraulic energy or vice versa. The conversion is done in the rotating impellers of pumps or in the runners of hydraulic turbines and the hydrodynamic force on the vanes of impellers or runners is the direct medium of the energy conversion. The effectiveness of this conversion process, i.e. efficiency, must be as high as possible and the hydrodynamic design of impellers or runners together with their surrounding stationary parts is the most important part of the whole design procedures.

Hydraulic machinery with an excellent hydrodynamic performance occasionally experiences a vibration or oscillation problem, which results in the fracture of a specific machine element or forces to shut down the operation. If such situation happens, the machine cannot continue the operation. To assure hydraulic machinery of vibration-free smooth operation is therefore as vital as to materialize a good hydrodynamic performance.

Many of vibration and oscillation troubles of hydraulic machinery are found or detected after the machine was constructed and started its operation at the site. This makes the situation quite unfavorable because the search for the true cause of the trouble is difficult due to limited possibility of measurement at the site and also because the choice of remedies applicable to the machine after the installation is quite restricted. If once such trouble happens, the damages are quite severe both for suppliers and users through the loss of operation and the cost of man power and additional reconstruction works at the site. Utmost precautions must be paid therefore to prevent vibration and oscillation troubles especially at off-design operating condition where most of unexpected vibration events take place.

The present volume consists of nine chapters: Chapter 1 and chapter 2 describe fundamental concepts and methods of modelling and mathematical analysis of vibration phenomena. Chapter 3 and chapter 4 are on the vibration of structure due to mechanical and hydraulic excitations of rotating machinery. Chapter 5 deals with the vibration of rotating shaft system, i.e. rotordynamics aspects. In chapter 5 the stability of the flow in the machine and examples of self-exciting mechanism of the fluid system are described. Chapter 7 deals with the noise problem, that is, generation, transmission and control of noise. The last chapter 8 summarizes the diagnosis and remedy of vibration problems.

The coordinator is responsible for the grand design and composition of the contents of this book. Each chapter was written by a single author or by a group of authors, totaling eleven authors from nine countries. The authors are all experts of global fame in the specific fields. The authors are in nature prone to select materials which they are most interested in and they have experienced often in the practice. Consequently some items, for example vibration caused by vortex core formation in the draft tube of Francis turbines, are referred to in plural chapters of this Volume. The coordinator has overlooked these duplications deliberately, since different authors describe the same phenomena in different ways and such duplication could help readers understand the phenomena better and from various view points.

The recent development of technology has been accelerated by the use of electronic computers to a great extent. The vibration analysis of large structures and the computation of fluid flow are the very fields which have been and are enjoying the powerful assistance of computers. This situation necessarily requires more mathematical modeling and expressions for the numerical analysis and simulation of the phenomena. The reader will find in the text rather heavy mathematical expressions, in many cases in matrix form, which are indispensable for computational purposes. In the era of computers the importance of mathematical modeling is steadily increasing even for an engineer in practice. The coordinator hope that the readers would appreciate these equations and formulations when they encounter the demands which require a quantitative prediction beyond a qualitative one.

None of the authors including the coordinator of this Volume is native English speaker. Though we are deeply obliged to Dr A P Boldy of the University of Warwick, UK, for his careful reading and polishing of the text, there remain still a lot of 'English as foreign language' in the book. We would like to ask the readers the forbearance in this respect.

Each author submitted the manuscript in floppy disc to the coordinator, who converted the files by different word processors into the format of LaTeX system and prepared the camera-ready copy for the publication. Appreciation is due to Miss Miho Yoshie who helped the conversion including expression of complicated mathematical formulae.

Hideo Ohashi,
Coordinator

Contributing Authors

Hideo Ohashi, Professor,
Department of Mechanical Engineering,
University of Tokyo, Japan.

Born in 1931, male, Japanese. BS in 1954 at Univ. of Tokyo, Dr.-Ing. (T.H. Braunschweig) in 1958 and Dr. Eng. (Univ. of Tokyo) in 1963. 5 years experience in compressor design. Education and research at Univ. of Tokyo since 1959 on dynamic characteristics of turbopumps, fluid forces on vibrating centrifugal impellers, unsteady and two-phase flow in cascades and ventilation of vehicle tunnels.

Hans Ingo Weber, Professor,
Faculty of Mechanical Engineering,
UNICAMP-State University of Campinas,
Brazil.

Born in 1943, Brazilian. Graduated Univ. of São Paulo in 1966, Dr.-Ing. (T.U. München) in 1971. Fellow of the Alexander von Humboldt Foundation. Active at UNICAMP since 1974, initiated area of dynamics for mechanical systems in lecturing and research. Main topic is rotordynamics, emphasizing hydraulic machinery and centrifuges. Identification and reduced order modeling are prioritary research subjects.

Hisao Tomita, Director,
Nuclear Engineering Laboratory,
Toshiba Corp., Japan.

Born in 1936, male, Japanese. BS in 1959 and Dr. Eng. in 1984 at Univ. of Tokyo. Specialized in analysis of dynamics of mechanical systems. Engaged in mechanical vibration problem studies for 10 years at Toshiba R&D Center. Directed the Toshiba project for the development of the gas centrifuge for the Japanese Uranium Enrichment Program. Deputy director of Toshiba R&D center in 1984.

Dung Yu-xin, Professor,
Department of Civil Engineering,
Dalian University of Technology, China.

Born in 1926, male, Chinese. BE in 1951 at Northeast Institute of Technology, Post graduate (Harbin Univ. of Tech.) in 1954 and Dr. Eng. (Moscow Energetic Inst.) in 1958. Education and research at Dalian Univ. of Tech. since 1959 on vibration of hydro-generator units, dynamic and static characteristics of hydropower structures and concrete dams.

Herbert Netsch, Consulting Engineer.

Born in 1921, male, Canadian. Dipl.-Ing. in 1943 at T.H. Stuttgart, Dr.techn. in 1948 at Univ. Wien and assistant at chair for turbomachinery and regulation. Industrial design of pumps and turbines. Senior Engineer, Snowy Mountains Hydro Elec. Authority, Sydney, Australia. Executive Engineer, Morse & Co., Fairbanks, USA. Calculation, design, model test, large mass flow axial pumps (watersupply to Chicago and Florida) and radial pumps. Prof. at Univ. Laval (until 1989), Québec, Canada. Design of pump-turbines and turbines, measurements of large power turbines; efficiency, vibrations and governor parameters. Development of autonomous turbines of small power. Consulting activities in various countries.

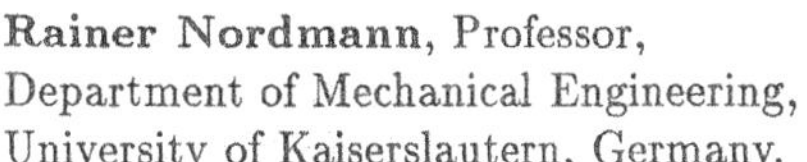

Rainer Nordmann, Professor,
Department of Mechanical Engineering,
University of Kaiserslautern, Germany.

Born in 1943, male, German. Dipl.-Ing. in 1970 at T.H. Darmstadt. Education and research at Univ. of Kaiserslautern since 1980 on rotordynamics, finite elements, modal analysis and parameter identification with applications to vibrations in turbomachinery. Actual research is in the field of fluid structure interactions (bearings, seals, balance pistons, impellers) of turbopumps and in mechatronics.

Werner Diewald, Design Department,
BASF AG, Ludwigshafen, Germany.

Born in 1958, male, German. Dipl.-Ing. in 1985 and Dr.-Ing. in 1989 at Univ. of Kaiserslautern. Education and research at Univ. of Kaiserslautern since 1985 on dynamic characteristics of turbopumps, machine dynamics and control theory. Since 1990 design department of BASF company for reactors and agitator vessels.

Michele Fanelli, Professor, Engineer,
Deputy Director of Centre for Hydraulic and Structural Research, ENEL (Italian National Power Board), Milano, Italy.

Born in 1931, male, Italian. Degree in hydraulic Engineering in 1954 at Univ. of Bologna (Italy); Diploma Ingénieur Hydraulicien de l'Université de Grenoble in 1956 at Grenoble (France); free-teaching Professorship in technique of constructions at Rome, 1970. Employed first by Edison Society and then by ENEL. Extended experience in dam analysis and design, hydraulic transients, physical hydraulic modeling, instationary behaviour of hydraulic machinery and systems.

Rafael Guarga, Professor, Director,
Institute of Fluid Mechanics and Environmental Engineering (IFMEE), University of Uruguay.

Born in 1940 in Uruguay. Mechanical engineer, graduate of Univ. of Uruguay. Dr.-Ing. in 1988 at National University of Mexico (UNAM). Assistant Prof. at Univ. of Uruguay until 1973. Researcher at Univ. of Mexico until 1985 and until today IFMEE Director. Research and education since 1980 on dynamic behaviour of hydraulic machinery, hydraulic transients in pressure conduits, stability of swirl flow in Francis and Kaplan units, combustion chambers and hydrocyclones.

Michihiro Nishi, Professor,
Department of Mechanical Engineering,
Kyushu Institute of Technology, Japan.

Born in 1943, male, Japanese. Graduated from School of Engineering Teachers, Kyushu Univ. in 1965. Dr. Eng. (Kyushu Univ.) in 1976. Studies on fluid dynamics at Res. Inst. of Industrial Sci., Kyushu Univ. from 1965 to 1971. Education and research at Kyushu Inst. Tech since 1971 on fluid mechanics of internal flow, boundary layer control, diffuser performance, draft tube surging and flow measurement.

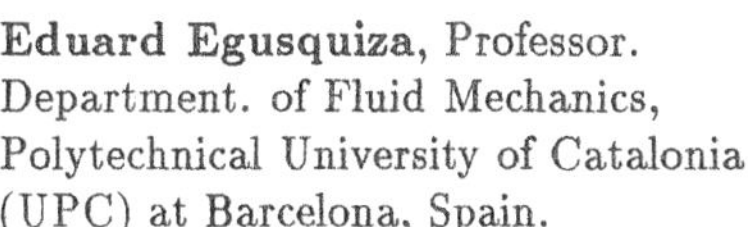

Eduard Egusquiza, Professor.
Department. of Fluid Mechanics,
Polytechnical University of Catalonia
(UPC) at Barcelona, Spain.

Born in Barcelona. Industrial Eng. and Dr. Eng. by the UPC. Assistant and Associate Prof. in 1977-83 at the Faculty of Industrial Eng. (ETSEI) in Terrassa. Prof. in 1983 at the Univ. of Oviedo. Prof. in 1988 at the ETSEI in Barcelona. Research on unsteady flows, flow induced vibrations and condition monitoring in turbomachinery (axial flow fans and hydraulic turbines). Researches also on molten metals.

Shrikant Bhave, Senior Deputy General Manager, Corporate R&D Division, Bharat Heavy Electricals Limited, Hyderabad, India.

Born in 1942, male, Indian. BE in 1964 at M.S. University, Baroda, M.E. in 1967 at M.S. University, Baroda and Ph.D. in 1975 at Univ. of Kanpur. 10 years teaching experience in Department of Mechanical Engineering, Univ. of Kanpur till 1976. Research in the area of vibration and stress analysis in rotating machinery, trouble shooting in power plant equipments, uprating and life extension studies of power plants.

Chapter 1

Fundamentals

H.I. Weber

1.1 Introduction

1.1.1 Hydraulic Machinery as a Mechanical System

This chapter provides the basic information and vocabulary to permit the understanding of vibration phenomena of hydraulic machinery and the methods used for it's analysis. As piping systems are dealt with in the Volume *Transient Phenomena of Hydraulic Machinery* of this Book Series, the scope of this volume is restricted to the presentation of the machine itself. Therefore, in trying to establish some categorization that is valid for several kinds of hydraulic machinery, the first concern relates to the definition of the mechanical system itself and its borders.

To explain phenomena related to the vibration of the machine, it is necessary to have the equations that describe its dynamic behaviour. The main purpose will therefore be the mathematical description of this system and the interaction with external perturbations to obtain information for the design on one hand, and to understand and to *survey* what the machine is doing on the other. All descriptions shall be necessarily simplified in such a way that they are mathematically feasible and also realistic.

Existing nomenclature such as IEC-TC 4, defines the hydraulic machine as a turbine or a pump that is interacting with fluid and is contained in a casing. The rotation is transmitted through a shaft, supported on bearings which interact with the structure, to a generator or a motor, which interact with the electric network. This whole assembly is called a hydraulic unit and if we wish to look at its dynamic behaviour, it will be rather difficult to consider the hydraulic machine separately. Some of the electric and magnetic phenomena are at much higher frequencies than that of hydraulic origin, but the mechanical connection of both parts results in a coupling that is important to be considered also at the turbine (pump) side. Therefore this chapter considers vibrations in the sense of the hydraulic unit, even if less emphasis is given to the electric part compared to the hydraulic one.

The chapter will further distinguish between *global vibrations* that affect the whole unit, from *local vibrations* where the energy is not - or damping is - sufficiently high to propagate it through the machine. In terms of these preliminary considerations the main point will be where to put the borders of the systems. Sometimes, for instance, when the fluid-structure interaction is considered (Chapter 2) the system will comprise also the fluid and the common behaviour is investigated. When considering the global behaviour of the unit, the mathematics would be extremely complex if the finite element modelling of this interaction is also to be included, therefore, it is more convenient to consider the fluid external to the system and try to describe its action as disturbing forces with an approximate mathematical expression. This will not have much use for the investigation of the vibration of the turbine blades but will probably be useful for the analysis of the overall machine, needed for instance in the diagnosis (Chapter 8) through measurements.

The concepts presented in the following sections will be valid for local vibrations (for instance guide vane vibrations) as well as for hydraulic unit vibrations in general. The explanations will be done, however, for the case of Francis and Kaplan Turbines, like the one represented in Figure 1.1. This case will be used repeatedly as examples for the technique and the definitions that will be presented.

1.1.2 Dynamics of a Hydraulic Unit

The first step in the dynamic analysis is to obtain the model of a real system. It is open for the engineers how precise to do this description but there are some compromises that must be made. A more complex formulation will demand more computing time, will make it more difficult to establish the dependences and the sensitivities and will transfer possible errors of simplifications to numerical errors in the computation.

The main measure of complexity is the number of equations that describe the problem, e.g. the number of chosen degrees of freedom for the system. The left side of these equations represents the dynamic properties of the machine whereas the right side represents the excitations to which it is responding. If the right side is made equal to zero then one obtains the eigenbehaviour; if not then one gets the resulting disturbed motion.

There are computer programs which enable the construction of these set of equations. Mainly these can be distinguished between the rigid body approach (MBS - Multi Body Systems) and the Finite Element Method (FEM). In the first case the structure is divided into rigid bodies and mass points connected by elastic elements (damping may be included) based on

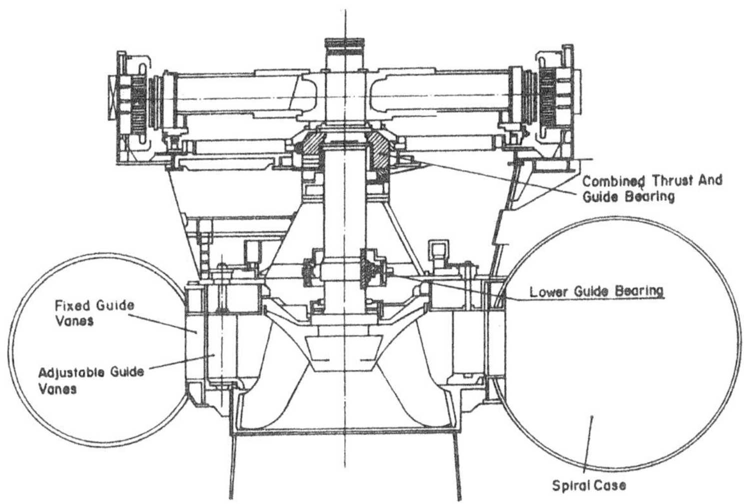

Figure 1.1 Francis turbine unit, umbrella type with combined guide and thrust bearings

engineering reasoning. The parameters may be obtained geometrically and by dynamic identification procedures, which will compensate for the errors introduced by the discretization. These are models with small number of degrees of freedom. Otherwise the FEM will use the geometrical configuration to make the most exact description possible. The huge amount of degrees of freedom can afterwards be reduced by a convenient technique. Normally the description of a system with continuously distributed mass and inertia (like a bar) with partial differential equations is avoided, which can be solved by approximation methods like Rayleigh-Ritz, but this can be done for the local vibration cases, for instance in hydroelastic vibrations.

The definition of the model is therefore always something that shall be done from the engineering point of view. Consequently it is also not unique. For instance to establish the model represented in Figure 1.2 (a) corresponding to the hydraulic unit represented in Figure 1.1 the following reasoning is required:

- the bearing supports may be modelled statically by FEM and the resulting stiffness will be represented directly in the final model; this means that the dynamics of the support is not interfering in the dynamics of the unit: this is a severe hypothesis that must be verified; the thrust bearing is simply substituted by an elastic restoring moment;

- the bearing stiffness (oil film) is much higher than the stiffness of the support and therefore it is assumed as a rigid connection; in that way we remove from the model all particularities which are typical of the bearings. This is not possible in all situations, but it will be possible in some;

- the rotor of the generator is considered rigid in itself, neglecting therefore its own dynamics; it is again a severe hypothesis;

- the negative stiffness characteristic to the generator is supposed to be included in the upper guide bearing;

- the shaft will be modelled by a convenient number of finite elements;

- the turbine is also considered rigid, eventually with some added mass due to the existing water; in that way all the dynamics of interactions between the water and the turbine blades are neglected;

- there is no consideration of the fact that the turbine is rotating in a casing that seals the flow of the water to the other side.

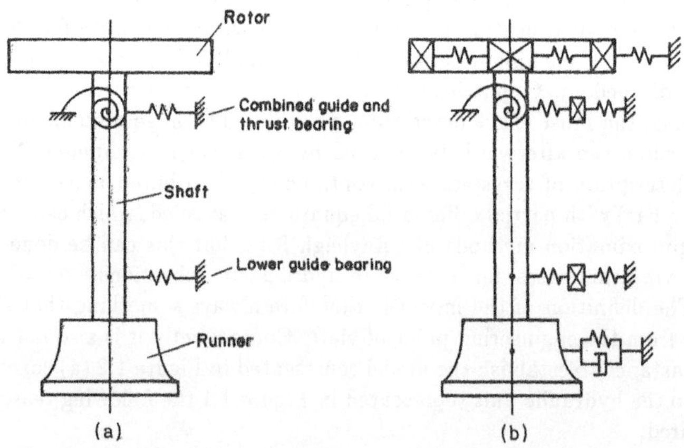

Figure 1.2 Models for the unit of Figure 1.1

This model may contain several imperfections. An attempt to remove some simplifications results in the model of Figure 1.2 (b). They allow a mathematical description, through any method, once we have chosen the coordinates (and therefore the number of degrees of freedom) that will define the dynamic behaviour. As the borders are established the excitations are also known that actuate on the system and which represent actions across the borders: the hydraulic forces through the turbine, the electric/magnetic forces through the generator, the perturbations coming from the foundation through the suspension, the weight due to the gravity and the unbalance excitation caused by the motion. Finally, in general due to non-linear effects, it is possible that some excitation interacts in such a way with the system that it is impossible to separate it, or consider it external to the system (mathematically, to put it on the right side of the equation): that is called self-excitation.

The usual mathematical model, for instance for a global vibration analysis based on a FEM, is restricted to linear effects and will comprise a symmetric mass matrix $\boldsymbol{M}$, that includes the mass and inertia properties of the generator, turbine and shaft, and a symmetric stiffness matrix $\boldsymbol{K}$, that includes the elastic properties of the suspension and shaft. Further the viscous damping can be considered through another symmetric matrix $\boldsymbol{C}$. Usually damping is neglected in a first approach as it is very difficult to establish values for the coefficients of the matrix. As the values cannot be well defined, the matrix $\boldsymbol{C}$ is assumed as a linear combination of $\boldsymbol{M}$ and $\boldsymbol{K}$ (proportional damping). In the model that is being considered, the damping in the metallic parts will probably be much less than that of the turbine vibrating in the water. Due to the rotation of a huge inertia, the generator, it is not possible to neglect the gyroscopic effects in big machines; they will be included in the model resulting in a skew-symmetric matrix $\boldsymbol{G}$. Mechanical laws of motion will combine these matrices in the form of an equation of motion, using the vector of displacement coordinates $\boldsymbol{x}$, defined according to the number of degrees of freedom n,

$$\boldsymbol{M}\ddot{\boldsymbol{x}} + (\boldsymbol{C} + \boldsymbol{G})\dot{\boldsymbol{x}} + \boldsymbol{K}\boldsymbol{x} = \boldsymbol{f} \tag{1.1}$$

This matrix equation of order n corresponds to a system of n second order coupled differential equations where $\boldsymbol{f}$ stands for the external forces. In this kind of problem, normally only $\boldsymbol{f}(t)$ and $\boldsymbol{x}(t)$ are functions of the time. Finally, there may be included in the model a more complex description of some specific phenomena like the journal bearings, where a general matrix $\boldsymbol{D}$ depending on velocity is substituted for the symmetric matrix $\boldsymbol{C}$ and the non-symmetric matrix $\boldsymbol{L}$, depending on displacement, is written instead of the matrix $\boldsymbol{K}$. A similar kind of analysis can also be done for seals or when one considers the motion of the turbine or pump in the casing.

1.1.3 System Parameters and Excitations

The mathematical model of a machine is identical to the equations of motion that describe its behaviour at the chosen coordinates, corresponding to the degrees of freedom, under the actions of forces which cross the borders of this system. The parameters of the system, corresponding to the coefficients of the matrices defined above, synthesize the physical properties in a way that allows model and reality to perform similarly.

It is not easy to give values to these parameters. If one uses the FEM, mass and stiffness matrices can be seen as a result of the modelled geometry (for instance using the design drawings). But this is not always possible, due to the extreme complexity resulting from a precise modelling. In general it is admitted that the mass parameters can be obtained from the geometry; then, with some lengthy calculations, the moments of inertia may be obtained. But it is also possible that the mass has to be previously discretized following some specific criteria which implies a good understanding of the physical system. The terms of the stiffness matrix are often obtained through static analysis (FEM or application of forces on the real structure); it is important to observe if this is possible and will not result in errors. Sometimes these values can only be used as a starting approach in a dynamic identification procedure. In this case for a model with a given (or a chosen) structure one will look for the *best values of parameters*, so that model and machine will behave similarly.

In several situations the definition of these matrices, matrix coefficients or system parameters, is not straight forward. Researchers are trying to identify physical phenomena and describe them adequately: for instance vibration in water will result in an *added mass* to be summed to the coefficients of the matrix M. There are solutions in the literature showing how to proceed for a plate in water or, maybe, a guide vane vibration. But the system is not yet so straightforward for the vibration of turbine blades that it could be used in the global vibration analysis as an extension of the FEM. It is probably still more practical to try to identify the values directly in the operation of the real machine. Another example is the bearings: there exists a huge number of papers on bearings and bearings properties, but often the values obtained for stiffness (and damping) can only be considered as an approximation; what the real machine is doing in its bearings can at best be identified from its own behaviour.

Gyroscopic terms are as easy to handle as the mass matrix coefficients, but what the real damping in the machine is, can not be exactly answered. It is difficult to establish a model (or to adapt the viscous model) to a numerical value based on experience and also to identify it (the identification procedures may present big errors when searching for damping coefficients).

The effects that cross the borders of the system will be called excitations.

They comprise the constant gravity force, which in most cases only defines an equilibrium position, time varying effects and transient perturbations. The time variation may present the basic harmonic or a multiple of the machine rotation velocity or of the net frequency, some other harmonic characteristic or as a (band limited) noise. The transient effects, of hydraulic or electric origin, excite the machine *suddenly* which begins vibrating in its proper characteristics. In this case the vibration is studied under eigendynamics and not under excited motion.

The excitations may have a hydraulic origin and are normally acting with low frequency. The perturbations on the water flow through the machine will act on the blades inducing them to vibrate. These hydraulic forces, which are very high, will in general not compensate only to a resulting torque, that is producing power, but will also give a radial rotating resultant and an axial component. These forces have been the subject of much research (see the literature review on pump impellers by Flack and Allaire, 1984 [1.7], and the analysis on small scale machines of Bachmann, 1980 [1.2]) but mainly in steady state conditions; there is not much knowledge on transients such as in the reversal of a turbo-pump. The periodic characteristic of these forces, derived from asymmetries of the flow supply from the spiral case and non-uniformity of flow in the machine, will depend on the rotation speed, the number of runner and guide vane blades, and so on.

The vortex formation may originate at the guide vanes, either the moving ones or the fixed ones. They may join a resulting vortex that is interrupted by the crossing of the runner blades: then there is a harmonic excitation with a basic frequency of $z_b\, n$ (Hz)(z_b - number of runner blades, n - rotational speed). It is also possible to have individual vortex formation at the lower part of the fixed guide vanes, resulting in a fundamental frequency $z_{gf}\, n$ (Hz) (z_{gf} - number of fixed guide vanes). More important is the vortex core in the draft tube and the huge hydraulic forces that it applies on the runner, outside of the nominal operation range of the unit. The investigation of this phenomena was instigated by Lord Kelvin but the dependence on the generated load was shown by Hosoi in 1965 [1.12]. Since then the number of attempts to obtain a good mathematical model have increased, see for instance Doerfler, 1980 [1.6], or Guarga, 1986 [1.9], showing the way to estimate the vortex core frequency, which is smaller than half of the rotating speed. There is also a further dependence on the water level at the tailrace.

The fluid-structure interaction can also form a mechanical system that gives rise to some changes in the way the mechanical part shall be analysed if the borders of the modelled system cut out the hydraulics. This may result in mass change (added mass) or it may induce a coupling between the vibration in flexion and torsion of the blade of a Kaplan turbine, for example, which can appear at very well defined flow velocity over the blades.

The excitations of a mechanical origin are somewhat easier to quantify. There are the unbalances acting at the rotating speed but there may be also a shaft deformation, an inclination of the shaft away from the vertical, or a generator pole that is not rigidly fixed, making things more difficult. It is certainly possible to compensate for the unbalance with additional masses but it will be difficult to compensate for a deformation. Also rubbing against some part may happen due to unskillful assembly making it difficult to balance the machine. There may exist some bearing problems; for instance a surface irregularity will produce a force each time it crosses one of the pads, having therefore a periodic characteristic. If there is a lateral misalignment at the coupling of the shaft parts, there will be an excitation at double the rotation speed.

Mechanical excitations can be of the self-excited type and in this case it is not possible to take them out of the borders of the system; they must be included in the eigendynamics of the system like the vibrations that a shaft may execute due to inner damping or the Coulomb damping in a bearing, due to bad lubrication or a clearance, producing position dependent forces demanding a non-linear way to treat the equations. Touching of the turbine at the housing may also produce strong vibrations of the self-excited type.

Some magnetic forces at the generator should also be mentioned. If there is a loss of circularity of the rotor then there will be a bigger force at smaller gaps; as this force is rotating it is possible to imagine the force as derived from a hypothetical negative spring. It is important to consider this magnetic unbalance in large machinery due to the dynamic consequences of this excitation. Another origin of a magnetic force may be a short circuit in the winding of a pole. Then there are again forces related to the harmonics of the rotating speed. Another effect comes from the electromagnetic forces between rotor and stator and the magnetic induction. This will result in an excitation at double the net frequency and is, in general, well separated in frequency from the above mentioned forces.

Each excitation will contribute to the dynamic behaviour of a machine and, in a linear hypothesis, the consequence is a superposition of the individual responses. As the analysis of the vibration of hydraulic machinery will often be concerned with the interpretation of measurements it is important to have an idea about all possible origins of vibrations. Some of them will be easily separated due to frequency characteristics. An analysis of the excitations in a hydraulic unit is worked out by A. H. Glattfelder, et. al., 1981 [1.8].

External forces acting at a certain frequency or in a frequency band may cause severe vibrations in the system that is being investigated. One speaks of resonances, meaning that some of the frequencies at which the system vibrates freely is coincident with one of these external frequencies.

It may have a global importance, for instance, when a system is operated at its critical velocity or it may be of only a local kind, being damped out when propagating through the machine. This is important to consider in the attempt to identify the sources of vibrations detected at some measurement points.

When trying to describe mathematically all the physical phenomena in a vibration problem, several engineering simplifications shall be made to keep the formulation at the easiest but still representative equations. One main decision in that sense is the linearization of the description of some inherently non-linear effects. For instance in a stable vibration with small amplitudes it is possible to linearize the mathematical relations around the working point as in the case of a hardening or softening spring characteristic, a non-linear damping characteristic, and so on. This usual assumption, which is also basic in this Chapter, may however not represent some typical behaviour of the system. Due to the resulting complications in this case one simplifies other aspects of the system and tries to describe fully the non-linearities, like the hysteresis loop of the damping at some point, the dry friction at some part, the sealing characteristics of a pump, etc. It may also be necessary to consider the time variation of a parameter, which may have its origin in a force non-linearly related to other parameters of the system.

1.2 Basic Dynamics of Mechanical Systems

1.2.1 Eigendynamics of Linear Systems

In this section a synthetic presentation of the structure and the properties of the linear model of a mechanical system is proposed, trying to make a bridge between linear algebra and engineering applications. Firstly the simplest case is considered, followed by additional considerations leading to more complex solutions. It was shown in section 1.1.2 that the resulting mathematical model, in the linear case, is the equation

$$\boldsymbol{M}\ddot{\boldsymbol{x}}(t) + (\boldsymbol{C} + \boldsymbol{G})\dot{\boldsymbol{x}}(t) + \boldsymbol{K}\boldsymbol{x}(t) = \boldsymbol{f}(t)$$

where $\boldsymbol{M}$ is the mass matrix, $\boldsymbol{C}$ is the viscous damping matrix, $\boldsymbol{G}$ is the gyroscopic matrix, $\boldsymbol{K}$ is the stiffness matrix, $\boldsymbol{x}(t)$ is the displacement vector, $\boldsymbol{f}(t)$ is the excitation and n, the order of the matrices, is the number of degrees of freedom.

The eigendynamics is mathematically expressed by the left hand side and, the external forces by the right hand side, therefore, the homogeneous equation with $\boldsymbol{f}(t) = \boldsymbol{o}$ indicates the absence of excitations on the system.

M K Systems

In a first approach the mechanical system is modelled as undamped and linear; its vibrational behaviour will be characterized by a time variation of its position around the equilibrium position. To describe this motion coordinates are required: the minimum number that can describe what the system is doing, will then be equal to the number of degrees of freedom n. It is obvious that this number is an engineering approximation. In this way a displacement vector $\boldsymbol{x}$ of order n is defined. The existence of an equilibrium position means that there is a structural stiffness, represented in a symmetric matrix $\boldsymbol{K}$ which, when multiplied with the displacements ($\boldsymbol{Kx}$), will result in the restoring forces. They are in equilibrium with the inertia forces which are associated with the symmetric mass matrix $\boldsymbol{M}$ multiplied by the acceleration, as it is expressed by Newton's law:

$$\boldsymbol{M}\ddot{\boldsymbol{x}} + \boldsymbol{K}\boldsymbol{x} = \mathbf{o}\,. \tag{1.2}$$

This equation contains not only the dynamical force equilibrium condition (D'Alembert's law) but also the constancy of the total energy, or an instantaneous balance of energy divided into energy of velocity (kinetic) and energy of displacement - or deformation (potential). Pre-multiplying the terms with the transpose of the velocity vector $\dot{\boldsymbol{x}}^T$ and integrating them we obtain:

$$\frac{1}{2}\dot{\boldsymbol{x}}^T\boldsymbol{M}\dot{\boldsymbol{x}} + \frac{1}{2}\boldsymbol{x}^T\boldsymbol{K}\boldsymbol{x} = E_k + E_p = E = \text{constant} \tag{1.3}$$

The value of this total energy E will be defined by the initial conditions $\boldsymbol{x}(0)$ and $\dot{\boldsymbol{x}}(0)$ and remains constant, circulating between the kinetic E_k and potential E_p parts: the system is called conservative, which again is an engineering approach.

Equation (1.2) is homogeneous which means that $\boldsymbol{x}(t) = \mathbf{o}$ is a solution, i.e., if the undisturbed system is in the equilibrium position, there it remains. It is of second order (order of the highest derivative, in mechanical systems usually the acceleration) therefore to define the motion $\boldsymbol{x}(t)$ two initial conditions are necessary, $\boldsymbol{x}(0)$ and $\dot{\boldsymbol{x}}(0)$, as can be seen from the energy equation (1.3). This is a matrix equation of order n, meaning that there are n coupled second order differential equations whose solution must be obtained. The couple of displacement and velocity can also be joined in a new vector $\boldsymbol{y}(t)$ of doubled order $2n$, called state of the system because its knowledge is necessary to compute a future state

$$\boldsymbol{y}(t) = \left\{ \begin{array}{c} \boldsymbol{x}(t) \\ \dot{\boldsymbol{x}}(t) \end{array} \right\}. \tag{1.4}$$

Equation (1.2) has two basic properties which are inherent to the system: the eigenfrequencies and the eigenmodes. They appear when we look for a solution. Independently from the initial state one may look for the solution that separates the amplitude $\boldsymbol{u}$ and a time function which is chosen as $e^{\lambda t}$:

$$\boldsymbol{x} = \boldsymbol{u}e^{\lambda t}\,. \tag{1.5}$$

Substituting in the equation one gets the so called eigenvalue problem:

$$(\boldsymbol{M}\lambda^2 + \boldsymbol{K})\boldsymbol{u} = \mathbf{o}, \tag{1.6}$$

where λ and $\boldsymbol{u}$ are unknown. This linear homogeneous system of equations has only a non-trivial solution if the lines of the matrix are linearly dependent, i.e., if the determinant

$$\det(\boldsymbol{M}\lambda^2 + \boldsymbol{K}) = 0. \tag{1.7}$$

This is a polynomial equation of order $2n$ in λ (but in this case without odd exponents) called the characteristic equation of the system. Further it can be shown that all the solutions for λ^2 are negative, since $\boldsymbol{M}$ and $\boldsymbol{K}$ are symmetrical positive definite matrices, which means that $\boldsymbol{M}$ has masses and/or inertias in all coordinates and that $\boldsymbol{K}$ corresponds to springs that directly or indirectly actuate in all coordinates (therefore all coordinates have an defined equilibrium value). Then

$$\lambda_1 = i\omega_1,\;\; \lambda_2 = i\omega_2, \ldots,\;\; \lambda_{n+1} = -i\omega_1,\;\; \lambda_{n+2} = -i\omega_2, \ldots,\;\; \lambda_{2n} = -i\omega_n \tag{1.8}$$

and the $2n$ roots of the characteristic equation result in only n values for ω. For each solution λ_p $(p = 1, \ldots, 2n)$ (1.6) produces a real vector $\boldsymbol{u}^p$ which, as it comes from a homogeneous system, has an undefined length, i.e., any $\alpha\boldsymbol{u}^p$ is also a solution. Further, as λ is squared in (1.6) the same solution $\boldsymbol{u}^r$ is obtained for both values $\lambda_r, \lambda_{n+r} = \pm i\omega_r$ $(r = 1, \ldots, n)$ and therefore there are only n distinct values of this vector. As the motion will be a linear superposition of these results, these couples may be combined to give

$$\begin{aligned} \boldsymbol{x}_r(t) &= \boldsymbol{u}^r(A_r e^{i\omega_r t} + A_{n+r}e^{-i\omega_r t}) \\ &= \boldsymbol{u}^r[C_r\cos(\omega_r t - \varphi_r)] \qquad r = 1, \ldots, n \end{aligned} \tag{1.9}$$

where A, C and φ are constants, C and φ may be written in terms of A.

Therefore as the system will only vibrate at n well defined frequencies, they are called the eigenfrequencies; also, the geometrical form $\boldsymbol{u}^r$ of vibration is well defined, which means that the relation between any two coordinates (or components of this vector) is constant in the motion at an eigenfrequency:

$$\frac{u_j^r}{u_{j+k}^r} = \text{constant} \quad j,k = 1,\ldots,n \tag{1.10}$$

for any $\alpha \boldsymbol{u}^r$ chosen as a solution for (1.6). This justifies the name eigenmode for the vector $\boldsymbol{u}^r$, which emphasizes the fact that there is a constancy in the relations between coordinates (describing a geometrical form or mode) disregarding the amplitude scalar factor.

The resulting motion of (1.2) for an initial disturbance $\boldsymbol{x}(0)$, $\dot{\boldsymbol{x}}(0)$ is a superposition of the eigensolutions (1.9):

$$\boldsymbol{x}(t) = \sum_{r=1}^{n} C_r \boldsymbol{u}^r cos(\omega_r t - \varphi_r) \tag{1.11}$$

where the $2n$ constants C_r and φ_r are defined through the $2n$ coordinates of the initial condition vectors. It is easy to visualize an eigenmode when the initial displacement assumes the form of the kth mode $\boldsymbol{x}(0) = \boldsymbol{u}^k$ and the motion is started without velocities $\dot{\boldsymbol{x}}(0) = \mathbf{o}$. Then $C_r = 0$ for $r \neq k$, C_k is an amplitude factor and the vibration keeps always the form of the mode, vibrating in the eigenfrequency ω_k with $\cos \omega_k t$. It is a basic property that a system without external perturbations is vibrating in the eigenfrequencies combining somehow the eigenmodes.

If the engineer would have chosen another set of coordinates $\bar{\boldsymbol{x}}$ to describe the motion of the system things would be very similar. The equation of motion would be changed to

$$\bar{\boldsymbol{M}}\ddot{\bar{\boldsymbol{x}}} + \bar{\boldsymbol{K}}\bar{\boldsymbol{x}} = \mathbf{o} \tag{1.12}$$

but with the transformation matrix $\boldsymbol{T}$, $\det \boldsymbol{T} \neq 0$

$$\bar{\boldsymbol{x}} = \boldsymbol{T}\boldsymbol{x} \quad , \quad \boldsymbol{M} = \boldsymbol{T}^T\bar{\boldsymbol{M}}\boldsymbol{T} \quad , \quad \boldsymbol{K} = \boldsymbol{T}^T\bar{\boldsymbol{K}}\boldsymbol{T} \tag{1.13}$$

again equation (1.2) is obtained. Also the characteristic equation is the same:

$$\begin{aligned} \det(\bar{\boldsymbol{M}}\bar{\lambda}^2 + \bar{\boldsymbol{K}}) &= \det(\boldsymbol{T}^{T^{-1}}\boldsymbol{M}\boldsymbol{T}^{-1}\bar{\lambda}^2 + \boldsymbol{T}^{T^{-1}}\boldsymbol{K}\boldsymbol{T}^{-1}) \\ &= \det(\boldsymbol{T}^{T^{-1}})\det(\boldsymbol{M}\bar{\lambda}^2 + \boldsymbol{K})\det(\boldsymbol{T}^{-1}) = 0\,. \end{aligned} \tag{1.14}$$

Therefore the eigenfrequencies are not dependent on the coordinates selected for the investigation. The eigenmodes are geometrically the same but will be expressed adequately in each of the coordinate systems:

$$\bar{\boldsymbol{u}} = \boldsymbol{T}\boldsymbol{u}. \tag{1.15}$$

There is a basic property of the eigenmodes that will be useful for the study of excited systems and the basis of modal analysis: two eigenmodes corresponding to different eigenfrequencies are orthogonal. Calling them $\boldsymbol{u}^r$ and $\boldsymbol{u}^s$, they satisfy equation (1.6), i.e.,

$$\left.\begin{aligned} \boldsymbol{M}\boldsymbol{u}^r\lambda_r^2+\boldsymbol{K}\boldsymbol{u}^r &= \mathbf{o}, \\ \boldsymbol{M}\boldsymbol{u}^s\lambda_s^2+\boldsymbol{K}\boldsymbol{u}^s &= \mathbf{o}, \end{aligned}\right\} \tag{1.16}$$

then pre-multiplying the first equation with $\boldsymbol{u}^{s^T}$ and the second with $\boldsymbol{u}^{r^T}$ and transposing the whole second equation, using the symmetry properties of the matrices, then subtracting both equations yields:

$$\boldsymbol{u}^{s^T}\boldsymbol{M}\boldsymbol{u}^r(\lambda_r^2-\lambda_s^2)=0. \tag{1.17}$$

It is assumed in this text that the eigenfrequencies are always different when $r \neq s$ and then orthogonality results with the mass matrix $\boldsymbol{M}$ as a weighting matrix

$$\boldsymbol{u}^{s^T}\boldsymbol{M}\boldsymbol{u}^r = 0 \quad ,\text{for} \quad s \neq r.$$

When $s = r$ the resulting value will be called the generalized mass μ_r and depends on the way the arbitrary length of the eigenmode was chosen (i.e., was scaled):

$$\boldsymbol{u}^{r^T}\boldsymbol{M}\boldsymbol{u}^r = \mu_r \quad r = 1,\ldots,n.$$

If the n eigenmodes are organized in a modal matrix

$$\boldsymbol{U} = [\boldsymbol{u}^1 \quad \ldots \quad \boldsymbol{u}^n] \tag{1.18}$$

then both informations are included in the equation

$$\boldsymbol{U}^T\boldsymbol{M}\boldsymbol{U} = \mathcal{M} \quad r = 1,\ldots,n. \tag{1.19}$$

where $\mathcal{M}$ is the diagonal matrix of the generalized masses. If this is the identity matrix then we speak of orthonormalized eigenvectors. This condition is convenient analytically and will be assumed in the following text: it may be established from a generic eigenvector $\widehat{\boldsymbol{u}}^r$ through

$$\boldsymbol{U} = \widehat{\boldsymbol{U}}\mathcal{M}^{-1/2}. \tag{1.20}$$

Now, let $\mathcal{M} = \boldsymbol{I}$ ($\boldsymbol{I}$ being the unity matrix) in (1.19). This property will also make this vector adequate to be used as a base of an orthogonal transformation. A geometrical example of a transformation of this kind is seen in the very simple case of a rotation Ωt where the representation in the

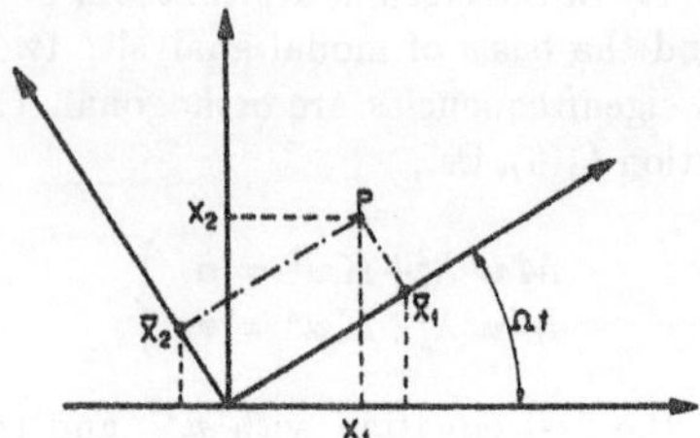

Figure 1.3 Plane rotation between two reference frames

system (x_1, x_2) shall be transformed into $(\bar{x}_1, \bar{x}_2)$ through a matrix $\boldsymbol{T}$,(see Figure 1.3) i.e.,

$$\left\{ \begin{array}{c} x_1 \\ x_2 \end{array} \right\} = \left[\begin{array}{cc} \cos \Omega t & -\sin \Omega t \\ \sin \Omega t & \cos \Omega t \end{array} \right] \left\{ \begin{array}{c} \bar{x}_1 \\ \bar{x}_2 \end{array} \right\} \quad , \quad \boldsymbol{x} = \boldsymbol{T}\bar{\boldsymbol{x}}$$

which is an orthogonal matrix because $\boldsymbol{T}^{-1} = \boldsymbol{T}^T$ and also normalized because $\det \boldsymbol{T} = 1$. In this way coordinates of equation (1.2) can be transformed to (1.12), for instance. It is possible to generalize these definitions with a weighting matrix and look at the modal matrix $\boldsymbol{U}$ at the place of the transformation matrix $\boldsymbol{T}$. The orthogonality will result from

$$\boldsymbol{U}^T \boldsymbol{M} \boldsymbol{U} = \boldsymbol{I} \tag{1.21}$$

and in this generalized form the following property is obtained:

$$\boldsymbol{U}^{-1} = \boldsymbol{U}^T \boldsymbol{M}. \tag{1.22}$$

Looking back to (1.16) it is observed that the orthogonality is also present with $\boldsymbol{K}$ as a weighting matrix and if the normalization is performed according to (1.21) then, considering (1.8)

$$\boldsymbol{u}^{s^T} \boldsymbol{K} \boldsymbol{u}^r = -\lambda_r^2 \delta_{rs} = \omega_r^2 \delta_{rs} \,, \quad r, s = 1, \ldots, n \quad \delta_{rs} = \left\{ \begin{array}{ll} 1 & r = s \\ 0 & r \neq s \end{array} \right. \tag{1.23}$$

resulting in

$$\boldsymbol{U}^T \boldsymbol{K} \boldsymbol{U} = -\boldsymbol{\Lambda}^2 \tag{1.24}$$

where $\boldsymbol{\Lambda}$ is the eigenvalue matrix (diagonal matrix, in the undamped case with each of the eigenfrequencies written only once, as one of the imaginary conjugates).

Finally using the modal matrix as a transformation and representing the coordinates of this spectral decomposition with $\boldsymbol{\eta}$ then

$$\boldsymbol{x} = \boldsymbol{U}\boldsymbol{\eta}, \tag{1.25}$$

which when substituted in (1.2) and using (1.21) and (1.24) results in an uncoupled system of equations:

$$\ddot{\eta}_r + \omega_r^2 \eta_r = 0\,, \quad r = 1, \ldots, n. \tag{1.26}$$

The complexity is transferred to the initial conditions which must be considered when searching for a solution in these coordinates:

$$\eta_r(t) = C_r \cos(\omega_r t - \varphi_r) \tag{1.27}$$

with $\boldsymbol{\eta}(0) = \boldsymbol{U}^T \boldsymbol{M}\boldsymbol{x}(0)$ and $\dot{\boldsymbol{\eta}}(0) = \boldsymbol{U}^T \boldsymbol{M}\dot{\boldsymbol{x}}(0)$.

The explicit solution for some initial condition to the undisturbed system is given in the original coordinates again by (1.11), as can be shown using (1.25), where

$$\boldsymbol{x}(t) = \boldsymbol{U}\boldsymbol{\eta}(t) = \sum_{r=1}^{n} \boldsymbol{u}^r \eta_r(t) = \sum_{r=1}^{n} C_r \boldsymbol{u}^r \cos(\omega_r t - \varphi_r) \tag{1.28}$$

$$C_r = \sqrt{\eta_r^2(0) + \dot{\eta}_r^2(0)/\omega_r^2}\,, \quad \varphi_r = \arctan \frac{\dot{\eta}_r(0)}{\omega_r \eta_r(0)}$$

Summarizing the eigendynamics of **M K** Systems results in two basic properties: the eigenfrequencies, which may be arranged in an *eigenfrequency* matrix and orthogonal eigenmodes that define a real modal matrix.

M C K Systems

This section demonstrates some aspects of what will be changed by the inclusion of viscous damping in the former analysis. The equation of motion will consider damping through a symmetric matrix $\boldsymbol{C}$:

$$\boldsymbol{M}\ddot{\boldsymbol{x}}(t) + \boldsymbol{C}\dot{\boldsymbol{x}}(t) + \boldsymbol{K}\boldsymbol{x}(t) = \mathbf{o} \tag{1.29}$$

and the new term $\boldsymbol{C}\dot{\boldsymbol{x}}$ will be responsible for the dissipation of energy, reducing the amplitudes of free motion with time. If there is some characteristic *period* T to be defined for the motion, then an integration similar to that performed in (1.3) results in the energy decay

$$E_d = \int_0^T \dot{\boldsymbol{x}}^T \boldsymbol{C}\dot{\boldsymbol{x}}\, dt. \tag{1.30}$$

Searching for the solution, proceeding similarly to the undamped case, again trying to separate amplitude $\boldsymbol{u}$ and a time function chosen as $e^{\lambda t}$ (like in (1.5)) the new eigenvalue problem is obtained:

$$(\boldsymbol{M}\lambda^2 + \boldsymbol{C}\lambda + \boldsymbol{K})\boldsymbol{u} = \mathbf{o}\,. \tag{1.31}$$

The determinant of this symmetric matrix has to be zero as in (1.7) resulting in a polynomial equation of order $2n$ with positive coefficients called the characteristic equation. If in addition to $\boldsymbol{M}$ and $\boldsymbol{K}$, $\boldsymbol{C}$ is also positive definite, meaning that the damping is reaching all the coordinates, then the roots λ, called the eigenvalues of the system, are:

$$\lambda_1 = -\sigma_1 + i\omega_1,\ \lambda_2 = -\sigma_2 + i\omega_2,\ \ldots,$$

$$\lambda_{n+1} = -\sigma_1 - i\omega_1,\ \lambda_{n+2} = -\sigma_2 - i\omega_2,\ \ldots \tag{1.32}$$

with $\sigma_r > 0$.

Therefore the time function of the eigenmotion is obtained by combining adequately the complex conjugate pairs:

$$e^{(-\sigma_r \pm i\omega_r)t} = e^{-\sigma_r t} e^{\pm i\omega_r t}$$

representing a decaying exponential function defined by the damping coefficients of the system that restrains a harmonic function at the eigenfrequency ω_r. If there is no damping, $\sigma_r = 0$, the former case is reproduced. If $\boldsymbol{C}$ is positive semi-definite then it is possible to have some of the $\sigma_r = 0$ resulting in undamped coordinates and the other $\sigma_r > 0$. If all $\sigma_r > 0$ then the damping again reaches all coordinates and the positive semi-definite matrix $\boldsymbol{C}$ is called pervasive. Otherwise, damping can be *negative*, as in self-excited situations, meaning that some $\sigma_r < 0$ and the harmonic behaviour will have increasing amplitudes due to the exponentially increasing envelope. Finally it is possible to think of a case where some $\omega_r = 0$, which means that this mode is *overdamped*, i.e., it looses its vibrational characteristic and is moving asymptotically.

Substituting the eigenvalues back in the homogeneous equation (1.31) $2n$ solutions will be obtained, which in this case are the complex eigenvectors $\boldsymbol{u}$. In this case there are two factors of arbitrariness corresponding to (1.10) since there is a real and an imaginary part resulting in two homogeneous systems of equations of order n, one for each part. This arbitrariness may be represented as a freedom in establishing the amplitude and the phase angle of the vector (or the amplitude of the real component and of the imaginary component of the eigenvector).

Further, as in the solution of the characteristic equation for each eigenvalue λ, its corresponding complex conjugate λ^* is also available, then in the solution of (1.31) if u^r is associated to $\lambda_r = -\sigma_r + i\omega_r \quad (r = 1, \ldots, n)$, u^{*^r} is associated to $\lambda_r^* = -\sigma_r - i\omega_r \quad (r = 1, \ldots, n)$. The resulting motion of (1.29) is given by the superposition of these solutions with adequate constants A_r and A_r^* (it is easy to show that they are also complex conjugate pairs), i.e.,

$$x(t) = \sum_{r=1}^{n} (A_r u^r e^{\lambda_r t} + A_r^* u^{*^r} e^{\lambda_r^* t})$$

or

$$x(t) = \sum_{r=1}^{n} C_r e^{-\sigma_r t} [u_R^r \cos(\omega_r t - \varphi_r) - u_I^r \sin(\omega_r t - \varphi_r)] \qquad (1.33)$$

where $u^r = u_R^r + i u_I^r$.

This shows how (1.11) is changed with damping. There are again $2n$ constants, C_r and φ_r, defined by the same number of initial conditions in displacement and velocity. The visualization of the eigenmode u^r is much more difficult: to make $x(t)$ dependent on only one of the n modes there must exit a specific initial condition in displacement and velocity. Also this vibrating mode would have its amplitudes decreasing with time due to the exponential envelope.

These eigenvectors may be put all together in a $(n \times 2n)$ modal matrix

$$[u^1 \quad u^2 \quad \ldots \quad u^{n+1} \quad u^{n+2} \quad \ldots] \qquad (1.34)$$

where n vectors are the complex conjugate of the other n, but there is another difficulty. If one tries to reproduce the orthogonality analysis of (1.16) and (1.17) in this case, it can be shown that only when the damping matrix is a linear combination of the mass matrix M and the stiffness matrix K it is possible to make an analogy in the deductions and therefore in the orthogonality conditions between the damped case and the undamped case. In this case C is called a proportional damping and relations (1.21) and (1.24) are valid as orthogonality definitions. Therefore also the real eigenmode of the undamped system may still be used for the modal uncoupling (1.25) resulting in a damped equation (1.26) for the mode. It is not unusual to do this simplification due to the imprecision in the knowledge of the exact damping itself. The reader may look in the specific literature for details.

If for some reason this simplification is not possible then the system must be analysed in the first order form by working with the state vector $y(t)$ (1.4) instead of the displacement $x(t)$. A way of doing it is transforming (1.29), for instance in:

$$\begin{bmatrix} C & M \\ M & 0 \end{bmatrix} \begin{Bmatrix} \dot{x} \\ \ddot{x} \end{Bmatrix} + \begin{bmatrix} K & 0 \\ 0 & -M \end{bmatrix} \begin{Bmatrix} x \\ \dot{x} \end{Bmatrix} = \begin{Bmatrix} 0 \\ 0 \end{Bmatrix} \qquad (1.35)$$

or

$$\tilde{M}\dot{y} + \tilde{K}y = o,$$

where $\tilde{M}$ and $\tilde{K}$ are symmetric matrices of order $2n$.

Again, looking for a solution that separates amplitudes $\tilde{u}$ from the time function:

$$y(t) = \tilde{u}e^{\lambda t}, \tag{1.36}$$

the eigenvalue problem is obtained:

$$(\tilde{M}\lambda + \tilde{K})\tilde{u} = o \tag{1.37}$$

which due to the physical reasoning that the system is still the same, must result in the same polynomial in order $2n$ for the characteristic equation, as in (1.31), and therefore also in the same eigenvalues (1.32). This may be easily checked in the equation above. Then again things may be done analogously to the undamped case. Substituting the eigenvalue λ_r or λ_r^* one gets the eigenvector $\tilde{u}^r$ or its complex conjugate $\tilde{u}^{*^r}$; the arbitrariness of $\tilde{u}^r$ will be in the real as well as in the imaginary part, each of which is result of a homogeneous system of order $2n$.

It was important to increase the order so that orthogonality relations could again be established. In this case, in a similar proof to (1.16) and (1.17) one establishes it as:

$$\tilde{u}^{p^T}\tilde{M}\tilde{u}^q = \delta_{pq}, \quad p, q = 1,.,2n \quad \delta_{pq} = \begin{cases} 1 & p = q \text{ (normalization)} \\ 0 & p \neq q \text{ (orthogonality)} \end{cases} \tag{1.38}$$

If one looks for the structure of the modal matrix it can be observed that for a certain solution

$$x(t) = ue^{\lambda t}, \quad \dot{x}(t) = \lambda ue^{\lambda t}$$

$$y(t) = \tilde{u}e^{\lambda t} = \begin{Bmatrix} x(t) \\ \dot{x}(t) \end{Bmatrix} = \begin{Bmatrix} u \\ \lambda u \end{Bmatrix} e^{\lambda t}$$

and rearranging conveniently the complex and its complex conjugates

$$\tilde{U} = [\,\tilde{u}^1 \ \dots \ \tilde{u}^{2n}\,] = [\tilde{u}^1 \ \dots \ \tilde{u}^n \ \vdots \ \tilde{u}^{*^1} \ \dots \ \tilde{u}^{*^n}\,]$$

$$= \left[\begin{Bmatrix} u^1 \\ \lambda_1 u^1 \end{Bmatrix} \cdots \begin{Bmatrix} u^n \\ \lambda_n u^n \end{Bmatrix} \ \vdots \ \begin{Bmatrix} u^{*^1} \\ \lambda_1^* u^{*^1} \end{Bmatrix} \cdots \begin{Bmatrix} u^{*^n} \\ \lambda_n^* u^{*^n} \end{Bmatrix} \right]$$

$$= \begin{bmatrix} U & U^* \\ U\Lambda & U^*\Lambda^* \end{bmatrix} \tag{1.39}$$

where Λ is the diagonal matrix with the eigenvalues with positive imaginary parts and Λ^* its complex conjugate. Then the $(n \times 2n)$ modal matrix $[\,U \;\; U^*\,]$ is the same from the second order system of differential equations (1.34) as can be seen from (1.37),

$$\begin{bmatrix} C\lambda_r + K & M\lambda_r \\ M\lambda_r & -M \end{bmatrix} \begin{Bmatrix} u^r \\ \lambda_r u^r \end{Bmatrix} = \begin{Bmatrix} o \\ o \end{Bmatrix} \tag{1.40}$$

which reproduces the solution of (1.31).

The complex $n \times n$ matrix U may also be called as a modal matrix remembering that associated to it we have always its complex conjugate. What has changed is the orthogonality definition, which analogous to (1.21) and (1.24) may be written as

$$\tilde{U}^T \tilde{M} \tilde{U} = \tilde{I} \quad \text{and} \quad \tilde{U}^T \tilde{K} \tilde{U} = -\tilde{\Lambda}\,. \tag{1.41}$$

Using (1.39) and (1.35) these conditions may be rearranged to $(n \times n)$ matrices:

$$U^T C U + \Lambda U^T M U + U^T M U \Lambda = I,$$

$$U^T K U - \Lambda U^T M U \Lambda = -\Lambda$$

and

$$U^{*T} C U + \Lambda^* U^{*T} M U + U^{*T} M U \Lambda = o,$$

$$U^{*T} K U - \Lambda^* U^{*T} M U \Lambda = o. \tag{1.42}$$

They uncouple the original system of equations (1.35) with the transformation

$$y = \tilde{U} \varsigma \tag{1.43}$$

and result in equations that may be arranged like:

$$\dot{\varsigma}_r - \lambda_r \varsigma_r = 0 \quad \text{and} \quad \dot{\varsigma}_{n+r} - \lambda_r^* \varsigma_{n+r} = 0\,, \quad r = 1, \ldots, n \tag{1.44}$$

where the solution is defined by the initial condition vector:

$$\varsigma(0) = \tilde{U}^T \tilde{M} y(0), \tag{1.45}$$

and therefore $\varsigma_{n+r} = \varsigma_r^*$.

The modal motion is defined by this couple of equations. They are combined according to

$$y(t) = \tilde{U}\varsigma(t) \quad , \quad \varsigma^T(t) = [\,\varsigma_1(t) \ \dots \ \varsigma_n(t) \ \varsigma_1^*(t) \ \dots \ \varsigma_n^*(t)\,] \tag{1.46}$$

to produce the following modal superposition for the displacement

$$x(t) = \sum_{r=1}^{n}(A_r u^r e^{\lambda_r t} + A_r^* u^{*r} e^{\lambda_r^* t})$$

which is again equation (1.33).

Damped Eigenproperties

We are frequently asking, what is damping doing to the system? It must be analysed by means of the eigenmodes, that is, one should take the solution for one specific undamped mode from (1.11), like

$$x(t) = u^r \cos\omega_r t \,, \qquad \text{some } r \text{ between } 1,\dots,n \tag{1.47}$$

then include some damping through a convenient matrix C resulting in a change of the undamped properties (natural frequency and eigenmodes) to that obtained by (1.33)

$$\bar{x}(t) = e^{-\sigma_r t}[\bar{u}_R^r \cos\bar{\omega}_r t - \bar{u}_I^r \sin\bar{\omega}_r t]. \tag{1.48}$$

Firstly there is the damping itself: the exponential envelope will define after how much time the original level of vibration will be reduced to a percentage of it. Therefore (σ_r) defines a damping measure in the time scale: it is also called the degree of stability. But it may be more adequate to compare damping on a frequency based way of thinking: after how many periods will the motion be damped to, for instance, 10%. One may introduce a non-dimensional time $\tau = \omega_r t$, but where ω_r is the undamped frequency, for the purpose of keeping the damping only in ξ_r, the modal damping ratio. The exponential envelope is then $e^{-\xi_r \omega_r t}$, and the most important modes for the dynamic behaviour may be selected.

Also the damped frequency of vibration will differ from the undamped one. In multi degree-of-freedom systems this relation is not easy to write analytically. It is useful to represent graphically the eigenvalues - also called in the nomenclature of System Theory the poles of the system - in the complex plane. Then considerable understanding can be obtained observing how these poles move, with the change of parameters of the problem. This is called root-locus diagram. The poles can move with increasing damping in a way that the imaginary part of one pair of poles become zero, which may finally result in a non-vibrating behaviour of that specific mode. In the one degree-of-freedom case the relations are easy to establish, the frequency $\bar{\omega}$ will be smaller than ω for increasing ξ, as can be seen in any undergraduate text on vibrations, we have

$$\bar{\omega} = \omega\sqrt{1-\xi^2}\,, \quad \text{with } \omega = \sqrt{\frac{k}{m}}\,, \quad \text{and } \xi = \frac{c}{2\sqrt{km}}\,.$$

A good understanding of the simple one degree-of-freedom case is essential for the understanding of the modal behaviour.

One also sees that the mode is changed by the inclusion of damping from the real mode in the undamped case: a complex mode makes the simple geometrical understanding of the motion much more difficult. The oscillating part of the mode (1.48) may be transformed using

$$\bar{u}_j^r = \sqrt{\bar{u}_{R_j}^{r^2} + \bar{u}_{I_j}^{r^2}} \quad \text{and} \quad \alpha_j^r = \arctan\frac{\bar{u}_{I_j}^r}{\bar{u}_{R_j}^r}$$

resulting in a modal motion

$$\bar{x}(t) = e^{-\xi_r\omega_r t}\bar{u}^r\cos(\omega_r t + \alpha^r) \tag{1.49}$$

showing that the maximum modal amplitude in each coordinate ($\bar{u}_i^r$) is also not reached simultaneously. In general one still tries to help understanding with the undamped case.

M G K Systems

When investigating the dynamic behavior of the rotor of a hydraulic machine, gyroscopic effects may appear mainly due to the inertia of the electric machine which is much bigger than that of the turbine. When rotating smoothly a system has an angular momentum around the axis, which is equal to the product of the polar inertia with the angular velocity. Newton-Euler's law says that the time variation of this value is equal to the applied moment. Then, if there exists a moment and the angular momentum is not changing in value, it must change in position, beginning to turn in the sense given by the moment. This means that the coordinates that describe vibration in the presence of rotating motion, for instance the two orthogonal deflections of a shaft, are coupled, the moment in one coordinate produces a velocity in the other and vice-versa (see Figure 1.4). The equations that include gyroscopic effects must therefore be necessarily derived in the 3-dimensional space. In the case of no rotation they may be decomposed in two orthogonal directions, uncoupled and if it is an axi-symmetric problem both systems of equations may have the same properties. If there is rotation then for a linear system with no damping the equations of motion are written as:

$$\bar{M}\ddot{v} + \bar{G}\dot{v} + \bar{K}v = o \tag{1.50}$$

where v is displacement vector and $\bar{G}$ is the skew symmetric gyroscopic matrix, that contains the mentioned dependence on the polar inertia times the velocity of rotation.

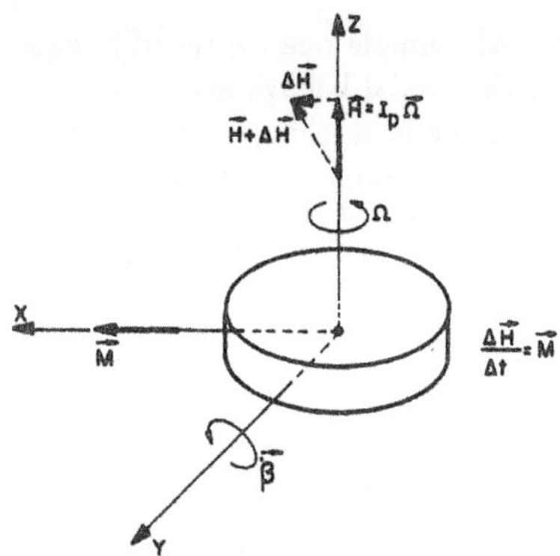

Figure 1.4 Coordinate coupling with rotating motion

Separating the vector v in coordinates (linear and angular, x related to the x-direction and y to the y-direction), a typical equation for the dynamic behaviour of an elastic shaft on elastic supports is:

$$\begin{bmatrix} M & o \\ o & M \end{bmatrix} \begin{Bmatrix} \ddot{x} \\ \ddot{y} \end{Bmatrix} + \begin{bmatrix} o & G \\ -G^T & o \end{bmatrix} \begin{Bmatrix} \dot{x} \\ \dot{y} \end{Bmatrix} + \begin{bmatrix} K_x & o \\ o & K_y \end{bmatrix} \begin{Bmatrix} x \\ y \end{Bmatrix} = \begin{Bmatrix} o \\ o \end{Bmatrix} \tag{1.51}$$

where M represents mass and inertia properties, which are independent from the direction, K_x and K_y represent the structural elastic properties of the shaft and its supports. This kind of equation is not valid for hydrodynamic journal bearings which also couple both directions, in the stiffness and damping matrices. The eigenvalue problem is then:

$$\begin{bmatrix} M\lambda^2 + K_x & G\lambda \\ -G^T\lambda & M\lambda^2 + K_y \end{bmatrix} \begin{Bmatrix} u_x \\ u_y \end{Bmatrix} = \begin{Bmatrix} o \\ o \end{Bmatrix}. \tag{1.52}$$

One may observe that the characteristic equation will only have even exponents for λ and, therefore, there are only pure imaginary eigenvalues, defining the frequencies of vibration. If n is the number of degrees of freedom ($n/2$ for each direction), then there are n frequencies of vibration, depending on the velocity of rotation. It is usual to represent this dependence in a diagram. With increasing rotation speed Ω the smallest frequency is normally tending to zero and the greatest going along an asymptote to infinite. Substituting this eigenvalues back in (1.52) to obtain the modes one sees that these are complex: for $\lambda_r = i\omega_r$ we will have an $\boldsymbol{u}^r$ and for $\lambda_r^* = -i\omega_r$ we will have $\boldsymbol{u}^{*^r}$. This result may be analysed as in the damped system case, resulting in a modal motion in analogy to (1.49)

$$v(t) = \left\{ \begin{array}{c} \bar{\boldsymbol{u}}_x^r \cos(\omega_r t + \alpha) \\ \bar{\boldsymbol{u}}_y^r \cos(\omega_r t + \alpha) \end{array} \right\} \tag{1.53}$$

but in this case corresponding coordinates that measure shaft deflections will define a point in space, with which we can define the modal motion of the shaft. Two fundamental possibilities may be distinguished: the resulting rotating motion of the center of the shaft is in the same sense of the velocity of rotation or it can also be in the opposite sense, resulting in the so called direct or backward whirl respectively.

The matrix in (1.52) is skew symmetric and therefore the orthogonality conditions cannot be derived as before. In this case the right eigenvectors (as defined) and the left eigenvectors (of the transposed system) are required in order to obtain an orthogonality condition.

M D L Systems

There are mathematical models of mechanical systems where, besides the symmetric mass matrix $\boldsymbol{M}$, there are general matrices for the forces depending on the velocities $\boldsymbol{D}\dot{\boldsymbol{x}}$ and on the displacements $\boldsymbol{L}\boldsymbol{x}$. That is the case when one analyses the behaviour of a rotor on journal bearings, for instance, also including the action of seals on the shaft. As any general matrix can be divided in a symmetric and a skew-symmetric part, it is possible to do it in the equation for a better understanding of the phenomena:

$$\boldsymbol{M}\ddot{\boldsymbol{x}}(t) + (\boldsymbol{C} + \boldsymbol{G})\dot{\boldsymbol{x}}(t) + (\boldsymbol{K} + \boldsymbol{N})\boldsymbol{x}(t) = \boldsymbol{o} \ . \tag{1.54}$$

Here $\boldsymbol{K}$ is symmetric and conservative, $\boldsymbol{C}$ is symmetric and corresponds to a viscous damping action. The eigenbehaviour will be found by looking for a solution of the type (1.5) and then defining the eigenvalue problem and the characteristic equation. It is very important in this general case to look at the real part of the eigenvalues and their dependence on the parameters of the systems. If some real part is positive then the system (through the corresponding eigenmodes) has an unstable behaviour.

Example

The association between a physical model and the numerical results through the developed mathematical tools will be shown for some strongly simplified situations. To begin with, consider the rigid body motion of the typical structure of a hydraulic unit as is shown in Figure 1.5 (a). The non-rotating system will have two degree-of-freedom, assuming that there is no motion along the shaft axis, i.e. in the vertical direction. This simplification means that the hydraulic lift force will act on an infinitely stiff support, or that the vertical vibration excited from this lift force on the machine supported by the thrust bearing, can be handled uncoupled from the transverse vibrations. The two coordinates that are necessary to describe this transverse motion can be chosen related to the centre of mass C of the unit, like the displacement x_C and the angle of rotation θ. This may be convenient analytically but disadvantageous experimentally, because x_G and x_T are immediately accessible measurements.

There will also be an approximation here because it will be difficult to get the shaft displacements in the bearings in an inertial way. Otherwise, to use relative shaft/bearing coordinates demands a more detailed model. For the equations of motion it is necesssary to know the whole mass of the unit, the transverse moment of inertia of the unit in rèspect to the center of mass C, the equivalent stiffness at the generator (k_G) and at the turbine (k_T) as well as the position of the supports and the center of mass (l_G, l_T, $l = l_G + l_T$). Damping c will be neglected at thè moment.

Newton's law for translation equates the mass, times the acceleration of the center of mass C to the elastic restoring forces:

$$m\ddot{x}_C + k_G x_G + k_T x_T = 0.$$

Newton-Euler's law of planar rotation equates the moment of inertia, times the angular acceleration to the moments of the restoring forces in respect to the centre of mass:

$$I\ddot{\theta} + k_G x_G l_G - k_T x_T l_T = 0.$$

The relation between the coordinates given by:

$$x_G = x_C + l_G\theta, \;\; x_T = x_C - l_T\theta, \;\; x_C = \frac{l_T}{l}x_G + \frac{l_G}{l}x_T, \;\; \theta = \frac{1}{l}(x_G - x_T)$$

allows the formulation of the equations of motion in both of the mentioned alternatives. By substitution:

$$\begin{bmatrix} m & 0 \\ 0 & I \end{bmatrix} \begin{Bmatrix} \ddot{x}_C \\ \ddot{\theta} \end{Bmatrix} + \begin{bmatrix} k_G + k_T & k_G l_G - k_T l_T \\ k_G l_G - k_T l_T & k_G l_G^2 + k_T l_T^2 \end{bmatrix} \begin{Bmatrix} x_C \\ \theta \end{Bmatrix} = \begin{Bmatrix} 0 \\ 0 \end{Bmatrix}$$

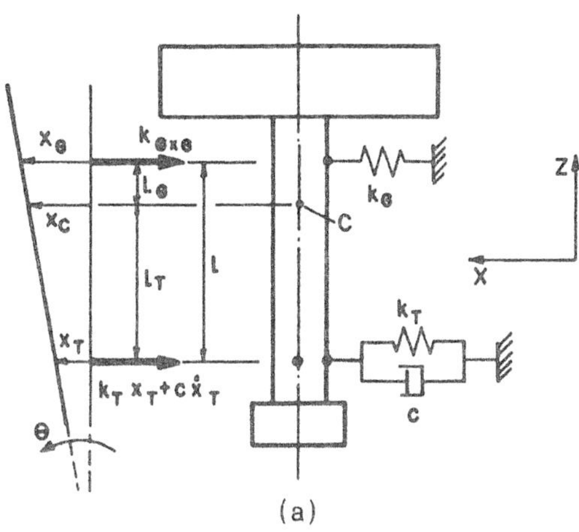

(a)

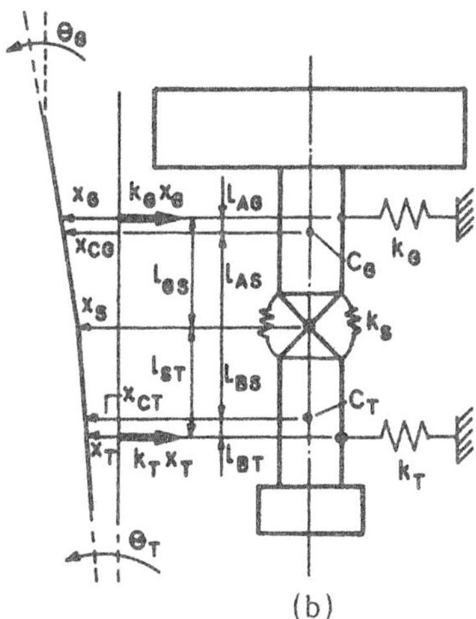

(b)

Figure 1.5 Models of hydraulic units

or

$$\begin{bmatrix} (I+ml_T^2)/l^2 & -(I-ml_Tl_G)/l^2 \\ -(I-ml_Tl_G)/l^2 & (I+ml_G^2)/l^2 \end{bmatrix} \begin{Bmatrix} \ddot{x}_G \\ \ddot{x}_T \end{Bmatrix} + \begin{bmatrix} k_G & 0 \\ 0 & k_T \end{bmatrix} \begin{Bmatrix} x_G \\ x_T \end{Bmatrix} = \begin{Bmatrix} 0 \\ 0 \end{Bmatrix}.$$

Observe that the transformation matrix $\boldsymbol{T}$

$$\bar{\boldsymbol{x}} = \begin{Bmatrix} x_C \\ \theta \end{Bmatrix} \ , \ \boldsymbol{x} = \begin{Bmatrix} x_G \\ x_T \end{Bmatrix} \ , \ \bar{\boldsymbol{x}} = \boldsymbol{T}\boldsymbol{x} \ , \ \bar{\boldsymbol{T}} = \begin{bmatrix} l_T/l & l_G/l \\ 1/l & -1/l \end{bmatrix}$$

simply substituted in the first equation will result in

$$\begin{bmatrix} ml_T/l & ml_G/l \\ I/l & -I/l \end{bmatrix} \begin{Bmatrix} \ddot{x}_G \\ \ddot{x}_T \end{Bmatrix} + \begin{bmatrix} k_G & k_T \\ k_Gl_G & -k_Tl_T \end{bmatrix} \begin{Bmatrix} x_G \\ x_T \end{Bmatrix} = \begin{Bmatrix} 0 \\ 0 \end{Bmatrix}$$

where both matrices are asymmetrical; only the pre-multiplication with $\boldsymbol{T}^T$ will bring symmetry again. If only numbers are being considered this shows that a symmetrical system may be hidden in very general-looking equations: always symmetrize (or diagonalize) the mass matrix before any further operation.

The integrated form of (1.3) defines kinetic and potential energy; the first one is easy to write with center of mass coordinates

$$E_k = \frac{1}{2}\dot{\boldsymbol{x}}^T\boldsymbol{M}\dot{\boldsymbol{x}} = \frac{1}{2}\begin{Bmatrix} \dot{x}_c & \dot{\theta} \end{Bmatrix} \begin{bmatrix} m & 0 \\ 0 & I \end{bmatrix} \begin{Bmatrix} \dot{x}_c \\ \dot{\theta} \end{Bmatrix} = \frac{1}{2}m\dot{x}_c^2 + \frac{1}{2}I\dot{\theta}^2$$

and the second one with bearing coordinates, where analogously

$$E_p = \frac{1}{2}k_Gx_G^2 + \frac{1}{2}k_Tx_T^2.$$

The numerical example will be performed with $m = 6.73\,\mathrm{E5\,kg}$, $I = 1.48\,\mathrm{E7\,kg.m^2}$, $l_G = 0.4\,\mathrm{m}$, $l_T = 3.6\,\mathrm{m}$, $l = 4.0\,\mathrm{m}$, $k_G = 3\,\mathrm{E9\,N/m}$, $k_T = 1.2\,\mathrm{E9\,N/m}$; giving

$$\boldsymbol{M} = \begin{bmatrix} 14.7013 & -8.6443 \\ -8.6443 & 9.3173 \end{bmatrix} \ , \ \boldsymbol{K} = \begin{bmatrix} 30000 & 0 \\ 0 & 12000 \end{bmatrix}.$$

A standard routine for mathematical manipulation like MATLAB will give

$$\boldsymbol{\Lambda} = \begin{bmatrix} 30.0040i & 0 \\ 0 & 80.1481i \end{bmatrix} \ , \ \boldsymbol{U} = \begin{bmatrix} -0.4642 & 0.8617 \\ 1.0 & 1.0 \end{bmatrix}.$$

The two eigenfrequencies of 4.78 Hz $(30/(2\pi))$ and 12.76 Hz characterize the motion of the system in its eigenmodes. In the lower one there is a nodal point inbetween the two coordinates, therefore it is mainly the rotational

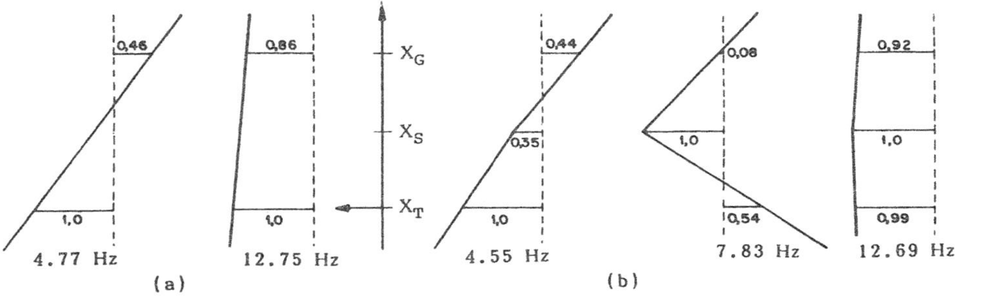

Figure 1.6 Models of the 2 and 3 degree-of-freedom models, showing the displacements at points G,S and T of Figure 1.5

mode, whereas the higher frequency corresponds to an essentially translatory mode. They are represented in Figure 1.6 (a).

The normalization of the eigenmodes may be checked for the generalized mass (1.19) and then the vectors may be re-normalized orthonormally:

$$\mathcal{M} = \begin{bmatrix} 20.5097 & 0 \\ 0 & 5.3362 \end{bmatrix}, \quad U_n = \begin{bmatrix} -0.1025 & 0.3730 \\ 0.2208 & 0.4329 \end{bmatrix}.$$

The matrix U_n satisfies (1.21) and (1.24) and can be used to uncouple the system following the orthogonality properties. Any motion derived from initial conditions is calculated through:

$$\left\{ \begin{array}{c} x_G(t) \\ x_T(t) \end{array} \right\} = C_1 \left\{ \begin{array}{c} -0.4642 \\ 1.0 \end{array} \right\} \cos(30t - \varphi_1) + C_2 \left\{ \begin{array}{c} 0.8617 \\ 1.0 \end{array} \right\} \cos(80.15t - \varphi_2),$$

where the former normalization of U was preferred.

Still looking for the undamped eigensolution, it is possible to include in the former model an approach for an elastic shaft through a stiffness k_S concentrated at a chosen point of the shaft, as in Figure 1.5 (b). Then, one degree-of-freedom more is included, keeping the restriction on the vertical motion: for instance if we need two coordinates to locate the turbine part, then the generator part has its position defined by one further coordinate. For instance, x_G, x_S, x_T may be chosen to describe the motion. Using this time Lagrange's Equation kinetic (E_k) and potential (E_p) energy expressions will be developed. It is necessary to know the position of the articulation (l_{GS}, l_{ST}, $l = l_{GS} + l_{ST}$), the mass, the position of the centres of mass, the transverse inertia in relation to this point of the upper (m_G, l_{AS}, I_G) and the lower part (m_T, l_{BS}, I_T) respectively, as well as the stiffnesses (k_G, k_T, k_S) and the position of the supports (l_{AG}, l_{BT}).

$$E_p = \frac{1}{2}k_G x_G^2 + \frac{1}{2}k_T x_T^2 + \frac{1}{2}k_S(\theta_G - \theta_T)^2,$$

$$E_k = \frac{1}{2}m_G \dot{x}_A^2 + \frac{1}{2}I_G \dot{\theta}_G^2 + \frac{1}{2}m_T \dot{x}_B^2 + \frac{1}{2}I_T \dot{\theta}_T^2.$$

Transforming the coordinates and making use of Lagrange's Equation for the displacement vector, $\boldsymbol{x} = \{x_G\ x_S\ x_T\}^T$, results in the following:

$$\boldsymbol{M} = \begin{bmatrix} m_{11} & m_{12} & m_{13} \\ m_{21} & m_{22} & m_{23} \\ m_{31} & m_{23} & m_{33} \end{bmatrix}, \text{and } \boldsymbol{K} = \begin{bmatrix} k_{11} & k_{12} & k_{13} \\ k_{21} & k_{22} & k_{23} \\ k_{31} & k_{32} & k_{33} \end{bmatrix}$$

where

$$
\begin{aligned}
m_{11} &= (I_G + m_G l_{AS}^2)/l_{GS}^2 \\
m_{12} &= -(I_G - m_G l_{AG} l_{AS})/l_{GS}^2 \\
m_{13} &= 0 \\
m_{21} &= -(I_G - m_G l_{AG} l_{AS})/l_{GS}^2 \\
m_{22} &= (I_G + m_G l_{AG}^2)/l_{GS}^2 + ({}_T + m_T l_{BT}^2)/l_{ST}^2 \\
m_{23} &= -(I_T - m_T l_{BS} l_{BT})/l_{ST}^2 \\
m_{31} &= 0 \\
m_{32} &= -(I_T - m_T l_{BS} l_{BT})/l_{ST}^2 \\
m_{33} &= (I_T - m_T l_{BS}^2)/l_{ST}^2
\end{aligned}
$$

and

$$
\begin{aligned}
k_{11} &= k_G + k_S/l_{GS}^2 \\
k_{12} &= -(1/l_{GS} + 1/l_{ST})k_S/l_{GS} \\
k_{13} &= k_S/(l_{GS} l_{ST}) \\
k_{21} &= -(1/l_{GS} + 1/l_{ST})k_S/l_{GS} \\
k_{22} &= (1/l_{GS} + 1/l_{ST})^2 k_S \\
k_{23} &= -(1/l_{GS} + 1/l_{ST})k_S/l_{ST} \\
k_{31} &= k_S/(l_{GS} l_{ST}) \\
k_{32} &= -(1/l_{GS} + 1/l_{ST})k_S/l_{ST} \\
k_{33} &= k_T + k_S/l_{ST}^2
\end{aligned}
$$

Using $m_G = 5.0475\,\mathrm{E5\,kg}$, $m_T = 1.6825\,\mathrm{E5\,kg}$, $I_G = 80.029\,\mathrm{E5\,kgm^2}$, $I_T = 57.372\,\mathrm{E5\,kgm^2}$, $l_{GS} = l_{ST} = l_{BS} = 2\,\mathrm{m}$, $l_{AS} = 2.5\,\mathrm{m}$, $l_{AG} = -0.5\,\mathrm{m}$, $l_{BT} = 0$ then

$$
\boldsymbol{M} = \begin{bmatrix} 27.8939 & -21.5845 & 0 \\ -21.5845 & 34.6655 & -14.3429 \\ 0 & -14.3429 & 16.0254 \end{bmatrix}
$$

$$
\boldsymbol{K} = \begin{bmatrix} 50000 & -40000 & 20000 \\ -40000 & 80000 & -40000 \\ 20000 & -40000 & 32000 \end{bmatrix}
$$

and again, using MATLAB for instance,

$$
\boldsymbol{\Lambda} = \begin{bmatrix} 28.5790i & 0 & 0 \\ 0 & 49.1862i & 0 \\ 0 & 0 & 79.7534i \end{bmatrix}
$$

$$U = \begin{bmatrix} -0.4424 & 0.0827 & 0.9198 \\ 0.3558 & 1 & 1 \\ 1 & -0.5387 & 0.9956 \end{bmatrix}.$$

The modes are represented together with the two degree-of-freedom model in Figure 1.6 showing how the inclusion of a new degree-of-freedom, due mainly to the elasticity of the shaft (as seen from the 7.83 Hz mode), changed the two existing modes, also slightly moving the eigenfrequencies.

Now, what happens when damping is considered? Imagine again the two degree-of-freedom model and that at the turbine support the damping of the vibration in the water is included; any other structural damping is neglected. Imagine also a numerical value of 1.0 E7 Ns/m for this damping, giving

$$C = \begin{bmatrix} 0 & 0 \\ 0 & 100 \end{bmatrix}$$

in the coordinates $\{ x_G \;\; x_T \}^T$. In spite of being semi-definite the damping is pervasive as will be shown. It is also non-proportional and therefore the generalized orthogonality condition must be considered. Again, the determination of eigenvalues and eigenvectors give:

$$\Lambda = \begin{bmatrix} -2.460 + 30.158i & 0 \\ 0 & -9.348 + 78.924i \end{bmatrix}$$

$$U = \begin{bmatrix} -0.4495 - 0.1353i & 0.8470 - 0.0932i \\ 1 & 1 \end{bmatrix}.$$

The generalized mass matrix is obtained, for this vectors of arbitrary length, calculating the left side of the first of equations (1.42), resulting in

$$\mathcal{M} = \begin{bmatrix} -0.2464 + 1.1732i & 0 \\ 0 & 0.1168 + 0.8172i \end{bmatrix} \text{E}3.$$

The normalization following (1.20) will give

$$U_n = \begin{bmatrix} -0.0112 + 0.0076i & 0.0201 - 0.0218i \\ 0.0182 - 0.0224i & 0.0263 - 0.0228i \end{bmatrix},$$

a matrix that satisfies all four equations of (1.42). The general solution in time of the homogeneous equation can again be written with any of the U's:

$$\left\{ \begin{matrix} x_G(t) \\ x_T(t) \end{matrix} \right\} = C_1 e^{-2.46t} \left[\left\{ \begin{matrix} -0.4495 \\ 1.0 \end{matrix} \right\} \cos(30.16t - \varphi_1) - \left\{ \begin{matrix} 0.1353 \\ 0 \end{matrix} \right\} \sin(30.16t - \varphi_1) \right] +$$

$$+C_2 e^{-9.35t}\left[\left\{\begin{array}{c}0.8470\\1.0\end{array}\right\}\cos(78.92t-\varphi_2)-\left\{\begin{array}{c}0.0932\\0\end{array}\right\}\sin(78.92t-\varphi_2)\right].$$

To sense what happened to the first mode compared to the undamped case, both solutions are compared with $C_2 = 0$ and $C_1 = 1$. It is observed that the first natural frequency increased a little bit and that a frozen position may be compared, looking at the equations written in the form:

$$x_G(t) = -0.4642\cos(30t); \quad x_{G_d}(t) = e^{-2.46t}0.4694\cos(30.16t - 16.75^o);$$

$$x_T(t) = 1.0\cos(30t)\,; \quad x_{T_d}(t) = e^{-2.46t}1.0\cos(30.16t)\,.$$

Therefore, the maximum amplitude is not reached simultaneously in both coordinates in the damped case and Figure 1.7 shows that as a consequence the node point from the first mode is *travelling*.

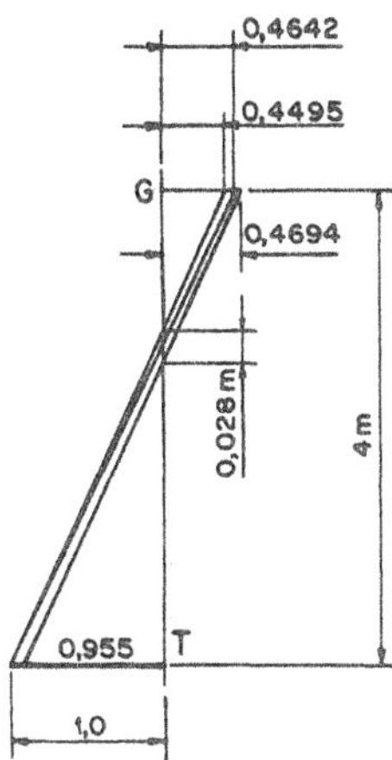

Figure 1.7 Mode of a damped system

Otherwise in the Pseudo Modal method, that will be presented in connection with excited systems (see 1.2.2), we have to take the orthonormal modes of the undamped case and make the coordinate transformation, diagonalizing $\boldsymbol{M}$ and $\boldsymbol{K}$ but not $\boldsymbol{C}$:

$$\begin{bmatrix}1 & 0\\0 & 1\end{bmatrix}\ddot{\eta}+\begin{bmatrix}4.88 & 9.56\\9.56 & 18.74\end{bmatrix}\dot{\eta}+\begin{bmatrix}900.24 & 0\\0 & 6423.71\end{bmatrix}\eta=\mathbf{o}\,.$$

As there is no order reduction this is still the same system with the same solutions. The idea of the method is to make further calculations with these small order system coupled only in the damping. Looking at the damping capacity in both modes and analysing the envelope, it is observed that to reduce vibrations to 1/10 of its original value it will take 0.93 s ($\Delta t = \ln 0.1/(-2.46)$) in the first mode and 0.25 s in the second mode. On the frequency side looking at how many cycles the system will vibrate in these times, results in 4.5 cycles at the lower frequency and 3.2 cycles at the higher frequency. The modal damping coefficients $\xi_1 = 0.082$ and $\xi_2 = 0.116$ are not very distinct, giving that result corresponding to a ***typical system***. If damping was induced through special bearings, seals, magnetic effects, etc. then some points of the approximations that were made have to be checked more carefully.

The inclusion of rotation in the undamped example will be made through a simple extension: imagine that the system has axi-symmetrical properties, i.e. that the y-direction orthogonal to the x-direction have both the same stiffness properties and that the shaft, as a vibrating system, is perfectly symmetric. Vibration in both directions is then coupled only through the gyroscopic effects. Figure 1.8 shows the shaft and the forces in space. The motion of the centre of mass C is in a plane and therefore Newton's equations are the same as before, but written for both directions. The rotation will be defined through Newton-Euler's equation in space, giving for the moments in respect to axes x and y (M_x , M_y):

$$M_x = I\ddot{\theta}_x + I_p\Omega\dot{\theta}_y$$

$$M_y = I\ddot{\theta}_y - I_p\Omega\dot{\theta}_x \, .$$

Keeping the same matrices $\boldsymbol{M}$ and $\boldsymbol{K}$ from the two degree-of-freedom undamped system, the equation (1.50) is written in the notation (1.51) with $\boldsymbol{v} = \{\, x_c \;\; \theta_y \;\; y_c \;\; \theta_x \,\}^T$ and

$$\boldsymbol{G} = \begin{bmatrix} 0 & 0 \\ 0 & -I_p\Omega \end{bmatrix} .$$

The change of coordinates to $\boldsymbol{v}_1 = \{\, x_G \;\; x_T \;\; y_G \;\; y_T \,\}^T$ will be made through

$$\boldsymbol{v} = \begin{bmatrix} \boldsymbol{T} & \mathbf{o} \\ \mathbf{o} & \boldsymbol{T} \end{bmatrix} \boldsymbol{v}_1$$

where $\boldsymbol{T}$ is known from the plane problem. Then

$$\boldsymbol{G} = \frac{I_p\Omega}{l^2} \begin{bmatrix} -1 & 1 \\ 1 & -1 \end{bmatrix} .$$

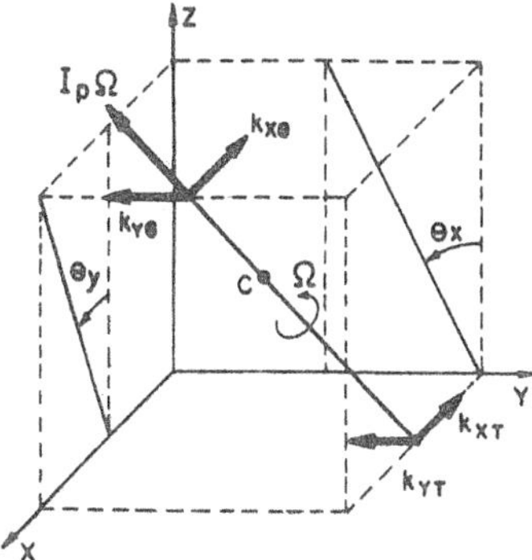

Figure 1.8 Position in space for the rotating shaft

Numerical solution will be looked for with $I_p = 1.4\,\mathrm{E7}\,\mathrm{kg.m^2}$ and $\Omega = 1.82\,\mathrm{Hz}$ ($I_p\Omega/l^2 = 1.0\mathrm{E7}\ \mathrm{kg.s^{-1}}$). Using again MATLAB for the numerical solution of the undamped system we get

$$\Lambda = \begin{bmatrix} 25.227i & & & \\ & 35.6670i & & \\ & & 79.9928i & \\ & & & 80.3593i \end{bmatrix}$$

showing that the gyroscopic effect splits the eigenfrequencies of the non-rotating system ($30\,\mathrm{s^{-1}}$ and $80.15\ \mathrm{s^{-1}}$) and also that this effect is more accentuated in the lower frequency. Looking at the non-rotating modes it is observed that the first one implies stronger angular motion of the huge generator inertia, justifying therefore what happens.

The corresponding modal matrix for the defined coordinates is:

$$U = \begin{bmatrix} -0.4426i & 0.4998i & 0.8785i & -0.8398i \\ i & -i & i & -i \\ -0.4426 & -0.4998 & 0.8785 & 0.8398 \\ 1 & 1 & 1 & 1 \end{bmatrix},$$

remembering that there is also its complex conjugate associated to it. Therefore x and y differ from 90°, i.e. whereas one is a pure imaginary number the other one is real. The symmetry of the structure is expressed by the fact that both values are equal. To analyse the motion for the two lower eigenvalues it is better to look at the modal motion representing it in time as in (1.53). Then:

$$\lambda_1 = 25.2227i \quad , \quad \boldsymbol{v}^1 = \left\{ \begin{array}{l} +0.4426\cos(25.2t + 90^o) \\ -1\cos(25.2t + 90^o) \\ -0.4426\cos(25.2t) \\ +1\cos(25.2t) \end{array} \right\}$$

and

$$\lambda_2 = 35.6670i \quad , \quad \boldsymbol{v}^2 = \left\{ \begin{array}{l} -0.4998\cos(35.7t + 90^o) \\ +1\cos(35.7t + 90^o) \\ -0.4998\cos(35.7t) \\ +1\cos(35.7t) \end{array} \right\}.$$

By looking carefully how the motion occurs as shown in Figure 1.9, it can be seen that for the first mode the motion is in the opposite sense of the rotation of the shaft: it is the so called backward whirl; whilst in the second mode it is in the same sense, called therefore forward whirl. Also it may be observed that in spite of the gyroscopic effects and the strong opening of the eigenvalues for increasing Ω, the modes resemble very much the modes of the non- rotating system. They are also normally used in place of the exact ones.

The inclusion of damping in the same manner as before gives eigenvalues with negative real parts and complex numbers as eigenvectors. The influence of damping and gyroscopic forces can be evaluated from

$$\boldsymbol{\Lambda} = \begin{bmatrix} \Lambda_1 & & & \\ & \Lambda_2 & & \\ & & \Lambda_3 & \\ & & & \Lambda_4 \end{bmatrix},$$

where

$$\begin{aligned} \Lambda_1 &= -2.1042 + 25.2870i \\ \Lambda_2 &= -2.7027 + 35.9581i \\ \Lambda_3 &= -9.0867 + 78.8707i \\ \Lambda_4 &= -9.7040 + 79.0104i\,. \end{aligned}$$

The real part is influenced by Ω, the backward whirl is showing in this case a decrease in damping and the forward whirl an increase.

1.2.2 Excited Dynamics of Linear Systems

Externally applied forces on a mechanical system will be in equilibrium with its inertia forces, gyroscopic forces and the already considered stiffness and

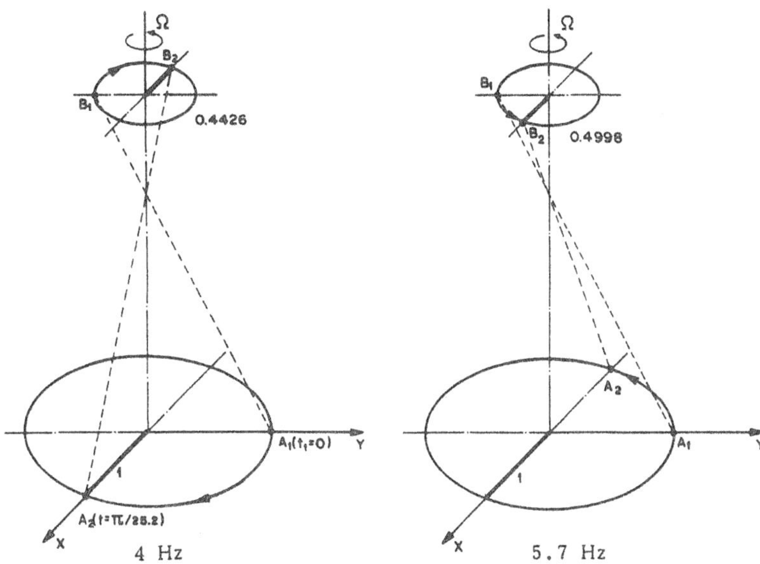

Figure 1.9 Modal motion in space of the center of shaft

damping forces. They will be written on the right hand side of the equation as a force $\boldsymbol{f}$ as in (1.1). Then the differential equation is not longer homogeneous and the solution will be a superposition of a particular solution for the excitation $\boldsymbol{f}$ with the homogeneous solution (including the constants to be defined through the initial conditions). As we are in general concerned with damped systems, even if the damping is only slight, after some time the homogeneous solution dies out, leaving only the excited behaviour.

Several kinds of excitations are common in hydraulic machinery. Transients will be caused by force pulses or impacts. They may be analysed considering the momentum that is transferred to the structure; this transforms the force pulse in a velocity initial condition and the solution is obtained through the analysis made above. Therefore these transients, which are frequent in the partial load operating condition of Francis turbines for instance, may be used to know something more regarding the eigendynamics of the machine.

Furthermore there are several periodic forces, in the frequency of rotation or multiple of it. It is possible to look for a general solution to the force

$f_o \cos(\Omega t)$ introducing it in the equation (1.2)

$$M\ddot{x} + Kx = f_o \cos(\Omega t). \tag{1.55}$$

The permanent solution will have the same structure of the force

$$x(t) = x_o \cos(\Omega t) \tag{1.56}$$

which, substituted in the equation, gives the amplitude of the resulting motion:

$$x_o = (-M\Omega^2 + K)^{-1} f_o. \tag{1.57}$$

The eigenvalues of the system (1.8) are known; therefore it is easy to conclude that if the frequency of excitation coincides with one of the eigenfrequencies, then the determinant (1.7) is zero and the amplitude in (1.57) goes to infinity. This is called a situation of resonance. It is interesting to observe that for $\Omega \to 0$ equation (1.57) reproduces the static deformation under the force ($x_o = K^{-1} f_o$) and that with increasing Ω the inertia effects become dominant.

Otherwise with increasing Ω there is also an increase of the amount of energy of the external force and due to the limited power supply of the usual phenomena f_o is consequently diminishing. This allows the restriction of the analysis to some frequency band where mechanical vibrations are important and can occur for a certain machine.

Due to the positive definiteness of M and K the resonance represents a point of change of signal for x_o. It is usual to keep the amplitude x_o always positive, then

$$x(t) = |(-M\Omega^2 + K)^{-1}| f_o \cos(\Omega t - \varphi) \tag{1.58}$$

where the phase $\varphi = 0$ up to the resonance and $\varphi = \pi$ afterwards.

In the analysis of the damped case the direct solution for the same excitation must be considered in the complex form:

$$f = f_o\left(\frac{1}{2}e^{i\Omega t} + \frac{1}{2}e^{-i\Omega t}\right), \quad x = \begin{cases} x_o \frac{1}{2} e^{i\Omega t} + x_o^* \frac{1}{2} e^{-i\Omega t} \\ x_{o_R} \cos \Omega t + x_{o_I} \sin \Omega t \\ \bar{x}_0 \cos(\Omega t - \varphi) \end{cases} \tag{1.59}$$

Therefore it is enough to solve

$$M\ddot{x} + C\dot{x} + Kx = f_o e^{i\Omega t}$$

through

$$\begin{bmatrix} -\Omega^2 M + K & -C\Omega \\ C\Omega & -\Omega^2 M + K \end{bmatrix} \begin{Bmatrix} x_{O_R} \\ x_{O_I} \end{Bmatrix} = \begin{Bmatrix} f_o \\ o \end{Bmatrix}. \tag{1.60}$$

The numerical solution may show several difficulties mainly in the neighbourhood of the resonances. In the investigation of physical background, analytical results may be obtained in the one degree-of-freedom case showing the limitation of amplitudes at the resonances and the phase change at these points, which is not so abrupt as in the undamped case. In the experimentation the change in 180° of the phase between applied force and consequent displacement is a good method to recognize the resonances, when one does a frequency sweep. The resonance situation itself will correspond to a phase of 90°.

A very particular excitation of rotating mechanical systems is originated in the unbalance: it is a force that is actuating always in the frequency of rotation and is proportional to the square of this velocity. Rotating unbalanced systems with damping may be investigated exactly like the former one, but due to the gyroscopic effects one must substitute the viscous damping matrix $\boldsymbol{C}$ by the general matrix $\boldsymbol{D}$. The eigenfrequencies of the system are changing with rotation Ω: at some rotation, if there is an impulsive action the response will show these eigenfrequencies and the frequency of excitation (Ω). There is a point where the frequency of excitation will be equal to some eigenfrequency: in this case a critical velocity is obtained for Ω. The diagram that represents eigenfrequencies and the excitation frequencies times the velocity of rotation, is known as Campbell's diagram. An example is presented in Figure 1.10.

Another kind of force shows strong stochastic character and is called noise. Looking for the frequency spectrum of this force usually produces a distribution over a range of frequencies. A theoretical approach to this force considers a linear, constant spectrum until infinite frequency and is called white noise. Full solutions need the theory of stochastic vibration, but some important points may be found out with the tools presented in this introduction. If there are eigenfrequencies in the range of frequency of the noise, then they will be resonant and present in the frequency analysis of the response. How strong this resonance will be, depends on damping and on the amount of energy the noise is able to inject into this eigenmode. Figure 1.11 shows how the resonance in the power spectral density (when modes are well separated) of a Francis turbine, excited by the stochastic forces of the partial load condition, may be integrated to give displacement:

$$x = \left(\int_{f_1}^{f_2} G_d(f_i)\, df_i \right)^{1/2}, \quad f_1 \leq f_i \leq f_2 \tag{1.61}$$

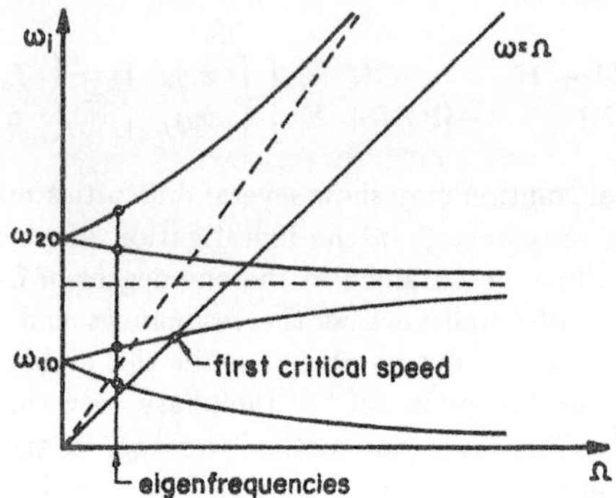

Figure 1.10 Dependence of eigenfrequencies with the speed of rotation of an axi-symmetric rotating system

where f are frequencies, G_d the represented power spectral density function of the displacement and x is the RMS displacement at the point of measurement.

Modal Approach

The orthogonality of the eigenmodes allows a general solution for any kind of excitation. This is the most usual way of treating the vibration problem of a mechanical system. Excited **MK** and $\mathbf{MC_pK}$ (where $\mathbf{C}_p$ is proportional damping) systems may be diagonalized from equation (1.1) changing to variable $\boldsymbol{\eta}$ (1.25), and pre-multiplying the equation by the transposed modal matrix $\boldsymbol{U}^T$ (1.20). Then if $\boldsymbol{C} = \alpha \boldsymbol{M} + \beta \boldsymbol{K}$, considering the orthogonality relation and a unitary normalization of the resulting mass matrix, one gets:

$$\ddot{\eta}_r + (\alpha + \beta\omega_r^2)\dot{\eta}_r + \omega_r^2\eta_r = \boldsymbol{u}^{r^T}\boldsymbol{f} \ , \quad r = 1, \ldots, n. \tag{1.62}$$

These n uncoupled second order equations may be solved with the basic theory of one degree-of-freedom systems. Afterwards it is again converted to the original coordinates with $\boldsymbol{x} = \boldsymbol{U\eta}$.

Even with non-proportional viscous damping it is possible to obtain a second order solution using the pseudo-modal method, see Lalanne, 1983 [1.14], Chapter 3. Here $\boldsymbol{U}$ is the modal matrix of the undamped system, n is very big because the equations may have been derived through the finite element method and $\boldsymbol{U}$ is used to reduce the system to the important modes. Therefore the modal matrix contains only the modes considered important

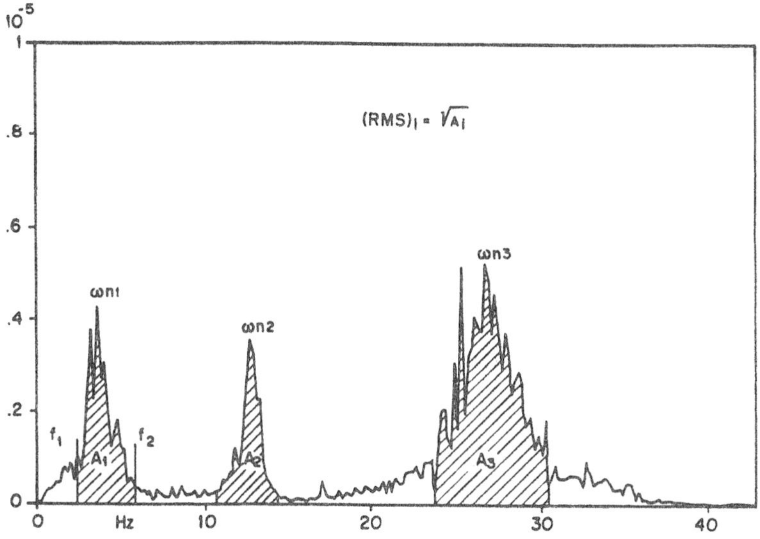

Figure 1.11 Power spectral density of the relative shaft bearing displacement of a Francis turbine in partial-load operation

for the analysis and is a $n \times m$ matrix, with $m << n$. Then the system will be reduced from order n, to order m ($\boldsymbol{U}^T \boldsymbol{M} \boldsymbol{U} \eta$, for instance), $\boldsymbol{C}$ will not be uncoupled but this will not cause problems because of the low order of the resulting system. In this case the equation will be

$$\ddot{\eta} + \boldsymbol{U}^T \boldsymbol{C} \boldsymbol{U} \dot{\eta} + \omega^2 \eta = \boldsymbol{U}^T \boldsymbol{f}. \tag{1.63}$$

Returning to the original displacements $\boldsymbol{x}$, it is shown that the method only works if all the important modes are actually obtained, but the conclusions are limited to a certain frequency range. Clearly one may do also the same including the gyroscopic matrix $\boldsymbol{G}$.

If all equations shall be uncoupled then again the formulation of (1.35) must be used, but one may also work directly with the second order equation (1.1) for **MCK** systems changing to the variable

$$\boldsymbol{x} = \boldsymbol{U} \varsigma + \boldsymbol{U}^* \varsigma^*, \tag{1.64}$$

$\boldsymbol{U}$ being the modal matrix defined in (1.39). Multiplying the equation with

U^T and using the orthogonality conditions (1.42) one gets instead of (1.62) couples of complex equations:

$$\left.\begin{array}{l}\dot{\varsigma}_r - \lambda_r \varsigma_r = \boldsymbol{u}^{r^T} \boldsymbol{f} \\ \dot{\varsigma}_{n+r} - \lambda_r^* \varsigma_{n+r} = \boldsymbol{u}^{*^{r^T}} \boldsymbol{f}\end{array}\right\} r = 1, \ldots, n. \tag{1.65}$$

In the particular case of the harmonic excitation, as in (1.55), one has

$$\begin{aligned} \boldsymbol{I}\dot{\varsigma} - \Omega\varsigma &= \boldsymbol{U}^T \boldsymbol{f}_o(\frac{1}{2}e^{i\Omega t} + \frac{1}{2}e^{-i\Omega t}) \\ \boldsymbol{I}\dot{\varsigma}^* - \Omega^*\varsigma^* &= \boldsymbol{U}^{*^T} \boldsymbol{f}_o(\frac{1}{2}e^{i\Omega t} + \frac{1}{2}e^{-i\Omega t}) \end{aligned} \tag{1.66}$$

where it can be shown that the second set of equations gives the complex conjugate to the first. Looking only for the particular solution

$$\varsigma = [i\Omega \boldsymbol{I} - \Lambda]^{-1} \boldsymbol{U}^T \boldsymbol{f}_o(\frac{1}{2}e^{i\Omega t}) + [-i\Omega \boldsymbol{I} - \Lambda]^{-1} \boldsymbol{U}^T \boldsymbol{f}_o(\frac{1}{2}e^{-i\Omega t}). \tag{1.67}$$

Using (1.64), the complex amplitude of response $\boldsymbol{x}_o$ from (1.56) and the decomposition (1.59) one has:

$$\boldsymbol{x}_o = (\boldsymbol{U}[i\Omega \boldsymbol{I} - \Lambda]^{-1} \boldsymbol{U}^T + \boldsymbol{U}^*[i\Omega \boldsymbol{I} - \Lambda^*]^{-1} \boldsymbol{U}^{*^T}) \boldsymbol{f}_o \,. \tag{1.68}$$

Only the inversion of diagonal matrices is necessary. The modal approach is very powerful and with few drawbacks: large damping or eigenvalues that are almost coinciding may introduce difficulties that shall not be detailed in this introduction.

Example

Looking again at the two degree-of-freedom damped system, the difference between the direct solution and the modal analysis solution of an excited mechanical system will be shown. Assuming that there is an external force of sinusoidal form acting at the lower coordinate, therefore

$$\{f\} = \left\{ \begin{array}{c} 0 \\ 1 \end{array} \right\} \cos \Omega t \quad \text{(N)}.$$

The direct solution demands the inversion of matrix (1.60). Numerical results will be checked at the resonance points which can be found out to be $\Omega_1 = 30.004\,\text{s}^{-1}$ and $\Omega_2 = 80.15\ \text{s}^{-1}$: they correspond to the situation when the displacements have zero real part $x_{GO_R} = x_{TO_R} = 0$. For Ω_1 it is obtained

$$x_G = -0.1547\text{E--}3 \sin \Omega_1 t \;, \quad x_T = 0.3333\text{E--}3 \sin \Omega_1 t \quad (\text{m})$$

showing that there is a phase angle $\varphi = \pi/2$ between excitation and response at the resonance and showing also that the motion corresponds to the conical mode. For Ω_2 we have

$$x_G = -0.1075\text{E--}3 \sin \Omega_2 t \;, \quad x_T = -0.1248\text{E--}3 \sin \Omega_2 t \quad (\text{m})$$

which is a motion in the nearly cylindrical mode. A general look at the response of the turbine displacement, for instance at $\Omega_3 = 20\,\text{s}^{-1}$ results in

$$x_{TO} = (0.1206 - 0.0310i)\,\text{E--}3$$

which can be written in real time dependent functions:

$$x_T = 0.1206\text{E--}3 \cos \Omega_3 t + 0.03\text{E--}3 \sin \Omega t = 0.1245\text{E--}3 \cos(\Omega_3 t - 14.42^\circ) \quad (\text{m})$$

The complete dependence of amplitude with excitation is shown in Figure 1.12.

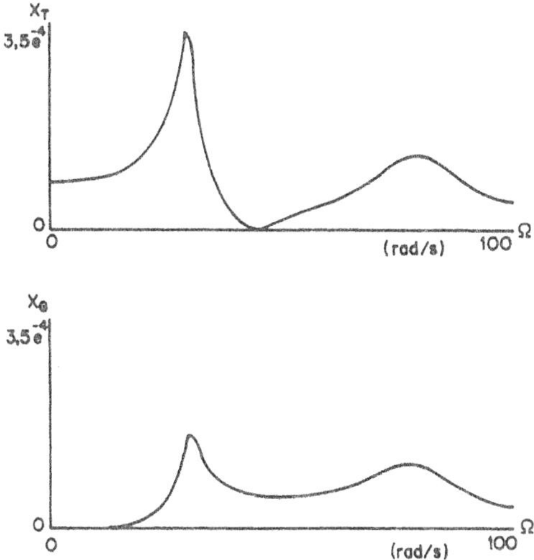

Figure 1.12 Amplitude response with excitation at the lower coordinate

The modal solution overcomes the difficulty of matrix inversion. It will be applied simply by calculating expression (1.68), using modal matrix and the eigenvalue matrix as well as their conjugates. The numerical results are in all situations coincident with the ones above.

1.2.3 Non-Linearities, Self-excitations

There are several possibilities of non-linearities in mechanical systems. They make solutions in multi degree-of-freedom systems extremely complex and therefore are analysed preferably in reduced order systems, trying to isolate the non-linear phenomena and understand what kind of consequences it may bring for the complex system. There are some analytical approaches that may be looked for in texts for non-linear vibrations, like Hagedorn, 1978 [1.11], but one may also try the numerical solution integrating the equations and looking for some general conclusions.

A common consequence of non-linearities in the equation of motion is the ability of the system to remove power from a source of energy and use it to perform a vibration. This power may come from a constant velocity fluid stream like in the case of a chimney, or from the constant velocity of a belt that supports an elastic restrained body and makes it vibrate always with the same characteristic, independent of the belt velocity. For self-excited systems, in general, one may define one or more limit cycles, which are orbits in the plane and represent the state (displacement × velocity) of a point of the system. This limit cycle vibration is a typical non-linear phenomena. It distinguishes from resonance because there is permanent vibration of a damped system without periodic excitation.

Mathematically the mechanism of interaction between the system and the surroundings will depend on position, velocity or acceleration, normally in a non-linear way. This interaction cannot therefore be considered at the right hand side of the equation of motion as an excitation, but must be included in the eigendynamics at the left hand side. If there is a linear dependence, then it is again a homogeneous equation and the big problem can be the instability resulting from an eigenvalue with a positive real part. If it is non-linear several situations can happen: for instance, it may be stable in the small (think of a pendulum clock), unstable in the large but again stable on the limit cycle. Several examples that clarify physically what is happening, can be refered to in the works of Magnus, 1961 [1.16], and of Bishop, 1965 [1.4].

Besides self-excitation it may happen that the restoring force is non-linear (it may for instance have a cubic character, like $k_0 x \pm k_1 x^3$), but in general the vibration is of small amplitude and linearization is possible. In this case what can significantly change physically is the frequency response: it can be calculated that the resonance peak is bent to the right (hardening spring) or to the left (softening spring) resulting in distinct resonances if one is increasing the excitation frequency or decreasing it. Figure 1.13 shows the hardening case, both frequencies and the unstable branch of the frequency response curve.

The damping is in general not purely viscous. Due to the convenience

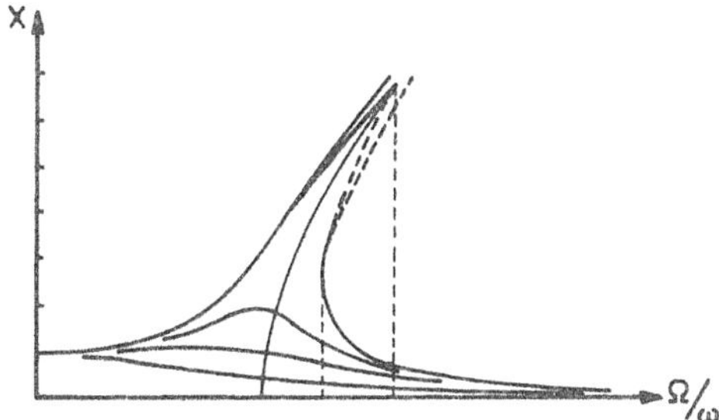

Figure 1.13 Amplitude response of a viscous damped system with a hardening spring as restoring force

of the mathematical treatment of viscous damping an equivalence is made, trying to reproduce a similar behaviour, for instance looking for a damping that is dissipating the same amount of energy. One may think of a material that has stiffness and damping properties, like rubber. The hysteresis loop of the material will indicate the amount of dissipated energy for certain amplitude, frequency and temperature conditions. Under these restrictions it is then possible to obtain some equivalent viscous value, through a simple formulation. The viscous energy dissipated in a cycle of a harmonic motion with frequency Ω is:

$$E_d = \int F\,dx = \int c\,\dot{x}^2\,dt = c\,\Omega^2 x_0^2 \int_T \sin^2 \Omega t\,dt = \pi c\,\Omega\,x_0^2. \tag{1.69}$$

This energy can be equated to the real dissipated energy and further calculations are made with this equivalent viscous damping. The non-linear damping problem of hysteresis loop is treated therefore in a linear form.

A special case for this equivalence derives from experiments for specimens of the materials and establishes a proportionality between the energy dissipated in a cycle of vibration and the square of the vibrating amplitude

$$E_d = \alpha x_0^2 = \pi \bar{c}\,\Omega\,x_0^2 \tag{1.70}$$

where α is a constant that is not depending on the frequency of the motion. This is called structural damping and it is usual to analyse it in the equation combined with the stiffness in a complex form. For a one degree-of-freedom system the equivalent viscous damping and the stiffness may be combined:

$$m\ddot{x} + k(1 + i\gamma)x = f_0 e^{i\Omega t}$$

where

$$\gamma = \frac{\alpha}{\pi k}$$

and this is exact for the permanent solutions of harmonically excited systems. It may also be generalized for several degree-of-freedom mainly if it is intended to use the approach of proportional damping.

1.3 Eigenproperties And Excitations of Hydraulic Machinery

1.3.1 Typical Structure of the Dynamical Model

There are some attempts, in the literature, to define a model for global vibrations of the hydraulic units, see Utecht, 1983 [1.23], Simon, 1982[1.22], Angehrn, 1980 [1.1]. In general they are developed for design purposes by the manufacturer of the hydraulic machine. Their drawback is that, as a rule, the model is not validated when the machine is commissioned. Therefore it looks questionable whether some conclusions are correct, also simple ones like the first critical speed for instance. This analytical approach has some rough simplifications, like in the value and in the action of the negative generator stiffness, or in neglecting the sealing effect at the turbine: Weber, 1986 [1.25], showed from measurements that this effect must be considered in the model, Dietzen, 1988 [1.19], calculated analytically its value for a given machine.

The experimental approach starts with a set of representative measurements, as in Figure 1.11, and the question of how many degree-of-freedom shall be considered relevant. So an engineering decision must be made as to the best method to smoothen the frequency response. As it was shown in the above theory, excitation frequencies and eigenfrequencies will be present in the response. For instance comparing the results in Figure 1.14 it may be concluded that the eigendynamics appears only in partial load condition. To define the model mainly this operation condition is helpful. But it is difficult to establish exact relations because the hydraulic excitation is not known, only its stochastic character. At least it seems reasonable to extract informations about resonances that may help in a better definition of the model. Therefore it is also not so important if one is measuring absolute displacements of the shaft, or relative shaft/bearing displacements. It is clear that frequency ranges where eigenfrequencies are expected have to be considered; therefore the main hydraulic frequencies are not present at these diagrams (they are lower), as well as electric generator frequencies (higher). Also the model being looked for will not *worry* about an eigendynamic outside of the specified range.

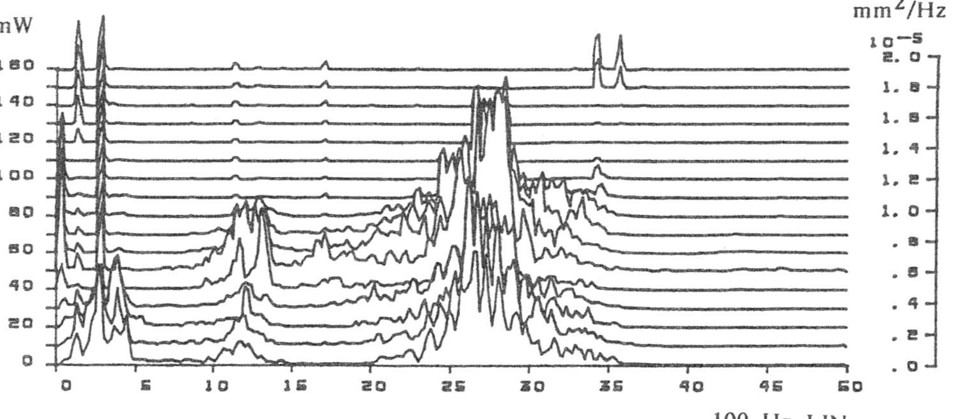

Figure 1.14 Power spectral density of the relative shaft/bearing displacement of a nominal 160 MW Francis turbine for several load conditions

It is important to remember that the machine is a unit in spite of being built by manufacturers of different technology and scope. The machine must normally be considered elastic, the number and characteristics of the supports is very important and, only as an example, the kind of sealing in a Francis turbine or the variation of the added mass in a Kaplan turbine will influence considerably the dynamic behaviour of the generator, as well as several of the generator frequencies, will be found in the turbine guide bearing.

Probably the best approach is to join the analytical and the experimental approaches. If the machine is in the design phase much of the former experience of the manufacturer must be included. The model will necessarily be bigger than that of the number of relevant degree-of- freedom: these may be obtained by the pseudo-modal transformation mentioned earlier and shall give resonances where they are expected from experiments. Observe also that modelling reduced scale hydraulic units, will in general not help very much to define a dynamical model for the real system.

For the experimental determination of the critical speed one should consider the gyroscopic effects. This means that the analysis shall be made with Campbell's diagram (Figure 1.10): from an analytical model of a large machine it was calculated an increase in 30% in the frequency of the critical speed compared to the non-rotating case; observe further that the eigenfrequencies of the system at its nominal speed will be between both these values (non-rotating and critical speed). For the determination of the response the full modal composition must be known, including that of the backward whirl corresponding to the eigenfrequencies that are decreasing with the increase of rotating speed.

To complete the knowledge of the model, the excitations must be known besides the knowledge on the mass matrix $\boldsymbol{M}$, the viscous damping matrix $\boldsymbol{C}$, the gyroscopic matrix $\boldsymbol{G}$ and the stiffness matrix $\boldsymbol{K}$. Vladislavlev, 1979 [1.24], presents a tabulation and a systematic analysis of the consequences of mechanical, hydraulic, electric and magnetic forces. Some may be estimated from the knowledge of the other machines like the unbalance, the vortex core, the magnetic forces at the electric machine, and so on. They may also be checked by measurements: then we need to know the transfer function between the point where the force is applied and where the measurements are being made; we also have to separate the effects at the same frequency to know for instance what is the contribution of the unbalance to the vibration at the frequency of the speed of rotation.

Finally it has to be clear whether the non-linear effects are to be included or not. Benkö and Holmen, 1966 [1.3], point out the possibility of combination or parametric resonances. There still seems to be not enough experience to decide when these phenomena really shall be included in the calculation.

1.3.2 Parameter Uncertainties

In any dynamical model there will be elements of the system matrices that are well known and others where the value is only roughly known. Looking again at the global vibration of the hydraulic unit we can exemplify the way of thinking.

The mass matrix terms will be well defined by a good finite element model but one has to take care of some points. The determination of the transversal inertia of the generator and turbine rotor is a good example of a well defined parameter but not included in the specification of a machine. The rotating part of the bearings on the shaft, i.e., the mass increase of the shaft at the bearings will be included but is also well defined. The *added mass* (and the consequent inertia) at the turbine or pump side must be considered. This is typically an uncertain parameter in the model and may also change with the generated power, as is clear when one imagines a Kaplan turbine. The mass distribution of the rotor ring in the electric machine and the way it is connected to the shaft (which is affected by velocity and load) is not an uncertain parameter, but it is not immediately clear how to simplify its analysis.

The terms of the stiffness matrix may be obtained by local static analysis (on design drawings or experimentally). Some will be well defined but others will not. When the system is in operation it may accommodate in a way that for instance clearances and dry friction terms are removed, changing some values of the elements of matrix $\boldsymbol{K}$. The shaft stiffness is certainly well defined but the bearings in general are not. In the analysis of the guide bearings one shall consider the support, the ring where the tilting pads are fixed, the pads itself and the oil film. The last one may be calculated with existing data through several theories to check whether it is in the same order of magnitude as the structural component or not. For large hydraulic machines analysed at UNICAMP, it was one or two powers of ten higher. The analysis of thrust bearings will be very similar. Finally, as already mentioned, the sealing effect on the turbine/pump rotor will be included.

The gyroscopic matrix depends on the polar inertia and is well known. The damping matrix is unknown and difficult to establish. It may be possible to neglect structural damping, mainly in a first approach, but what about the damping induced by the water - for instance at the sealing of the hydraulic machine? There is still not enough experience that would enable a detailed analysis to be performed. Engineering sense has to be used to establish the consequences of neglecting damping in the behaviour of the model.

1.3.3 Identification Problems

The first eigenfrequencies of the real system can be determined when transients or stochastic forces excite the eigendynamics of the hydraulic unit. They may be used to check some of the uncertain parameters. It would be very useful to know also the eigenmodes associated with those frequencies, but they are very difficult to measure. Some points of the machine that may be accessed may be used to validate the analytical model used to obtain the modes. Again, the problem is that the excitations are not well known and, therefore, the modes must be extracted from a very complex response that includes eigendynamics and external forces. Also some parameters depend on load conditions, others on the speed of rotation; the excitation of the hydraulic forces depends on several external operating conditions. Due to all the particularities of hydraulic machinery it is very difficult to use straightforward identification procedures; it will be convenient to determine approximately the matrices off-line and then improve these matrices with on-line identification procedures from measurements, such as Natke, 1988 [1.17]. The resulting model will be useful for diagnostic purposes and for future designs.

1.3.4 Measurement Difficulties

For the full understanding of the vibration of hydraulic units there must be a strong interaction between the analytical model and the information from experiments. The efforts of the investigations are directed to the determination of the excitations, i.e. the right hand side of the equation and the understanding of the eigendynamics, the left hand side. The determination of the excitations in nominal conditions is also well known, whereas investigations on the forces in transients and stochastic effects are still under investigation, see Schneider, 1988 [1.20]. The study of models and the extrapolation of the results to full scale machines is a common practice. Unfortunately this will not always work with the dynamic effects, since the eigendynamics of the full scale machine is completely different and, as mentioned, there are phenomena that cannot be isolated from the machine itself.

Difficulties are of several kinds. For instance there is usually no machine availability time to make deeper investigations; measurements are seldom made on good machines; in general they are only measured on crash situations trying to find out the reasons for some greater failure. Also, the information has not been collated in a comprehensive data bank. All this means is that a longer time is necessary to formulate good mathematical models that are the base of modern engineering.

Modern hydraulic units are being produced with eddy current relative shaft/bearing sensors. Two orthogonally placed sensors at a bearing can produce orbit plots when the electric signals are conveniently filtered, they can also provide peak values to actuate an alarm or they can be used to look at the frequency spectra for diagnostic purposes. For the dynamic analysis of the unit, the sensors at the generator guide bearing and at the turbine guide bearing should be in the same vertical plane. This will help to verify the global model. In this measurement the vibrations of the bearing support and the motion of the shaft will be presented. The bearing support should be itself measured with a velocity transducer or, may be, with an accelerometer. The velocity transducer is more appropriate to obtain the low frequency response due to the hydraulic excitation. There are also attempts to measure the absolute motion of the shaft in space: the signal will contain only the dynamics of the rotating shaft and its response to all external inputs.

An important role in measurement is played by the strain gauges; a typical example of such application is the vibration of guide vanes and modelling of its dynamic behaviour in the stream of water. The torsional behaviour of shafts must also be measured by strain gauges: the signal will then be transmitted from the rotating shaft to an inertial station through FM or another convenient method.

The measurements have to be treated before further processing: this means cleaning it from known information that does not contribute to the dynamic analysis, such as the run out of the shaft or from unwanted d.c. components or noise inputs. These time signals may be used as inputs to the mathematical model in the time domain identification procedures. Otherwise they may be Fourier-transformed and viewed in the frequency domain giving information on eigenfrequencies, amplitudes of response or damping. Figure 1.15 shows a measurement made with an accelerometer in the draft tube of a model Francis turbine in order to look at the frequency of the vortex core. It is a common practice to clean the harmonics (for instance of the velocity of rotation) out of the diagram. Good skills are always required to identify what is found in a measurement of that kind. For instance, it is still a research problem how to interpret the measurements at the turbine guide bearing of a Francis machine in order to recognize when there is rubbing through the water seal. It appears that a signal for it will be the sudden change or appearance of a lot of harmonics of the rotation speed.

Important open questions still exist relating to how best to optimize the sensor position and minimize the number of sensors. As global vibrations will cross the whole machine it is expectable that the number of measuring points can be kept small.

Information on Standards

International Norms and Guidelines will give a valuable introduction for the

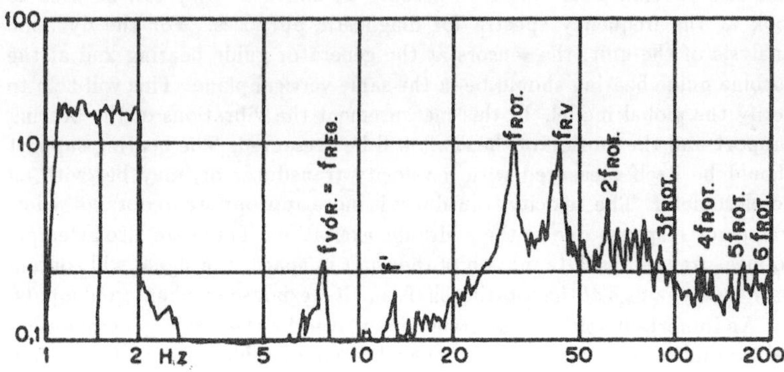

Figure 1.15 Draft tube response of a model Francis turbine with vortex core

understanding of the behaviour of the hydraulic machinery. They sum up years of experience of manufacturers and customers. It is not our intention to make any ranking of norms but to mention only some examples of the actual standard. At the International Electrotechnical Commission, the Technical Committee No.4 on Hydraulic Turbines made a valuable contribution to the measurement techniques through the Guide for Measurement of Vibration in Hydraulic Turbines and Reversible Pump-Turbines or the Guide for Field Measurement of Vibrations and Pulsations in Hydraulic Machines. Also the VDI (Verein Deutscher Ingenieure) Guideline 2059 on Shaft Vibrations of Turbosets gives very important informations on measurements and interpretations. Part 5 is concerned only with hydraulic machines: roughly speaking it establishes, in the orbit plot at the bearing, a maximum tolerable displacement depending on the rotation. But there is no information on very slow machines (i.e. large hydraulic machinery) and on working conditions outside of the guaranteed region. On the other hand the norm brings a table trying to correlate frequencies of measured signal with possible causes.

1.3.5 Excitations and System Parameters of Hydraulic Origin

Flow induced forces and vibrations constitute an important field of research of mechanical and civil engineering. There are several text books and survey articles, see for instance Naudascher, 1980 [1.18], where the classification of the phenomena is made, the excitation mechanisms are described and the behaviour of the structure is analysed. In this item however we will restrict the scope to only some of the aspects.

Naudascher divides the excitation mechanisms in three types. The *extraneous* (or external) excitation caused by a flow or pressure pulsation that is not produced by interaction with the system itself. The force is in general deterministic, an example is the periodic vortex shedding from the guide vanes when the considered system is the rotating turbine runner. The *instability excitation* is caused by a flow instability and, in general, it is intrinsic to the vibrating system itself, as in the example above when the vibrating system is the guide vanes. In this case the excitation mechanism comes from a more or less periodic energy transfer from the main flow to the fluctuating flow, therefore it also belongs to the category of self-excitation. However as this is a phenomenon in the flow equations, consequently the structure can be treated merely as externally excited. The *movement excitation* is due to forces arising from the motion of the body in the fluid. A half cylinder vibrating in a fluid stream will generate a damping-like force, but with negative coefficient, that induces instability through a self-excited mechanism in the equations of motion of the structure.

Considering the linear system analysis presented in this chapter the hydraulic action may be included only on the right hand side of the equation as the extraneous and instability excitations or it may be included on the left hand side as the movement excitation. In this case, hydraulic phenomena may produce negative damping terms that can be dangerous due to the loss in stability, but they may produce also acceleration proportional terms (added mass) or displacement proportional terms (added stiffness). As a matter of fact, the vibrating motion in a fluid, even if it is stagnant, will induce an unsteady flow resulting in forces on the body. Linearly, these forces will have acceleration proportional terms that may be included in the mass matrix of the system and therefore we will speak of added mass; similarly there is also added damping and added stiffness for the velocity and the displacement proportional terms.

The value of the added mass may produce considerable changes in the eigendynamics of the system. It can be imagined as a part of the fluid mass surrounding the body and vibrating with it. As a consequence the eigenfrequencies tend to be reduced. This phenomenon is well studied for several conventional configurations: for a quasi two dimensional structure (for instance flat and long plate) an added mass coefficient is introduced which, when multiplied to the fluid mass of a circular cylinder circumscribing the body, gives the effective added mass. This coefficient is 1 for a long circular cylinder and for a long flat plate and is reduced with the increase of the width. The added mass coefficient depends not only on the body shape, but also on amplitude, frequency, submergence and confinement as well as the flow conditions. For larger amplitudes or flow velocities the hydraulic force will change from a inertia dominated force to a drag force, influencing the damping term of the system.

In a more complex situation as in the case of the runner blades of a turbine one has to use an analysis with the finite element technique. D.K. Gupta, 1984 [1.10], developed a numerical method for this case. Equations for the structure and for the fluid must be solved in a coupled form. But neglecting damping and considering the fluid to be incompressible then the structure equation can be uncoupled when an added mass term is included. When the fluid is compressible, in an approximate way, the structure equation may again be uncoupled after including a secondary added mass matrix. The solution is surely complex and there are still some steps to do to transfer the result for one blade to the complete action on a Francis or a Kaplan runner, but this will produce an analytical estimate of the values of the required parameters.

It is also important to observe that when describing the global vibration of the hydraulic unit, several terms of the mass matrix must include the added mass terms. Special care is required not to forget the corresponding added inertia of the added mass at the runner.

At present it seems more convenient to apply identification procedures to obtain the parameters of the mass matrix from the real machine, as it is done by Bolleter et al., 1987 [1.5], and several other authors. Liess, 1984 [1.15], measured transfer functions at model machines that allow conclusions to be obtained on added mass parameters. Further, the model results cannot be extrapolated in this case to the full scale: hydraulic and dynamical similarity do not scale in the same way. Looking at the hydraulic unit there is probably no added stiffness and damping is still a very unknown subject.

At the right hand side of the equation of motion the hydraulic forces of extraneous kind have to be integrated over the runner. The resulting forces and moments as defined in the paper of Bachmann, 1980 [1.2], have a permanent part and a variable one, which may be of deterministic (periodic) or stochastic nature, see for instance Schwirzer, 1977 [1.21]. There is a component from the force parallel to the shaft, representing the lift of the machine and exciting the axial/vertical vibration on the thrust bearing. The other component is a rotating radial force, not necessarily with the rotation speed of the machine, and an out of vertical plane moment which excite the radial/flexure vibration. Finally there is the torsional moment that will be transformed in power, which due to its variable parts, may also excite torsional vibrations of the hydraulic unit.

References

1.1 Angehrn, R. (1980). Berechnungsverfahren für das Querschwingungsverhalten von Rotoren, Escher Wyss Mitt., 1/2, pp. 173-178.

1.2 Bachmann, P. (1980). Fortschritte im Erfassen und Auswerten von Kräften und Momenten an Rotoren hydraulischer Modellturbomaschinen, Escher Wyss Mitt., 1/2, pp. 69-81.

1.3 Benkö, G.B. and Holmen, E.K. (1966). 'Parametric resonances in umbrella-type generating units', Symp. Vibr. in Hydr. Pumps and Turbines., Manchester, Sept., Paper(3).

1.4 Bishop, R.E.D. (1965). Vibration, Cambridge Press.

1.5 Bolleter, et al. (1987). 'Measurement of hydrodynamic interaction matrices of boiler feed impellers', J. of Vibr. Acoust. Stress and Rel. in Design, Vol. 109, pp. 144-151.

1.6 Doerfler, P. (1980). Mathematical model of the pulsations in Francis turbines, Escher Wyss Mitt., 1/2, pp. 101-106.

1.7 Flack, R.D. and Allaire, P.E. (1984). 'Lateral forces on pumps and impellers: a literature review', Shock and Vibration Bulletin, pp. 10.

1.8 Glattfelder, A.H., Grein, H. and Doerfler, P.K. (1981). 'Intense system vibrations in hydroplants', Water Power & Dam Constr., March.

1.9 Guarga, R. (1986). 'Modelo bidimensional del vortice sin simetria axial creado por el flujo helicoidal dentro de un tubo cilindrico', Proceeding XII Congresso Latino-Americano de Hidraulica, S. Paulo, September.

1.10 Gupta, D.K. (1984). 'Vibration analysis of fluid submerged blades', Ph.D. Thesis, I.I.T., Delhi.

1.11 Hagedorn, P. (1978). Nichtlineare Schwingungen, Akad. Verlagsg. Wiesbaden.

1.12 Hosoi, Y. (1965). 'Experimental investigation of pressure surge in draft tubes of Francis water turbines', Hitachi Rev., Tokyo, 14 (12).

1.13 IEC-TC4: International Electrotechnical Commission-Technical Committee 4: 'Hydraulic turbines: International guide for field measurements of vibrations and pulsations in hydraulic machines'.

1.14 Lalanne, M., Berthier, P. and DerHagopian, J. (1983). Mechanical vibrations for engineers, J. Wiley.

1.15 Liess, C., Jaeger, E.U. and Klemm, D. (1984). 'Hydraulically induced radial forces on Francis turbines: measurement, evaluation and results', IAHR-Symposium, Stirling.

1.16 Magnus, K. (1961). Schwingungen, Teubner Verl.

1.17 Natke, H.G. (1988). 'Updating computational models in the frequency domain based on measured data: a survey', Prob. Engineering Mech., Vol. 3, pp. 28-35.

1.18 Naudascher, E. (1980). 'Engineering for structures subject to flow-induced forces and vibrations', Intensive Course Monograph at Univ. of São Paulo.

1.19 Dietzen, F.J. (1988). Bestimmung der dynamischen Koeffizienten von Dichtspalten mit Finite-Differenzen-Verfahren, VDI-Fortschritt-Berichte, Reihe 11, Schwingtechnik, Nr. 103.

1.20 Schneider, K. (1988). 'Measuring the dynamic forces and moments acting on the impellers of hydraulic turbomachines' (in German), Pump Congress, Karlsruhe, October.

1.21 Schwirzer, T. (1977). 'Dynamic stressing of hydroelectric units by stochastic hydraulic forces on the turbine runner', Water Power, Vol. 29. pp. 39-44.

1.22 Simon, F. (1982). Berechnung des dynamischen Verhaltens von Wellensystemen bei Wasserkraftanlagen, Voith Forschung und Konstruktion, Heft 28.

1.23 Utecht, M. (1983). 'Vibrations analysis for the Itaipu generator turbine units', Proceedings Cigré Symposium, Rio de Janeiro, Brazil, November.

1.24 Vladislavlev, L.A. (1979). Vibration of hydro units in hydroelectric power plants, Amerind Publ. Co., New Delhi.

1.25 Weber, H.I. et al. (1986). 'Flexural vibrations of Francis turbines in partial load operation', Proceedings of IAHR Symposium on hydraulic machinery, Montreal.

Chapter 2

Modelling and Analysis

H.Tomita

2.1 Fluid-Structure Interaction

Fluid-structure interaction should be taken into account in vibration analysis of a hydraulic machine, since there are many cases in which the interaction between a vibrating machine structure and a contiguous fluid has a profound influence upon the magnitude and frequencies of the structural vibration. A typical effect of the fluid on the structural vibration consists in *added mass* to the structure. The basic mechanism of fluid-structure interaction is described in this section.

2.1.1 Basic Equations

The equation of motion for inviscid fluid in the absence of external forces is given by

$$\frac{\partial \boldsymbol{v}}{\partial t} = \boldsymbol{v} \times \operatorname{rot} \boldsymbol{v} - \frac{1}{2}\operatorname{grad} |\boldsymbol{v}|^2 - \frac{1}{\rho}\operatorname{grad} p \, , \tag{2.1}$$

where ρ, $\boldsymbol{v}$ and p are the density, velocity vector and pressure of the fluid. In discussing the small perturbation of fluid motion, the convective acceleration terms in (2.1) may be omitted resulting in

$$\frac{\partial \boldsymbol{v}}{\partial t} = -\frac{1}{\rho}\operatorname{grad} p \, . \tag{2.2}$$

In addition to (2.1) the continuity relation given by

$$\frac{\partial \rho}{\partial t} + \operatorname{div}(\rho \boldsymbol{v}) = 0 \tag{2.3}$$

must be satisfied. The relation between the pressure and density of the fluid may be assumed to take the form

$$\delta p = -B\frac{\delta V}{V} \quad \text{or} \quad \frac{dp}{d\rho} = \frac{B}{\rho} = c^2 \, , \tag{2.4}$$

in which δ denotes small variation. V and B are the specific volume and bulk modulus of the fluid, and c is the sound velocity in the fluid.

When the fluid motion is assumed to be irrotational, the velocity potential Φ can be defined by

$$\boldsymbol{v} = \operatorname{grad} \Phi \, . \tag{2.5}$$

From equations (2.2), (2.3), (2.4) and (2.5), the equation which determines Φ can be obtained as

$$\nabla^2 \Phi - \frac{1}{c^2} \frac{\partial^2 \Phi}{\partial t^2} = 0 \, . \tag{2.6}$$

Using the acoustic assumption

$$p = -\rho \frac{\partial \Phi}{\partial t} \tag{2.7}$$

and substituting p for Φ in (2.6),

$$\nabla^2 p - \frac{1}{c^2} \frac{\partial^2 p}{\partial t^2} = 0 \, . \tag{2.8}$$

Equations (2.6) and (2.8) are the linearized wave equations. When the compressibility of the fluid need not be considered, these equations will be further simplified as

$$\nabla^2 \Phi = 0 \quad \text{and} \quad \nabla^2 p = 0 \, . \tag{2.9}$$

Now the governing equations for the structural deformation should be defined. Assuming;

(1) elastic deformation

(2) small displacement

for the motion of the structure, then in the tensor description the motion of the structure can be expressed as

$$\rho_S \frac{\partial^2 w_i}{\partial t^2} = \frac{\partial \sigma_{ij}}{\partial x_j} \, , \tag{2.10}$$

where ρ_S is the density of the structure; w_i is a component of the displacement vector of the structure; σ_{ij} is a component of stress tensor of the structure. In the tensor description a repeated suffix implies summation with regard to that suffix. Equation (2.10) is the basic equation describing the motion of the structure.

The conditions at the boundary of the fluid and the structure play an important role in fluid-structure interaction, because the motions of both will be coupled there. The pressure of the fluid at the surface of the structure will impose an external force onto the structure. This external force must be balanced with the internal stresses of the structure. Thus,

$$\sigma_{ij} \, n_j = p \, n_i \, , \tag{2.11}$$

where n_i is the direction cosine of the outward normal vector at the surface of the structure.

When viscosity of the fluid is not taken into consideration, there will be no constraint conditions on the surface in the tangential direction and the following condition must be satisfied at the boundary for the fluid velocity $\boldsymbol{v}$ and displacement $\boldsymbol{w}$:

$$\boldsymbol{v} \cdot \boldsymbol{n} = \frac{\partial \boldsymbol{w}}{\partial t} \cdot \boldsymbol{n} . \tag{2.12}$$

By introducing the fluid pressure instead of the velocity, this condition may be expressed in a different form,

$$- \operatorname{grad} p \cdot \boldsymbol{n} = \rho \frac{\partial^2 \boldsymbol{w}}{\partial t^2} \cdot \boldsymbol{n} . \tag{2.13}$$

2.1.2 Basic Aspects of Fluid-Structure Interaction

A variety of practical methods have been developed to analyze fluid-structure system dynamics based on the formulation of the problem discussed in the previous paragraph. Detailed discussion of these methods will not be presented here. Some simple analyses will be made to give the reader an understanding of the basic aspects of fluid-structure interaction.

As shown in Figure 2.1, part of an elastic shell is immersed in an inviscid and incompressible fluid which is contained in a rigid vessel. The following section will study how the vibration characteristics of the immersed structure are affected by the fluid in contact with it.

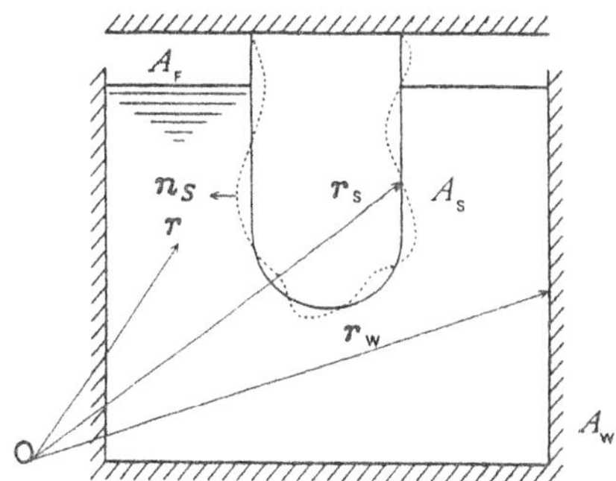

Figure 2.1 Simplified fluid-structure model

The displacement of the structure can be expressed as the superposition of the normal modes of vibration,

$$w_S = \sum q_n(t)\, \Psi_n(\boldsymbol{r}) , \tag{2.14}$$

in which w_S is the structural displacement in the normal direction to the surface of the structure, and $q_n(t)$ and $\Psi_n(\boldsymbol{r})$ are the nth generalized coordinate and the nth modal function.

Once the displacement at the boundary of the fluid is defined, the motion of the fluid can be determined by solving the following set of equations:

$$\begin{aligned} \nabla^2\Phi &= 0 \ : \boldsymbol{r} \in V\,, \quad \operatorname{grad}\Phi \cdot \boldsymbol{n}_S = \dot{w}_S \ \ : \boldsymbol{r} \in A_S\,, \\ \Phi &= 0 \ : \boldsymbol{r} \in A_F\,, \quad \operatorname{grad}\Phi \cdot \boldsymbol{n}_W = 0 \ \ : \boldsymbol{r} \in A_W\,, \end{aligned} \tag{2.15}$$

in which A_S and A_W are the surfaces where the fluid makes contact with the structure and the vessel respectively; A_F is the free surface of the fluid. The region of the fluid is denoted by V, and $\boldsymbol{r}$ is the vector which designates location in the volume. The over dot on w_S indicates time differentiation. As the effect of the gravity is usually small, it is not considered in (2.15). Using the same series of generalized coordinates q_n's, the solution of the problem can be written in the following form:

$$\Phi = \sum \dot{q}_n(t)\, \Phi_n(\boldsymbol{r})\,, \tag{2.16}$$

where Φ_n is the velocity potential corresponding to the nth normal mode of vibration of the structure, satisfying

$$\begin{aligned} \nabla^2\Phi_n &= 0 \ : \boldsymbol{r} \in V\,, \quad \operatorname{grad}\Phi_n \cdot \boldsymbol{n}_S = \Psi_n \ \ : \boldsymbol{r} \in A_S\,, \\ \Phi_n &= 0 \ : \boldsymbol{r} \in A_F\,, \quad \operatorname{grad}\Phi_n \cdot \boldsymbol{n}_W = 0 \ \ : \boldsymbol{r} \in A_W\,. \end{aligned} \tag{2.17}$$

Now the kinetic and potential energy of the total system can be calculated. The kinetic energy of the fluid is

$$E_{kF} = \int_V \frac{1}{2}\rho\, \boldsymbol{v} \cdot \boldsymbol{v}\, dV = \int_V \frac{1}{2}\rho\, (\operatorname{grad}\Phi)^2\, dV\,.$$

Application of the first form of Green's theorem and assumption of the irrotational flow will lead to

$$\begin{aligned} E_{kF} &= \frac{1}{2}\rho \int_{A_S+A_W+A_F} \Phi \operatorname{grad}, \Phi \cdot \boldsymbol{n}\, dA - \frac{1}{2}\rho \int_V \Phi\, \nabla^2\Phi\, dV \\ &= \frac{1}{2}\rho \int_{A_S+A_W+A_F} \Phi \operatorname{grad}\Phi\ \cdot \boldsymbol{n}\, dA\,. \end{aligned}$$

Using (2.17) and (2.16), and applying the orthogonality relation for Φ_n, we obtain

$$E_{kF} = \frac{1}{2}\rho \sum \dot{q}_n^2 \int_{A_S} \Phi_n \operatorname{grad}\Phi_n \cdot \boldsymbol{n}_S\, dA = \frac{1}{2}\rho \sum \dot{q}_n^2 \int_{A_S} \Phi_n\, \Psi_n\, dA\,. \tag{2.18}$$

The potential energy of the fluid, E_{pF}, due to the displacement of the free surface is estimated to be small and will not be considered here.

The kinetic energy and the potential energy of the structural system will be denoted as E_{kS} and E_{pS}. These can be expressed as

$$E_{kS} = \frac{1}{2} \sum M_n \, \dot{q}_n^2 \tag{2.19}$$

and

$$E_{pS} = \frac{1}{2} \sum K_n \, q_n^2 \, , \tag{2.20}$$

where M_n and K_n are the nth modal stiffness and mass of the structural system, respectively. Thus the Lagrangian function for the fluid- structure system is

$$L = E_{kF} + E_{kS} - E_{pS} = \frac{1}{2} \sum (M_n + m_n) \, \dot{q}_n^2 - \frac{1}{2} \sum K_n \, q_n^2 \, , \tag{2.21}$$

where m_n is defined by

$$m_n = \rho \int_{A_S} \Phi_n \, \Psi_n \, dA \, . \tag{2.22}$$

The equations of motion can be derived by using Lagrange's equation:

$$(M_n + m_n) \, \ddot{q}_n + K_n \, q_n = 0 \, . \tag{2.23}$$

As evident from this result, the fluid in contact with the structure causes an increase in the inertia of the structure. Hence m_n is called *added mass* to the nth mode of the vibration of the structure.

The study of a simple fluid-structure problem may be helpful in deepening the understanding of the interaction between fluid and structure. An elastic annular disc vibrates in water in a circular rigid vessel as shown in Figure 2.2. It will be assumed that the clearances between the disc and the walls of the vessel are so small that no flow can occur through the clearances. The width of the disc is small compared with its average radius $r_0 = (r_1 + r_2)/2$, so that it may be assumed that the deflection of the disc is uniform in the radial direction. The motion of the fluid caused by the vibration of the disc can be assumed to be two-dimensional in the circumferential and vertical directions.

The flexural vibration of the disc in its nth mode can be written as

$$q_n(t) \, \Psi_n(\theta) = q_n(t) \cos n\theta \, . \tag{2.24}$$

Then the velocity potential of fluid corresponding to the nth mode of vibration can be obtained as

$$\Phi_n = \frac{\cosh\{(n/r_0)(z - z_0)\}}{(n/r_0) \sinh\{(n/r_0) H_1\}} \cos n\theta \tag{2.25}$$

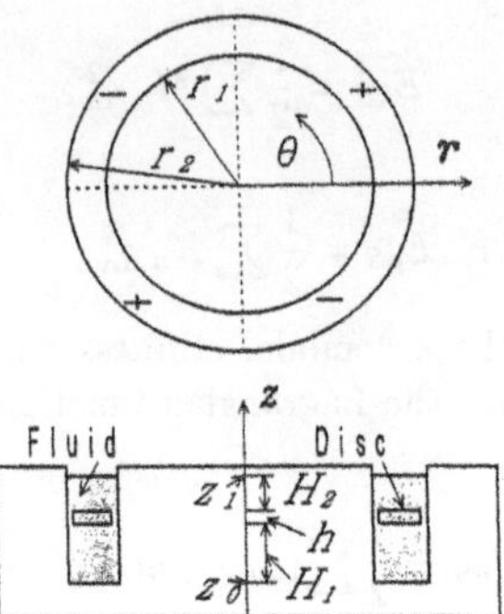

Figure 2.2 Vibration mode of disc in fluid

for the fluid volume below the disc. Similarly for the volume above the disc,

$$\Phi_n = \frac{\sinh\{(n/r_0)(z - z_1)\}}{(n/r_0)\cosh\{(n/r_0)H_2\}} \cos n\theta \, . \tag{2.26}$$

Arriving at these results, $r \approx r_0$ is assumed through the fluid volumes. The hydraulic mass added to the disc is

$$m_n = \frac{\pi\rho r_0(r_2^2 - r_1^2)}{2n}\{\coth(\frac{nH_1}{r_0}) + \tanh(\frac{nH_2}{r_0})\} \, . \tag{2.27}$$

The modal mass of the disc in the absence of the fluid can be calculated as

$$M_n = \frac{1}{2}\pi\rho_S h(r_2^2 - r_1^2) \, , \tag{2.28}$$

where h is the thickness of the disc. The ratio of the effective modal mass of the disc to M_n is

$$\frac{M_n + m_n}{M_n} = 1 + \frac{\rho\, r_0}{n\,\rho_S\, h}\{\coth(\frac{nH_1}{r_0}) + \tanh(\frac{nH_2}{r_0})\} \, . \tag{2.29}$$

In Figure 2.3 the analytical solution is compared with the experimental results of Kubota, 1979 [2.1]. As recognized in Figure 2.3 (a), the smaller the modal number is and the smaller the depth of the confined fluid, the stronger is the effect of the fluid on the vibration of the disc. Comparing the effect of the variation of H_2 on m_n, shown in Figure 2.3 (b), with that of H_1 in Figure 2.3 (a), it can be understood that the influence of a fluid on a structure is relatively small when the fluid has a free surface.

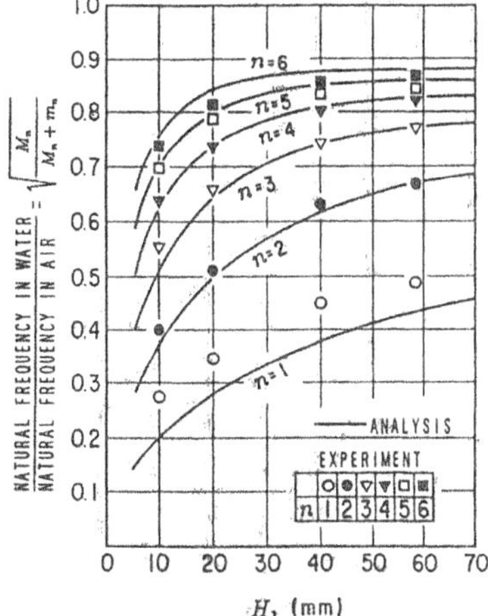

(a) $(H_2 = 0\,,\ h = 10\,\text{mm})$

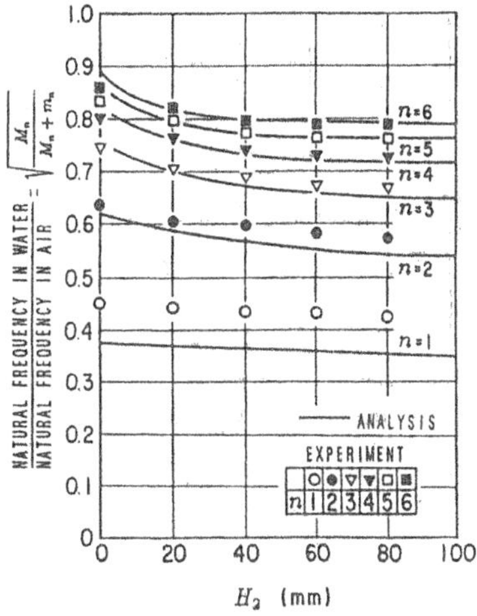

(b) $(H_1 = 40\,\text{mm},\ h = 10\,\text{mm})$

Figure 2.3 Influence of fluid on natural frequencies of submerged disc

2.1.3 Influence of Flexibility of Outer Confinement

In the system shown in Figure 2.1 flexibility of the vessel has not been taken into consideration. If there is the flexibility of the vessel, its flexural vibration will be coupled with that of the inner structure via the fluid. The change in the vibration characteristics of the system due to the coupling may not be negligibly small. This effect should be kept in mind in designing a model for vibration testing.

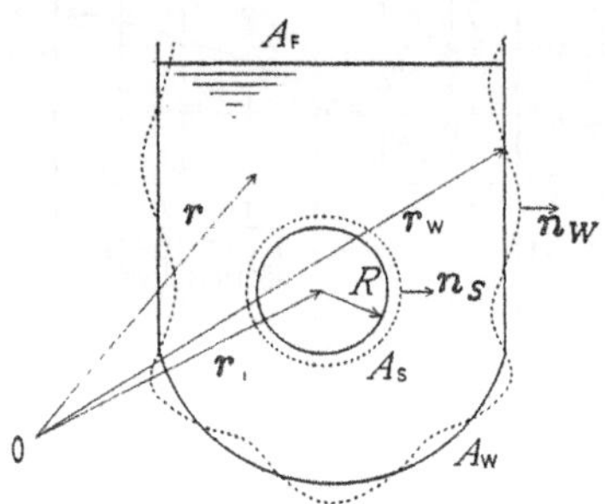

Figure 2.4 Influence of flexible boundary

For simplicity the inner structure in Figure 2.1 is substituted with a small elastic sphere which is vibrating in the breathing mode as shown in Figure 2.4. The displacement of the surface of the sphere is one-dimensional in the normal direction to its surface. The displacement of the wall of the outer vessel in the normal direction to its surface can be expressed in the form similar to that for w_S in (2.14):

$$w_W = \sum q_n(t)\, \Psi_n(\boldsymbol{r})\,. \tag{2.30}$$

The velocity potential Φ is determined by the following equations:

$$\left.\begin{array}{l} \nabla^2\Phi = -4\pi R^2 \dot{R}\, \delta(\boldsymbol{r}-\boldsymbol{r}_i) \ : \ \boldsymbol{r} \in V\,, \\ \operatorname{grad}\Phi\cdot \boldsymbol{n}_S = \dot{R} \ \text{ or } \ \int_{A_S} \operatorname{grad}\Phi \cdot \boldsymbol{n}_S\, dA = 4\pi R^2 \dot{R} \ : \ \boldsymbol{r}\in A_S\,, \\ \Phi = 0 \ : \ \boldsymbol{r} \in A_F\,, \ \operatorname{grad}\Phi\cdot\boldsymbol{n}_W = \dot{w}_W \ : \ \boldsymbol{r}\in A_W\,, \end{array}\right\} \tag{2.31}$$

in which the fluid motion induced by the small radial vibration of the sphere is approximated by the fluid motion induced by a source at the centre of the sphere; δ represents Dirac's delta function.

Let us assume the velocity potential is the sum of two components Φ_S and Φ_W:

$$\begin{array}{rcl} \Phi & = & \Phi_S + \Phi_W\,, \\ \Phi_S & = & R^2\dot{R}\,\Phi_S^*(\boldsymbol{r})\,, \qquad \Phi_W = \sum \dot{q}_n\, \Phi_n(\boldsymbol{r})\,. \end{array} \tag{2.32}$$

The velocity potential Φ_S is chosen to satisfy the following equations:

$$\left.\begin{array}{l}\nabla^2\Phi_S = -4\pi R^2\dot{R}\,\delta(\boldsymbol{r}-\boldsymbol{r}_i)\ :\ \boldsymbol{r}\in V\ ,\\ \operatorname{grad}\Phi_S\cdot\boldsymbol{n}_S = \dot{R}\ \text{ or } \int_{A_S}\operatorname{grad}\Phi_S\cdot\boldsymbol{n}_S\,dA = 4\pi R^2\dot{R}\ :\ \boldsymbol{r}\in A_S,\\ \Phi_S = 0\ :\ \boldsymbol{r}\in A_F\ ,\ \operatorname{grad}\Phi_S\cdot\boldsymbol{n}_W = 0\ :\ \boldsymbol{r}\in A_W\ ,\end{array}\right\} \tag{2.33}$$

and Φ_W is a function which satisfies

$$\left.\begin{array}{l}\nabla^2\Phi_W = 0\ :\ \boldsymbol{r}\in V\ ,\\ \operatorname{grad}\Phi_W\cdot\boldsymbol{n}_W = \dot{w}_W\ :\ \boldsymbol{r}\in A_W\ ,\\ \Phi_W = 0\ :\ \boldsymbol{r}\in A_F\ ,\ \operatorname{grad}\Phi_W\cdot\boldsymbol{n}_S = 0\ :\ \boldsymbol{r}\in A_S\ .\end{array}\right\} \tag{2.34}$$

The kinetic energy of the fluid is obtained as

$$\begin{aligned} E_{kF} &= \frac{1}{2}\rho\int_V \boldsymbol{v}\cdot\boldsymbol{v}\,dV = \frac{1}{2}\rho\int_{A_S+A_W+A_F}\Phi\operatorname{grad}\Phi\cdot\boldsymbol{n}\,dA\\ &= \frac{1}{2}\rho\Big(\int_{A_S}\Phi_S\operatorname{grad}\Phi_S\cdot\boldsymbol{n}_S\,dA + \int_{A_W}\Phi_W\operatorname{grad}\Phi_W\cdot\boldsymbol{n}_W\,dA\Big) + E_{SW}\\ &= 2\pi\rho\Phi_R R^4\dot{R}^2 + \frac{1}{2}\sum m_n\dot{q}_n^2 + E_{SW}\ , \end{aligned} \tag{2.35}$$

where the boundary conditions for Φ_S and Φ_W are taken into account. The definition of m_n is given as

$$m_n = \rho\int_{A_S}\Phi_n\,\Psi_n\,dA\ , \tag{2.36}$$

and $\Phi_R = \Phi_S^*(R)$. E_{SW} in the equation is given by

$$\begin{aligned} E_{SW} &= \frac{1}{2}\rho\Big(\int_{A_W}\Phi_S\operatorname{grad}\Phi_W\cdot\boldsymbol{n}_W\,dA + \int_{A_S}\Phi_W\operatorname{grad}\Phi_S\cdot\boldsymbol{n}_S\,dA\Big)\\ &= \frac{1}{2}\rho R^2\dot{R}\sum\dot{q}_n\Big(\int_{A_W}\Phi_S^*\,\Psi_n\,dA + \int\Phi_n\,d\Omega\Big)\\ &= \frac{1}{2}\rho R^2\dot{R}\sum\dot{q}_n\,g_n\ , \end{aligned} \tag{2.37}$$

where

$$g_n = \int_{A_W}\Phi_S^*\,\Psi_n\,dA + \int\Phi_n\,d\Omega\ , \tag{2.38}$$

in which $d\Omega$ is the solid angle of the sphere surface element for its centre.

The potential energy of the deformation of the sphere can be calculated as $\frac{1}{2}k(R-R_0)^2$, where k is a constant of the elasticity of the sphere and R_0 is the radius of the sphere in its equilibrium state. Taking the thickness of the sphere as h, the kinetic energy of the sphere can be expressed as $\frac{1}{2}\rho_S h\,4\pi R_0^2\dot{R}^2 = \frac{1}{2}M_S\dot{R}^2$. The kinetic and potential energy of the vessel can be expressed in the form of (2.19) and (2.20) respectively. Then the Lagrangian function of the total system is

$$L = 2\pi\rho\Phi_R R^4 \dot{R}^2 + \tfrac{1}{2}M_S\dot{R}^2 + \tfrac{1}{2}\sum(M_n + m_n)\dot{q}_n^2 + \tfrac{1}{2}\rho R^2\dot{R}\sum q_n\, g_n$$
$$-\tfrac{1}{2}k(R - R_0)^2 - \tfrac{1}{2}\sum K_n\, q_n^2 \qquad (2.39)$$

and the equations of motion of the system can be determined as

$$(M_S + 4\pi\rho\Phi_R R^4)\ddot{R} + 2\pi\rho R^3(4\Phi_R + R\tfrac{d\Phi_R}{dR})\dot{R}^2 + k(R - R_0)$$
$$+\tfrac{1}{2}\rho R^2\sum\ddot{q}_n\, g_n + \rho R\dot{R}\sum\dot{q}_n\, g_n = 0\,, \qquad (2.40)$$
$$(M_n + m_n)\,\ddot{q}_n + K_n\, q_n = -\tfrac{1}{2}\rho(R^2\ddot{R} + 2R\dot{R}^2)\, g_n\,, (n = 1, 2, \ldots). \qquad (2.41)$$

Now let $R = R_0(1 + \lambda)$ and assume λ to be sufficiently small compared with unity. Then expanding R in (2.40) and (2.41) with regard to λ and omitting the terms including the second and higher power of λ, we have the first order approximation of the equations as

$$(M_S + 4\pi\rho\Phi_R R_0^4)\ddot{\lambda} + k\lambda = -\tfrac{1}{2}\rho R_0\sum\ddot{q}_n\, g_n\,, \qquad (2.42)$$
$$(M_n + m_n)\ddot{q}_n + K_n q_n = -\tfrac{1}{2}\rho R_0^3\, g_n\,\ddot{\lambda}\,,\ (n = 1, 2, \ldots)\,. \qquad (2.43)$$

Equations (2.42) and (2.43) show that vibration of the sphere and the vessel are coupled by the inertia of the fluid. For a further study of the vibration characteristics of the system, it will be assumed that only the $\tilde{n}^{\text{th}}$ mode dominates the vibration of the vessel. Then the original system will be reduced to a two degree-of-freedom system which is described by the following equations:

$$(M_S + m_S)\ddot{\lambda} + k\lambda = -\frac{1}{2}\rho R_0\,\ddot{q}_{\tilde{n}}\, g_{\tilde{n}}\,, \qquad (2.44)$$
$$(M_{\tilde{n}} + m_{\tilde{n}})\ddot{q}_{\tilde{n}} + K_{\tilde{n}}q_{\tilde{n}} = -\frac{1}{2}\rho R_0^3\, g_{\tilde{n}}\,\ddot{\lambda}\,, \qquad (2.45)$$

where $m_S = 4\pi\rho\Phi_R R_0^4$.

For the determination of the natural frequencies of the reduced system, a simple harmonic motion with a circular frequency ω is assumed. Then, the following equation which determines the natural frequencies of the system is obtained:

$$\det\begin{bmatrix} -(M_S + m_S)\omega^2 + k & -\tfrac{1}{2}\rho R_0\, g_{\tilde{n}}\omega^2 \\ -\tfrac{1}{2}\rho R_0^3\, g_{\tilde{n}}\omega^2 & -(M_{\tilde{n}} + m_{\tilde{n}})\omega^2 + K_{\tilde{n}} \end{bmatrix} = 0 \qquad (2.46)$$

or

$$-(M_S + m_S)\,\omega^2 + k = \frac{\tfrac{1}{4}\rho^2\, R_0^4\, g_{\tilde{n}}^2\,\omega^4}{-(M_{\tilde{n}} + m_{\tilde{n}})\omega^2 + K_{\tilde{n}}} \qquad (2.47)$$

The equation shows that unless $K_{\bar{n}}$ is relatively large, the natural frequency of the sphere in the fluid is affected by the flexibility of the wall of the outer vessel.

In the case in which the natural frequency of the vessel in the absence of the elastic sphere is near to the natural frequency of the sphere, the resulting change in the latter frequency due to the flexibility of the wall of the vessel may not be small.

2.2 Modal Analysis Based On The Finite Element Method

Numerous studies have been carried out since the late 1970s on the practical methods of solving the fluid-structure problems. The method which will be introduced in this section is a so-called hybrid method. Since this method consolidates both equations for fluid and structure, established structural-analysis programs can be used for the numerical calculation involved in this method. The formulation of the hybrid approach reviewed in this section is according to McNeal et al., 1979 [2.2].

2.2.1 Consolidated Equations for Fluid-Structure Analysis

Expressing (2.8) in Cartesian coordinate, the equation for the fluid may be written as

$$\frac{\partial^2 p}{\partial x^2} + \frac{\partial^2 p}{\partial y^2} + \frac{\partial^2 p}{\partial z^2} = \frac{\rho}{B}\frac{\partial^2 p}{\partial t^2} \,. \tag{2.48}$$

The corresponding equation for structure is (2.10). The equation for the equilibrium of the stresses in a particular direction is

$$\frac{\partial \sigma_{xx}}{\partial x} + \frac{\partial \tau_{xy}}{\partial y} + \frac{\partial \tau_{zx}}{\partial z} = \rho_S \frac{\partial^2 w_x}{\partial t^2} \,, \tag{2.49}$$

where σ_{xx}, τ_{xy} and τ_{zx} are the components of stress. These two equations will become the same equations, if

$$w_x = p \,, \tag{2.50}$$

$$\rho_S = \tfrac{1}{B} \,, \tag{2.51}$$

$$\left.\begin{array}{rcccl} \sigma_{xx} & = & \frac{1}{\rho}\frac{\partial p}{\partial x} & = & -\dot{v}_x \,, \\ \tau_{xy} & = & \frac{1}{\rho}\frac{\partial p}{\partial y} & = & -\dot{v}_y \,, \\ \tau_{zx} & = & \frac{1}{\rho}\frac{\partial p}{\partial z} & = & -\dot{v}_z \,, \end{array}\right\} \tag{2.52}$$

where $\dot{v}_x$, $\dot{v}_y$ and $\dot{v}_z$ are the components of fluid acceleration. The second parts of (2.52) is given by (2.2).

It is necessary to establish an analogous stress-strain relation for the fluid. By setting w_y and w_z equal to zero, the stress-strain equation for the structure is

$$\begin{Bmatrix} \sigma_{xx} \\ \tau_{xy} \\ \tau_{zx} \end{Bmatrix} = \begin{bmatrix} G_{11} & G_{14} & G_{16} \\ & G_{44} & G_{46} \\ \text{sym.} & & G_{66} \end{bmatrix} = \begin{Bmatrix} \epsilon_{xx} \\ \gamma_{xy} \\ \gamma_{zx} \end{Bmatrix} , \tag{2.53}$$

where ϵ_{xx}, γ_{xy} and γ_{zx} are the components of strain, and G_{ij} is an element of 6×6 anisotropic matrix which relates $(\sigma_{xx}, \sigma_{yy}, \sigma_{zz}, \tau_{xy}, \tau_{yz}, \tau_{zx})$ to $(\epsilon_{xx}, \epsilon_{yy}, \epsilon_{zz}, \gamma_{xy}, \gamma_{yz}, \gamma_{zx})$. The relation between the displacements and strains are

$$\epsilon_{xx} = \frac{\partial w_x}{\partial x} , \quad \gamma_{xy} = \frac{\partial w_x}{\partial y} , \quad \gamma_{zx} = \frac{\partial w_x}{\partial z} . \tag{2.54}$$

Hence, setting

$$G_{11} = G_{44} = G_{66} = \frac{1}{\rho} , \quad G_{14} = G_{16} = G_{46} = 0 , \tag{2.55}$$

and using equations (2.53), (2.54) and (2.50), equation (2.52) is obtained. The other components of $[G_{ij}]$ can take any values, since ϵ_{yy}, ϵ_{zz} and γ_{yz} are all zero. The transformation introduced above makes it possible to perform the analysis of the motion of fluid by computer programs for structural analysis. Treatment of boundary conditions remains to be shown. At the free surface it should be put

$$w_x = p = 0 . \tag{2.56}$$

The boundary condition for the fluid at the rigid wall boundary is

$$\frac{\partial p}{\partial n} = n_x \frac{\partial p}{\partial x} + n_y \frac{\partial p}{\partial y} + n_z \frac{\partial p}{\partial z} = 0 , \tag{2.57}$$

where n_x, n_y and n_z are the direction cosines. Transforming the equation into the corresponding equation for the structure, we have

$$f_x = n_x \sigma_{xx} + n_y \tau_{xy} + n_z \tau_{zx} = 0 , \tag{2.58}$$

where f_x is the x-component of traction at the boundary. Comparison of these two equations shows that no external force needs to be applied onto the fluid at the boundary.

At the boundary between the fluid and the structure the following pseudo-force should be applied to the fluid:

$$f_F = -A\dot{v}_n , \tag{2.59}$$

where $\dot{v}_n$ is the outward component of the fluid acceleration and A is the area associated with the fluid element. Equation (2.57) follows from (2.58) and (2.52). Substituting $\ddot{w}_n$ for $\dot{v}_n$, the normal acceleration of the structure, gives

$$f_F = -A\ddot{w}_n \, . \tag{2.60}$$

The force which should be applied to the structure at the fluid-structure interface is

$$f_S = A\,p \, . \tag{2.61}$$

2.2.2 Modal Analysis

Since fluid and structure are continuous media, they have infinite degrees-of-freedom of motion, and accordingly, an infinite number of natural modes of vibration. It is usually difficult to take all these modes into consideration in the analysis of the vibration problem of a continuous system. Therefore, in many cases of vibration analysis, the dynamic characteristics of the system are approximated by considering only a few dominant modes of vibration of the system. This is the main reason why the technique of modal analysis, which is introduced in this section, is useful in engineering analysis of vibration problems.

The equations of motion of a continuous body are first discretized by the standard finite element technique as

$$\boldsymbol{M}\ddot{\boldsymbol{x}} + \boldsymbol{C}\dot{\boldsymbol{x}} + \boldsymbol{K}\boldsymbol{x} = \boldsymbol{f} \, , \tag{2.62}$$

where $\boldsymbol{x}$ is the vector of displacements, $\boldsymbol{f}$ is the vector of externally applied forces, and $\boldsymbol{M}$, $\boldsymbol{C}$ and $\boldsymbol{K}$ are mass, damping and stiffness matrices respectively.

The next step is the determination of the eigenvalues, or the natural circular frequencies, of the system. Omission of the damping and external force terms from the equation of the system leads to

$$\boldsymbol{M}\ddot{\boldsymbol{x}} + \boldsymbol{K}\boldsymbol{x} = \boldsymbol{o} \, . \tag{2.63}$$

Assuming the solution can be separated to the amplitude and a time function, we put $\boldsymbol{x} = \boldsymbol{u}^{\lambda t}$ in the equation and obtain

$$(\boldsymbol{M}\lambda^2 + \boldsymbol{K})\boldsymbol{u} = \boldsymbol{o} \, . \tag{2.64}$$

In order that $\boldsymbol{u}$ in (2.64) has a value not equal to zero, the following condition must be satisfied:

$$\det\,(\boldsymbol{M}\lambda^2 + \boldsymbol{K}) = 0 \, . \tag{2.65}$$

Solving this equation for λ, we obtain the eigenvalue $\lambda_n = i\omega_n$ of each mode of vibration of the system. Any vector which satisfies (2.64), when λ

equals λ_n , is called the eigenvector of the nth mode of vibration and will be expressed by $\boldsymbol{u}^n$.

Using the matrix which is defined by

$$\boldsymbol{U} = [\boldsymbol{u}^1 , \boldsymbol{u}^2 , ..., \boldsymbol{u}^n] , \tag{2.66}$$

we can express the displacements as

$$\boldsymbol{x} = \boldsymbol{U}\boldsymbol{q} , \tag{2.67}$$

where $\boldsymbol{q}$ is the generalized coordinates. Substituting (2.67) into (2.62) and premultiplying both sides of the equation by $\boldsymbol{U}^T$, we obtain

$$\boldsymbol{U}^T\boldsymbol{M}\boldsymbol{U}\ddot{\boldsymbol{q}} + \boldsymbol{U}^T\boldsymbol{C}\boldsymbol{U}\dot{\boldsymbol{q}} + \boldsymbol{U}^T\boldsymbol{K}\boldsymbol{U}\boldsymbol{q} = \boldsymbol{f}_q , \tag{2.68}$$

where $\boldsymbol{f}_q = \boldsymbol{U}^T \boldsymbol{f}$.

Damping is often an important consideration in the vibration analysis. A popular definition of damping matrix in the modal analysis is of the type

$$\boldsymbol{C} = \alpha\boldsymbol{M} + \beta\boldsymbol{K} , \tag{2.69}$$

where α and β are constants determined experimentally. For the more generalized treatment of damping the reader should refer to a text by Hurty and Rubinstein, 1964 [2.3]. Since the eigenvectors are orthogonal to each other, $\boldsymbol{U}^T\boldsymbol{M}\boldsymbol{U}$ and $\boldsymbol{U}^T\boldsymbol{K}\boldsymbol{U}$ are diagonal matrices. We normalize the $\boldsymbol{u}^n$'s, so that $\boldsymbol{U}^T\boldsymbol{M}\boldsymbol{U}$ becomes unity matrix $\boldsymbol{I}$, and we express $\boldsymbol{U}^T\boldsymbol{K}\boldsymbol{U}$ as

$$\boldsymbol{U}^T\boldsymbol{K}\boldsymbol{U} = \boldsymbol{\Omega} = \begin{bmatrix} \omega_1^2 & & 0 \\ & \ddots & \\ 0 & & \omega_n^2 \end{bmatrix} . \tag{2.70}$$

Then (2.68) can be written as

$$\ddot{\boldsymbol{q}} + (\alpha\boldsymbol{I} + \beta\boldsymbol{\Omega})\dot{\boldsymbol{q}} + \boldsymbol{\Omega}\boldsymbol{q} = \boldsymbol{f}_q . \tag{2.71}$$

As $\boldsymbol{\Omega}$ is diagonal, (2.71) can be resolved into a set of the following independent equations of motion for individual generalized coordinate:

$$\ddot{q}_n + 2\zeta_n\omega_n\dot{q}_n + \omega_n^2 q_n = f_{qn} , \tag{2.72}$$

where

$$\zeta_n = \frac{1}{2}\left(\frac{\alpha}{\omega_n} + \beta\omega_n\right) . \tag{2.73}$$

Solving these equations with proper initial conditions and combining these solutions by (2.67), the displacement of the original system can be reconstructed.

2.2.3 Analysis of Fluid-Structure System

The equations for fluid and structure are discretized by the application of the standard finite element technique (Zienkiewicz, 1971 [2.4]) and can be written in the form

$$\boldsymbol{M}_S\,\ddot{\boldsymbol{x}} + \boldsymbol{K}_S\,\boldsymbol{x} = \boldsymbol{f}_S + \boldsymbol{f}\,, \tag{2.74}$$

$$\boldsymbol{M}_F\,\ddot{\boldsymbol{p}} + \boldsymbol{K}_F\,\boldsymbol{p} = \boldsymbol{f}_F \tag{2.75}$$

where $\boldsymbol{x}$ and $\boldsymbol{p}$ are the vectors of structural displacements and fluid pressures, respectively. $\boldsymbol{M}_S$ and $\boldsymbol{K}_S$ are the mass and stiffness matrices of the structure. $\boldsymbol{M}_F(=1/B)$ and $\boldsymbol{K}_F(=1/\rho_F)$ are those of the fluid; $\boldsymbol{f}$ is the vector of externally applied forces; $\boldsymbol{f}_F$ and $\boldsymbol{f}_S$ are the vectors of forces applied onto the fluid and the structure respectively, at the boundary between the fluid and the structure. It can be proved that (2.60) and (2.61) lead to

$$\boldsymbol{f}_S = \boldsymbol{A}^T\boldsymbol{p}\,, \tag{2.76}$$

$$\boldsymbol{f}_F = -\boldsymbol{A}\,\boldsymbol{x}\,, \tag{2.77}$$

where $\boldsymbol{A}$ is the area matrix. Substituting these into (2.74) and (2.75) and expressing these equations in a matrix form, we obtain

$$\begin{bmatrix} -\omega^2\boldsymbol{M}_S + \boldsymbol{K}_S & -\boldsymbol{A}^T \\ -\omega^2\boldsymbol{A} & -\omega^2\boldsymbol{M}_F + \boldsymbol{K}_F \end{bmatrix} \begin{Bmatrix} \boldsymbol{x} \\ \boldsymbol{p} \end{Bmatrix} = \begin{Bmatrix} \boldsymbol{f} \\ \mathbf{o} \end{Bmatrix}, \tag{2.78}$$

where $\boldsymbol{x}$ and $\boldsymbol{p}$ are assumed to be harmonic functions. This equation is difficult to solve directly, because the coefficient matrix in the equation is asymmetric. The coefficient matrix can be symmetrized by the following procedure. By solving the equation without the coupling, the eigenvectors for the fluid and the structure can be derived. Then $\boldsymbol{x}$ and $\boldsymbol{p}$ can be written in the modal expression as

$$\boldsymbol{x} = \boldsymbol{U}_S\,\boldsymbol{q}_S\,, \tag{2.79}$$

$$\boldsymbol{p} = \boldsymbol{U}_F\,\boldsymbol{q}_F\,, \tag{2.80}$$

where $\boldsymbol{U}_S$ and $\boldsymbol{U}_F$ are the eigenvector matrices of the structure and the fluid respectively, without interaction. And $\boldsymbol{q}_S$ and $\boldsymbol{q}_F$ are generalized coordinates of the structure and the fluid, respectively.

The resulting equation of motion is

$$\begin{bmatrix} -\omega^2\boldsymbol{m}_S + \boldsymbol{k}_S & -\boldsymbol{s}^T \\ -\omega^2\boldsymbol{s} & -\omega^2\boldsymbol{m}_F + \boldsymbol{k}_F \end{bmatrix} \begin{Bmatrix} \boldsymbol{q}_S \\ \boldsymbol{q}_F \end{Bmatrix} = \begin{Bmatrix} \boldsymbol{U}_S^T\boldsymbol{f} \\ \mathbf{o} \end{Bmatrix}, \tag{2.81}$$

where

$$\begin{aligned} \boldsymbol{s} &= \boldsymbol{U}_F^T\boldsymbol{A}\boldsymbol{U}_S\,, \\ \boldsymbol{m}_S &= \boldsymbol{U}_S^T\boldsymbol{M}_S\boldsymbol{U}_S\,, \quad \boldsymbol{k}_S = \boldsymbol{U}_S^T\boldsymbol{K}_S\boldsymbol{U}_S\,, \\ \boldsymbol{m}_F &= \boldsymbol{U}_F^T\boldsymbol{M}_F\boldsymbol{U}_F\,, \quad \boldsymbol{k}_F = \boldsymbol{U}_F^T\boldsymbol{K}_F\boldsymbol{U}_F \end{aligned} \tag{2.82}$$

are diagonal matrices. From the bottom row of (2.81) we have

$$q_F = \omega^2 k_F^{-1}(s\, q_S + m_F\, q_F)\,, \tag{2.83}$$

which is substituted into the top row of (2.81). By multiplying both sides of (2.83) by m_F, we obtain the new bottom row and the coefficient matrix in (2.81) is symmetrized as

$$\begin{bmatrix} -\omega^2(m_S + s^T k_F s) + k_S & -\omega^2 s^T k_F^{-1} m_F \\ -\omega^2 m_F k_F^{-1} s & \omega^2 m_F k_F^{-1} m_F - m_F \end{bmatrix} \begin{Bmatrix} q_S \\ q_F \end{Bmatrix} = \begin{Bmatrix} U_S^T f \\ o \end{Bmatrix} \tag{2.84}$$

Since the coefficient matrix of this equation is symmetrical, eigenvalues can easily be found and the response analysis can be made by the method introduced in the previous paragraph. As an example of application of the theory, the eigenvalues of a water-filled cylindrical shell structure made of steel were calculated. Figure 2.5 shows only one fourth of the structure, since the structure is symmetric with respect to its central axis. The shell is simply supported at both ends and its thickness is 14 mm. The vibration analysis of this type of fluid-structure system may be necessary in the design of a large diameter pipe structure. The first several mode shapes are shown in Figure 2.6. None of these modes have any nodes between the supporting points in the longitudinal direction.

In Figure 2.7 the results of numerical calculations of the natural frequencies for these modes are compared with the analytical solution, Kondo, 1989 [2.5]. Agreement between the two solutions is practically acceptable. The influence of the fluid is relatively large on the modes of simpler shapes. Note that lower circumferential wave numbers do not correspond to lower natural frequencies. Special attention should be paid to this fact in this type of analysis, since the standard programmes of finite element analysis will find the natural frequencies in their order of magnitude from the smallest. It is important not to drop these modes of low circumferential wave numbers in the analysis, because the omission of these important modes may badly damage the response analysis of the fluid-structure system.

2.3 Model Test Of Hydroelastic Vibration

In the study of a vibration problem in hydraulic machinery, a theoretical analysis of the problem can be a good guide for the understanding of the basic nature of the phenomena concerned. However, as the analysis cannot always give the quantitative information on the phenomena, which is needed for the actual settlement of the vibration trouble, some experiments will become necessary. The basic consideration which should be given in planning the experiments on fluid-structure problems will be summarized in this section.

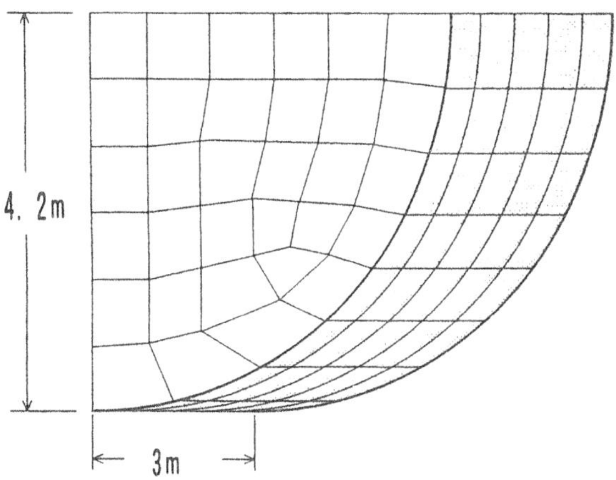

Figure 2.5 Model for FEM calculation

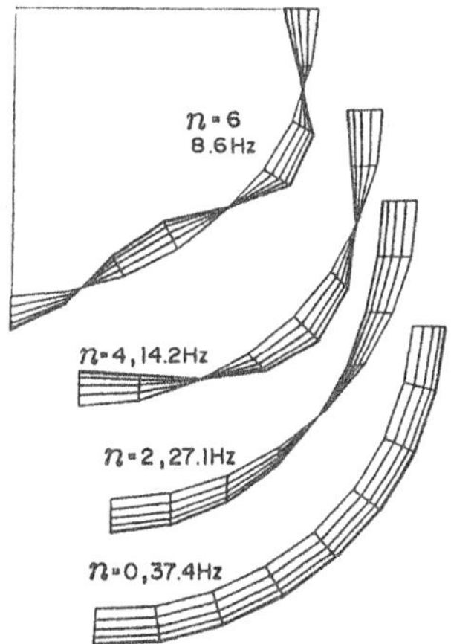

Figure 2.6 Example of mode shapes

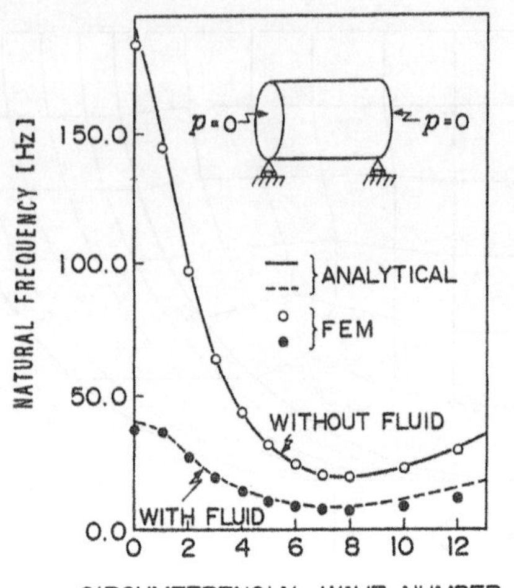

Figure 2.7 Natural frequencies of fluid-contained vessel

2.3.1 Similarity Law for Hydroelastic Vibration

It is a common practice to test the performance of large hydraulic machines using reduced-scale models. It is sufficient for the purpose to keep similarity of only the flow conditions for a prototype machine and its model. However, when the hydroelastic vibration of a machine is the object of the study, the similarity must be maintained not only on flow parameters but also on parameters related to vibration. Ohashi et al., 1970 [2.6], show the similarity conditions necessary for testing models of hydroelastic vibration. The following discussion is based on their work.

There are ten hydrodynamic and structural parameters which are involved in setting the running condition of a machine. These parameters are: representative length D (m, outer diameter of runner, for example), rotational speed n (Hz), flow rate Q (m^3/s), density of fluid ρ (kg/m^3), kinetic viscosity of fluid ν (m^2/s), bulk modulus of fluid B (Pa), net positive suction head (NPSH) (m, difference between total head at the runner position and vapor pressure head of the fluid), density of structural material ρ_S (kg/m^3), modulus of longitudinal elasticity E (Pa) and modulus of transverse elasticity G (Pa). According to dimensional analysis, there are seven non-dimensional numbers which uniquely identify the running condition of a machine. They are;

(1) flow coefficient

$$\phi = \frac{Q}{D^3 n}, \tag{2.85}$$

(2) cavitation coefficient

$$\sigma_c = \frac{2g(\mathrm{NPSH})}{(\pi D n)^2} \tag{2.86}$$

(3) Reynolds number

$$R_e = \frac{\pi D^2 n}{\nu} \tag{2.87}$$

(4) Mach number in fluid

$$M_K = \frac{\pi D n}{\sqrt{B/\rho}} \tag{2.88}$$

(5) Mach number for longitudinal elastic wave

$$M_E = \frac{\pi D n}{\sqrt{E/\rho_S}} \tag{2.89}$$

(6) Mach number for transverse elastic wave

$$M_G = \frac{\pi D n}{\sqrt{G/\rho_S}} \tag{2.90}$$

(7) density ratio

$$\rho/\rho_S \tag{2.91}$$

When these seven numbers coincide for two hydraulic machines with similar form and different sizes, they can be said to be running under identical conditions. Then coincidence can be assured for the non- dimensional numbers of other combinations of parameters, such as efficiency η, head coefficient $\psi = 2gH/(\pi Dn)^2$ (H = total head), non- dimensionalized stress $\sigma_{ij}/\rho g H$, coefficient of pressure fluctuation $c_p = \Delta p/\rho g H$, non-dimensionalized frequency f/n, and strain ϵ.

Of the seven numbers above, flow and cavitation coefficients are relatively easy to set equally for the two machines with different sizes. So model tests are usually run by keeping the similarity of these conditions. However, it is usually very difficult to make all the seven similarity numbers identical at the same time for the two machines. So, the numbers which dominate the phenomenon of concern are deliberately chosen, and the model tests are performed with the remaining similarity conditions unsatisfied.

2.3.2 Model Test Under Actual Head Condition

When a model machine is constructed with the material used in the prototype machine and it is run at the same peripheral speed and with the same operating fluid as the prototype machine, all other similarity conditions except the condition on the Reynolds number can be met. The Reynolds number for the reduced-scale machine will be reduced to the scale. However,this type of test operation can be satisfactory for very large hydraulic machines, because they are usually operated at very high values of Reynolds number and the variation of the Reynolds number will not cause any significant changes in their internal flow patterns. In this type of model test the head can be set approximately equal for the two machines. The magnitude of the static stresses or their dynamic variation will also be the same for both machines. The periods of vibration of the reduced-scale model will reduce in proportion to the reduction of the scale.

An example of an actual head test was performed by Tanaka et al., 1985 [2.7]. A high-head pump-turbine was suspected of suffering from vibration trouble in its runner. Thus a model test for the hydroelastic vibration of the runner was planned and conducted. In this test the main concern was in the hydrodynamic interaction between the runner vanes and the guide vanes.

At the instant a runner vane passes by a guide vane, a hydrodynamic impulse is transmitted onto the runner vane. The transmitted impulses to the runner vane are periodical with the frequency of

$$f = z_g \, n \; , \tag{2.92}$$

where n (Hz) is the rotational speed of the runner and z_g is the number of the guide vanes set around the runner. These periodical impulses are applied to every runner vane and may contribute to the excitation of the structural vibration of the runner, when the following condition is met;

$$\tilde{n} \, z_g \pm k = \tilde{m} \, z_r \; , \tag{2.93}$$

where k is the number of nodal diameters of the structural vibration of the runner as shown in Figure 2.8; z_r the number of the runner vanes; *tilden* and $\tilde{m}$ arbitrary integers. $\tilde{n}$ and $\tilde{m}$ correspond to the order of harmonics of the excitation to rotating and stationary structures, respectively. The second and higher harmonics of the excitation ($\tilde{n} \geq 2$) are usually unimportant. When unfortunately (2.93) is met for a certain mode of the structural vibration of the runner and the natural frequency of the structural vibration coincides with the exciting frequency $\tilde{n} z_g n$, resonance occurs.

A schematic view of the model pump-turbine is shown in Figure 2.9, and the test conditions are given in Table 2.1. Some typical results of the test are observed in Figure 2.10, in which $\Delta\sigma$ is the variation of stress observed on

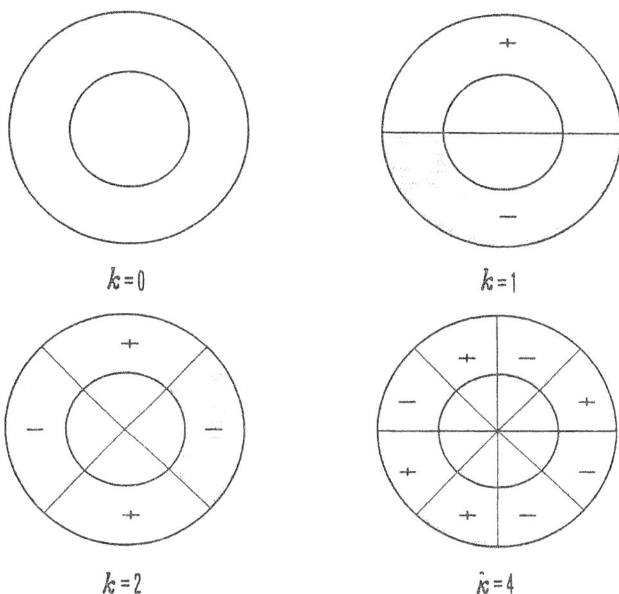

Figure 2.8 Vibration mode of runner

Table 2.1 Ratings of model tested

Ratings	Single-stage	Multi-stage
Test head (m)	1,200	2,000
Test speed (rpm)	8,000	5,000
Max. runner dia. (mm)	500	450
Max. input power (kW)	7,000	5,000

the surface of the runner and n_0 is the rated speed of the runner. Resonance can be clearly observed at the speed a little smaller than the rated value, $n = 0.85n_0$.

Since the flow patterns at the periphery of the runners are similar for most of the pump-turbines of over 500 m head, we can assume $Dn \propto \sqrt{H}$ for them. Then, as the natural frequencies of structural vibration of a certain vibration mode for a similarly shaped pump-turbine, f_0, can be assumed to vary inversely to its diameter, we obtain

$$\frac{f}{f_0} \propto z_g\sqrt{H} . \tag{2.94}$$

Equation (2.94) predicts that the value of the relative speed n/n_0, at which resonance occurs, will almost be the same for the prototype turbine and its reduced-scale model, when z_g and H are set equal for the two. The equation also tells us that this kind of resonance is more likely to occur in high head pump-turbines.

The field measurements of the stress fluctuation of the prototype turbine were made afterwards to check the test results on the model turbine. The results are shown in Figure 2.11. Comparison of the results of the two tests, shown in Figure 2.12, confirms that similarity conditions had been properly satisfied in the model test.

2.3.3 Model Test Under Reduced Head Condition

Even when the model test cannot be carried out under the actual head condition, the similarity conditions can be partly satisfied by choosing the proper material for the model. In the following discussion the suffices M and P denote the model and prototype respectively.

Suppose that it is required to make Mach number for longitudinal elastic wave of a model equal to that of the prototype. Then,

$$M_E = \frac{\pi D_P n_P}{\sqrt{E_P/\rho_{SP}}} = \frac{\pi D_M n_M}{\sqrt{E_M/\rho_{SM}}} . \tag{2.95}$$

This requirement can be met, if the following condition is satisfied:

$$\frac{H_M}{H_P} = \frac{(D_M n_M)^2}{(D_P n_P)^2} = \frac{E_M \rho_{SP}}{E_P \rho_{SM}} . \tag{2.96}$$

Thus, equation (2.95) holds, even when it is taken as $H_M/H_P < 1$, by the proper choice of material for the model.

Then, the other similarity conditions will be checked. First, similarity of flow will be studied. As long as the same fluid is used for the model

① SUPPORTING PLATE
② SPIRAL CASE
③ BOTTOM COVER
④ DRAFT TUBE
⑤ MECHANICAL SEAL
⑥ GUIDE VANES
⑦ HEAD COVER
⑧ SHAFT
⑨ RUNNER

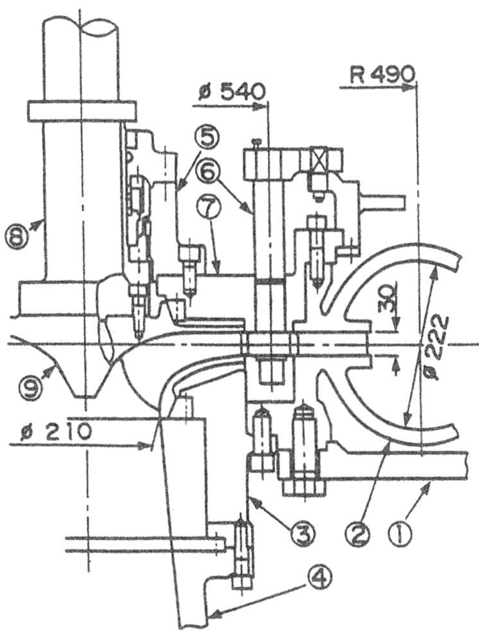

Figure 2.9 Schematic view of model turbine

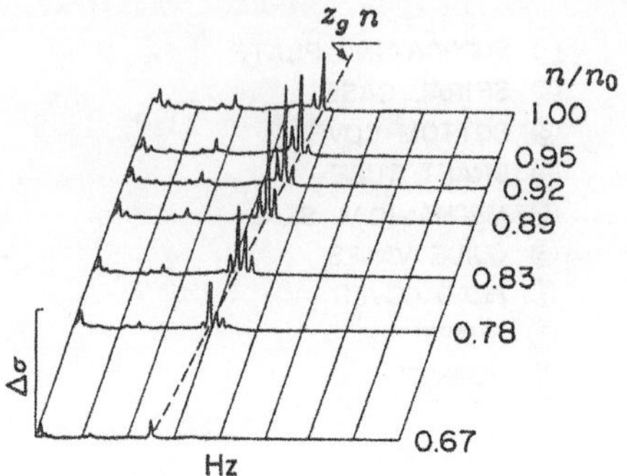

Figure 2.10 Typical results of model test

and prototype machine, it is impossible to keep the similarity of Reynolds number R_e and Mach number M_K, because

$$\frac{R_{eM}}{R_{eP}} = \frac{D_M^2 n_M \nu_P}{D_P^2 n_P \nu_M} = \sqrt{\frac{H_M}{H_P}} \frac{D_M \nu_P}{D_P \nu_M} , \tag{2.97}$$

$$\frac{M_{KM}}{M_{KP}} = \frac{D_M n_M}{D_P n_P} \sqrt{\frac{\rho_M B_P}{\rho_P B_M}} = \sqrt{\frac{H_M \rho_M B_P}{H_P \rho_P B_M}} . \tag{2.98}$$

It is usually not practical to use different fluids for both machines. Therefore, the possible effects of the difference of the Reynolds number in the results of the test should be studied carefully. Fortunately, in most hydroelastic tests, the Reynolds number is not the dominant parameter.

In the experiments in which such effects as the propagation of pressure waves or a rapid transient phenomenon is involved, the compressibility of fluid plays an important role and Mach number in the fluid can be the dominant parameter. So, in these experiments, the model cannot simulate the actual phenomena exactly.

In regard to the Mach number for transverse elastic wave, the following relation can be derived from (2.89) and (2.90):

$$\frac{M_{GM}}{M_{GP}} = \frac{M_{EM}}{M_{EP}} \sqrt{\frac{E_M G_P}{E_P G_M}} = \frac{M_{EM}}{M_{EP}} \sqrt{\frac{1+(1/m)_M}{1+(1/m)_P}} , \tag{2.99}$$

in which $(1/m)_M$ and $(1/m)_P$ are Poisson's ratio of the material used for model and prototype. As Poisson's ratio is in the range of $0.3 \sim 0.4$ for ordinary materials, the coincidence in M_G can be achieved approximately, if the coincidence in M_E is achieved.

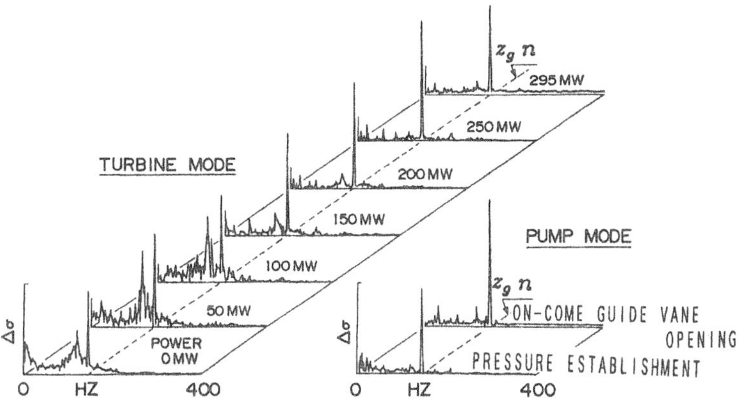

Figure 2.11 Results of prototype test

The density ratios do not coincide, when the fluid used in the model test is the same as that used in the prototype machine and the structural material used in the model is different from that used in the prototype machine. As the change in the natural frequencies of a structure due to the surrounding fluid depends on the density ratio, the frequencies of excitation must be modified so as to compensate for the effect of the difference of the density ratio between the model and prototype machines, especially in the test for the study of resonant vibration.

2.3.4 Self-Excited Vibration of Francis Turbine

In planning a test for the study of a self-excited vibration caused by the fluid-structure coupling, special care must be taken so that the mechanism of excitation will be simulated correctly. This point is illustrated by a test which was performed by Tomita and Kawamura, 1965 [2.8], and will be introduced in this paragraph.

Severe flexural vibration of the rotor shaft of a Francis turbine generator shown in Figure 2.13 was experienced in its test operation. The vibration always occurred, while the load of the generator was being increased. Once started, the vibration grew by itself to a very large amplitude. After thorough investigation of the phenomenon, it was suspected that the water which leaked through the sealing gap between the runner and the casing, shown in Figure 2.14 (a), was the cause of the vibration.

As long as the runner remains in the centre of the casing, the leak flow is uniform around the runner. Once the runner deviates from the centre of

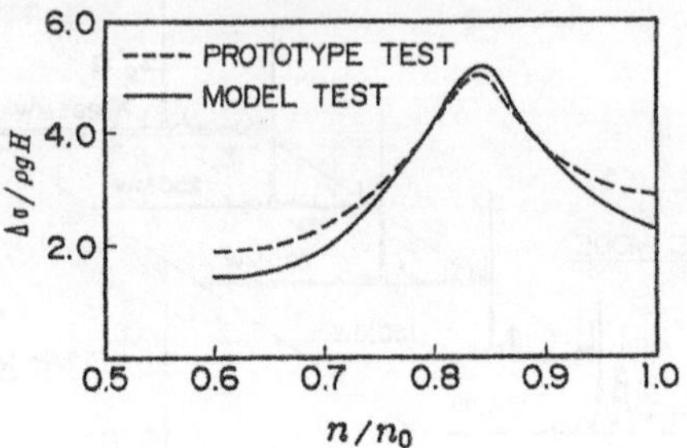

Figure 2.12 Measured stress

the casing by y_1, the uniformity will be broken and the resulting change of the leak flow, represented by q in Figure 2.14 (b), will occur in the back chamber. When the runner vibrates, the variational flow must change its direction accordingly, thus causing the variation of pressure in the chamber. If we see the chamber as a kind of duct whose area is hD, the space gradient of the variational pressure p can be expressed as

$$\frac{\partial p}{\partial y} = -\rho \frac{1}{hD} \dot{q} , \tag{2.100}$$

which follows directly from (2.2). The integration of (2.100) gives

$$p = -\rho \frac{y}{hD} \dot{q} . \tag{2.101}$$

The assumption of

$$q \propto \bar{v} D \, y_1 , \tag{2.102}$$

in which $\bar{v}$ is the mean flow velocity in the sealing gap, leads to

$$p \propto -\rho \frac{y}{hD} \bar{v} D \, \dot{y}_1 = -\rho \frac{y}{h} \bar{v} \, \dot{y}_1 . \tag{2.103}$$

The variational pressure which is applied to a small surface element of the runner will give the runner a hydraulic moment. Integration of the moments acting on the individual surface elements of the runner gives the total hydraulic moment acting on the runner:

$$\mathcal{M}_H = -\int p \, dA \, (y - y_1) \approx -\int p \, dA \, y = a_H \, \dot{y}_1 , \tag{2.104}$$

where

$$a_H = c_H \, \rho \, \frac{\bar{v}}{h} \, D^4 \tag{2.105}$$

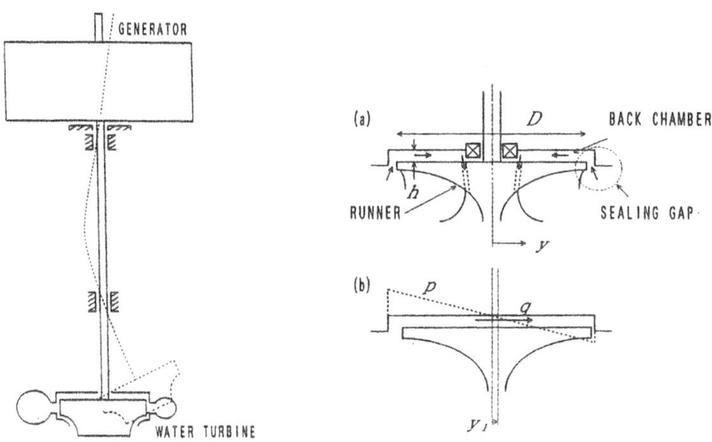

Figure 2.13 Flexural vibration of Francis turbine

Figure 2.14 Leak flow model

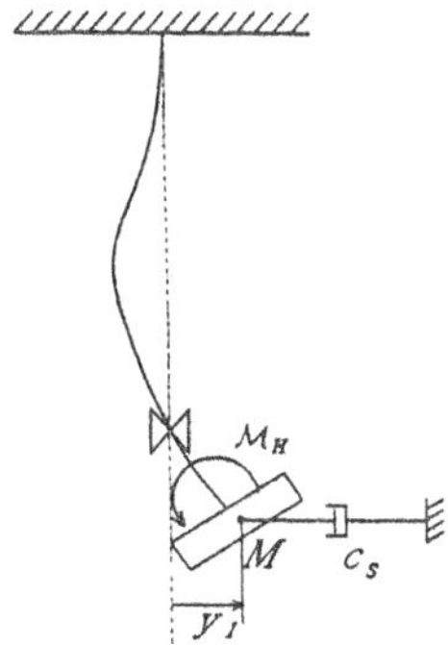

Figure 2.15 Model of lateral vibration

and dA is the area of a small surface element; c_H is a non-dimensional constant of positive value.

Now, in order to write the equation of motion of the rotor system, we will take a simplified model of the system shown in Figure 2.15, in which M is the inertia of the runner and c_s is the equivalent coefficient of the structural damping of the system. Actually the energy of vibration is dissipated mainly in the supporting structures of the rotor. The equation of the motion can be written down as

$$y_1 = \xi(-M\ddot{y}_1 - c_s\dot{y}_1) + \eta(a_H\dot{y}_1)\,, \tag{2.106}$$

where ξ (η) is the influence number of the system for force (moment), and is defined as the displacement of the shaft at the runner, when unit force (moment) is applied to the runner. Rearranging (2.106) gives

$$\xi M\ddot{y}_1 + (\xi c_s - \eta a_H)\dot{y}_1 + y_1 = 0\,. \tag{2.107}$$

The system is stable, if the following condition is satisfied:

$$\xi c_s > \eta a_H\,. \tag{2.108}$$

And at the boundary of stability

$$\bar{v}_{critical} = \frac{c_s}{\rho c_H(D^4/h)}\cdot\frac{\xi}{\eta}\,. \tag{2.109}$$

Now the dependence of $\bar{v}_{critical}$ on the scale of the machine will be studied. Taking D as the representative parameter of the scale of the machine, it can easily be proved that

$$\frac{\xi}{\eta} \propto D\,. \tag{2.110}$$

To identify the dependence of structural damping c_s on the scale of the machine, we need an additional assumption. If the energy lost per cycle of vibration in the system is supposed to be proportional to the maximum potential energy stored in the rotor shaft, then,

$$\pi c_s\omega_0\xi^2 \propto \frac{1}{2\xi}\xi^2 = \frac{1}{2}\xi\,, \tag{2.111}$$

where ω_0 is the circular natural frequency of the system.

As ξ and the natural frequency of the system ω_0 vary inversely to D, it is concluded that

$$c_s \propto \frac{1}{\omega_0\xi} \propto D^2\,. \tag{2.112}$$

Thus, assuming that (2.111) holds, it follows that $\bar{v}_{critical}$ does not depend on D, the scale of the machine.

According to the conclusion thus obtained, a model turbine-generator which was similarly reduced to the scale of 1/20 of actual size was constructed and tested under the actual head condition. The self-excited vibration was reproduced on the model machine with exactly the same characteristics of the phenomenon observed in the field. The test was quite satisfactory.

References

2.1 Kubota, Y. et al. (1979). Trans. ASME, Ser. C ,Vol. 50, No. 449, pp. 243-248.

2.2 McNeal, R.H. et al. (1979). Trans. 5th SMiRT, B4/9.

2.3 Hurty, W.C. and Rubinstein, M.F. (1964). *Dynamics of Structures*, Prentice Hall.

2.4 Zienkiewicz, O.C. (1971). *The Finite Element Method in Engineering Science*, McGraw-Hill.

2.5 Kondo, H. et al. (1989). Proc. PVP-89, ASME.

2.6 Hironaka, K. and Ohashi, H. (1970). J. JSME, Vol. 73, No. 623, pp. 1638-1644.

2.7 Tanaka, H. et al. (1985). *Turbomachinery*, Vol. 13, No. 15, pp. 607-612.

2.8 Tomita, H. and Kawamura, M. (1965). *Toshiba Review*, Vol. 20, No. 8, pp. 787-791.

Chapter 3

Structural Vibration due to Mechanical Excitation

Dung Yu-Xin

3.1 Excitation By Rotating Shaft

3.1.1 Introduction

Mechanical excitations resulting in the vibration of the hydrogenerator shaft, its supporting elements and foundation may be produced by a number of factors such as; defects in the construction and erection (for instance bent hydrogenerator shaft); unbalanced rotor or turbine runner; friction or impact between rotating and fixed parts; faulty guide and thrust bearings; internal hysteresis of shaft.

3.1.2 Response to Unbalance

The unbalance of generator rotors or turbine runners means that there is unbalanced force, or unbalanced force and moment. The response to this unbalance is one of mechanical excitations caused by the rotating shaft, which includes the generator rotor, exciter rotor and turbine runner. The unbalanced rotor is shown in Figure 3.1. The centre of gravity lies at a distance of e from the shaft centre. In operation the rotor is rotating about the shaft axis, producing a rotating centrifugal force F perpendicular to the shaft as shown in the figure, then

$$F = M\,e\,\omega^2 \quad (\mathrm{N})\,, \quad \omega = 2\,\pi\,n \tag{3.1}$$

where M represents the mass of rotor (kg), e the eccentricity (m) and n the rotating speed (Hz).

The unbalance of the generator rotor and turbine runner can be divided into static and dynamic unbalance. To solve the problem of static unbalance,

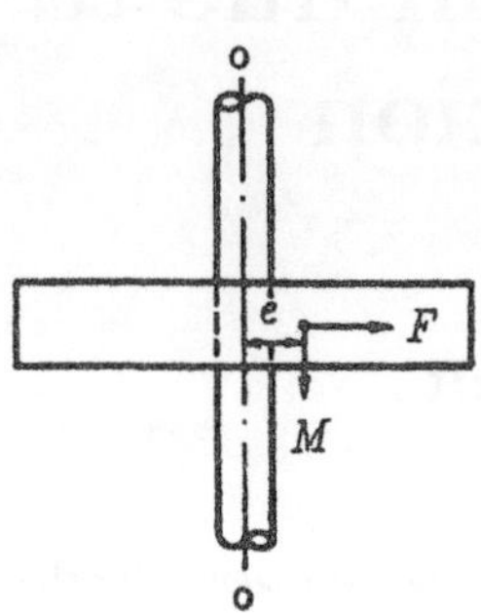

Figure 3.1 Unbalanced rotor

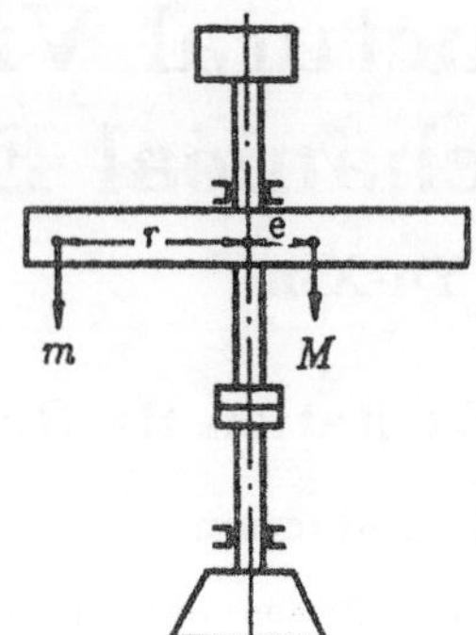

Figure 3.2 Correction of statically unbalanced rotor

a balance weight is added to the lighter side of the rotor as shown in Figure 3.2, making the balance moment and unbalance moment equal about the shaft axis. For the original eccentricity e and the mass M, the balance mass m is applied at a radial distance r from shaft axis, then

$$m\,r = e\,M \qquad \text{or} \qquad m = \frac{e}{r}\,M.$$

Generally speaking, static unbalance of the generator rotor takes place when $n_s \geq 200$, where $n_s = n\sqrt{P}/H^{5/4}$is the specific speed of the turbine (rpm, kW and m units are used), while static unbalance of the turbine runner occurs when $n_s < 200$. Inversely, dynamic unbalance of the generator rotor takes place when $n_s \leq 200$ and dynamic unbalance of the turbine runner occurs when $n_s > 200$.

When a generator has a disk-like form with relatively small axial length, static balance is the only one required to ensure quiet operation at all speeds. In case the rotor has an elongated axial length, static balance alone will not be sufficient. Any unbalance of a rigid rotor (static, dynamic or combined) can be corrected by placing appropriate correction masses in two planes AA and BB. For example, let the existing unbalance eM be a moment of 12 kg m at $\ell/4$ (ℓ is the length of the rotor) from x axis as shown in Figure 3.3. The 12 unit unbalance will cause a 12 unit rotating centrifugal force, which can be held in static equilibrium by a 9 unit force at A and by a 3 unit force at B. In this case, the moment about point zero due to the centrifugal forces on two correction masses is equal and opposite to that of original unbalance mass and, therefore, dynamic balance satisfies the quiet operation at all speeds. Static and dynamic unbalance of generator rotor usually exist side by side. Dynamic unbalance is adjusted after regulation of static unbalance rotor.

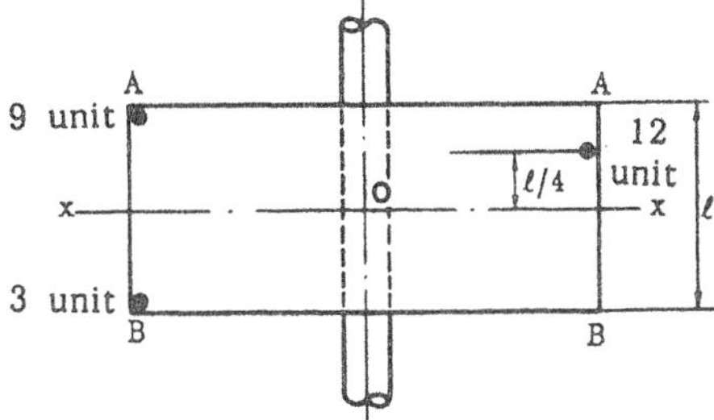

Figure 3.3 Dynamic unbalance can be corrected by placing one correction mass in each of two planes, AA and BB

Vibration of a generator unit produced in operation indicates that rotor unbalance often exists together with electric and hydraulic unbalance. For example, a shaft system of pump-turbine with its four bearings arrangement has to be designed with respect to the dynamics of the system, taking into account predominantly the residual unbalanced forces and the unbalanced magnetic pull in the air gap of the electrical machines. Both produce force vectors rotating at the frequency of the rotation. Their amplitudes depend on the mechanical stability of the rotors and their residual eccentricity and can be reduced to the necessary levels by appropriate design, improved erection and balanced quality.

3.1.3 Dry Friction Excitation

An important case of dry friction excitation is caused by a loose guide bearing or by a poorly lubricated bearing with excessive clearance. Let the circle A shown in Figure 3.4 designate the inside of guide bearing and B the cross section of a vertical shaft. The shaft is rotating clockwise and is temporarily deflected from its equilibrium position at the centre of A, so that it strikes A at point C.

On account of its rotation the shaft sets up friction forces F and F', of which F is the force on the shaft, while $F' = -F$ acts on the guide bearing. The force F can be replaced by a parallel force of equal magnitude through the centre of the shaft B and a couple Fr, where r is radius of the shaft. The couple acts merely as a brake on the shaft which is supposed to be driven at uniform speed, so that the only effect of the couple is to require some increase in the driving torque which is of little consequence.

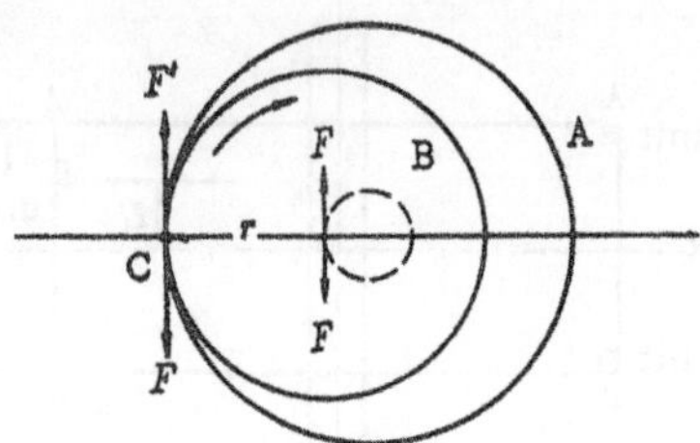

Figure 3.4 Shaft whirl caused by dry friction

The force F through the centre, however, drives it downward or rather to the direction tangent to circle A. The direction of F changes with the position of the shaft B in A, so that the shaft will be driven around as indicated by the dotted circle. It should be noted that the shaft is driven around the clearance to the direction opposite to that of its own rotation. If the shaft rotates in the centre of the guide bearing without touching it, the shaft is stable. As soon as it strikes the guide bearing for any reason, the shaft will be set into a violent whirling vibration.

An experience at a hydroelectric plant indicated that in such a case vibration did not take place when the hydrogenerator unit operated at a normal speed without load. When the speed rose 10% above the normal speed, strong vibration did take place on the shaft and upper supporting elements, and when the speed was 30% below the normal speed strong vibration did occur again. Therefore it was difficult to put the hydrogenerator unit into operation. Studies indicated that the vibration of shaft system was caused by protrusion of runner seal gum. In operation the rotating part brushed against the fixed part and the excited frequency was equal to the rotational frequency in this case. The calculated lateral natural frequency of the shaft system was equal to 2.2 times the rotational frequency at the normal speed. When the speed was equal to 110% or 70% of normal speed, resonance of double or triple frequency did occur.

3.1.4 Vibration Severity

Excited forces on the hydrogenerator unit include hydraulic, mechanical and electrical forces. The vibration state of the unit is usually appraised by a field test. The test value is compared with the standard or the permitted value. Some nations, firms or academic organizations put forward standards, which are based upon test material. Although the theoretical basis of those standards has not yet been sufficiently verified, practice indicates that, when the hydrogenerator vibration satisfies these standards, normal operation and safety of hydrogenerator can usually be ensured.

At the present time there is not any unified international vibration standards of hydrogenerator unit. Although the vibration standards of general machines include the vibration severity of various kinds of rotating machines, the vibration severity of hydrogenerator unit must, in addition, consider the following characteristics:

1. Low-speed machines. The general standard (for example VDI- 2056, ISO-2372) is used for a rotating machine with a speed greater than 600 rpm, but for hydrogenerator, the speed is mostly lower than 600 rpm.

2. Vertical shaft. The vibration standard of general machines is, however, given by experiment with machines of horizontal shaft.

3. Machines of heavy type with low speed and large diameter.

4. Irregular vibration sources (hydraulic vibration forces). The vibration standard of hydrogenerator unit takes account not only vibration severity, but also measuring equipment and measuring methods.

Several representative proposals for vibration appraisal of hydrogenerator unit are introduced here for reference:

1. USSR proposal for IEC TC4 (turbine) in 1963. A monogram of vibration criteria, see [3.11], is put forward based on experiments on domestic hydrogenerator vibration in USSR. This proposal includes generator upper/lower guide bearings and turbine guide bearings. It is applicable to vertical vibration and horizontal vibration in two orthogonal directions. The vibration standard is based on runner diameter $D = 5$ m. For an arbitrary diameter of D_r the vibration standard can be calculated by the following equation

$$A = A_5\sqrt{D_r/5}$$

In Figure 3.5 four zones are classified: Excellent, Good, Satisfactory and Bad. Points A_1 and A_2 indicate the case of a runner diameter $D = 3.5$ m, speed $n = 187.5$ rpm and double vibration amplitude $2A = 0.08$ mm. From the location of point A_2 the vibration severity is appraised as satisfactory. Points B_1 and B_2 indicate the case of $D = 4.5$ m, $n = 214$ rpm and $2A = 0.13$ mm, where point B_1 is located on the border of satisfactory and bad zones.

The critical value (double amplitude) of turbine shaft vibration is

$$\delta = \Delta + 2A_{\max}$$

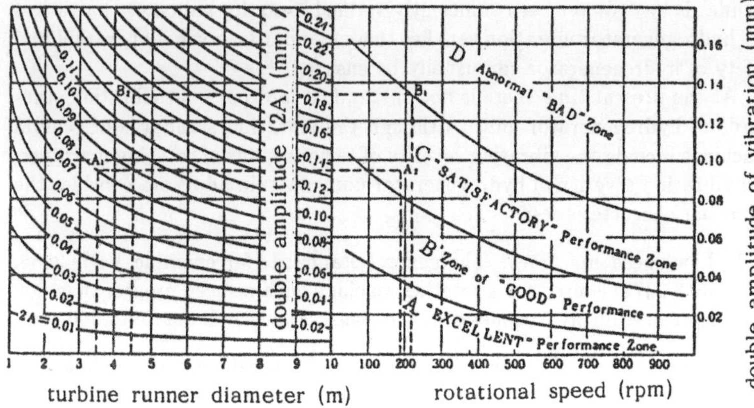

Figure 3.5 Vibration standard of hydrogenerator unit
A-excellent, B-good, C-satisfactory, D-bad

where Δ is the gap between shaft and guide bearing in the radial direction and $2A_{max}$ is double amplitude of bearing vibration.

2. VDI-2059. In 1979 the German Engineering Association, VDI, provided important recommendations [3.8] for the transient state of turbosets. Turbines whose vibration magnitudes are below the line A of Figure 3.6 may be operated without restriction even when its probability of appearance is 100%. Turbines whose vibration magnitudes fall on line B are considered acceptable for the transient state of hydroturbine provided its probability of appearance is $1/100 \sim 1/1000$. Turbines whose vibration magnitudes fall on line C are considered acceptable for very short term operation provided its probability is 10^{-5}. S_{max} is the maximum vibration amplitude of the shaft vibration track.

3. Rund's evaluation. This proposal (Rund, 1980 [3.12]) suggested the vibration amplitude of normal operation and transient state. Recently vibration of transient state has become very important. In addition to the vibration value in normal operation, the probability of vibration appearance must also be considered. In Figure 3.7 the probability of vibration appearance is also given.

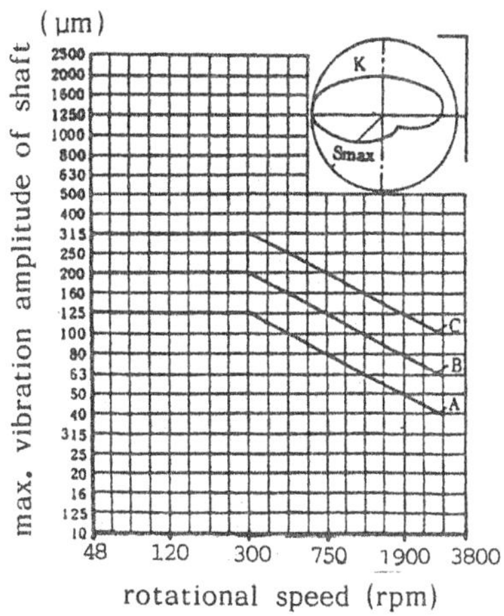

Figure 3.6 Evaluation of shaft vibration of turbosets

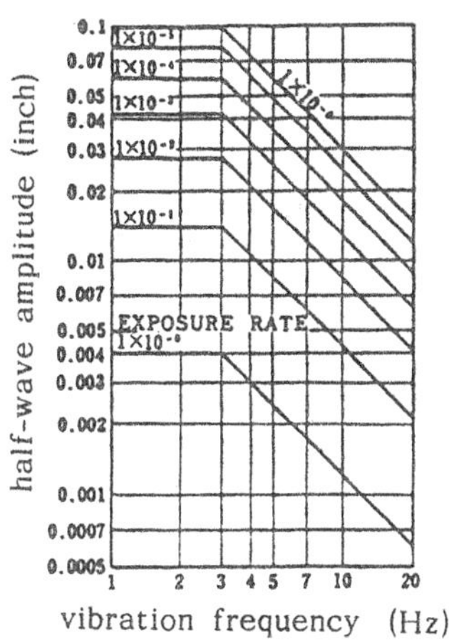

Figure 3.7 Vibration amplitude of normal operation and transient state

Probability of appearance:

1×10^0 –continuous operation
1×10^{-3} –time of permissible appearance is 9 hours in one year
1×10^{-6} –time of permissible appearance is 9 seconds in one year

Other appreciated standards:

1×10^0 –Normal
1×10^{-3} –Strong
1×10^{-6} –Severe

4. USSR proposal for IEC (1972). The vibration standard of turbine and generator is given in Figure 3.8 (a) and (b). These figures show that the allowable value of double amplitude is inversely proportional to vibration frequency. The vibration standard is divided into five zones: A-excellent, B-good, C- satisfactory, D-unsatisfactory and E-impermissible.

In equation 3.1 the unbalance mass M and eccentricity e of the test unit are constant, and hence the value of double amplitude $2A$ is proportional to the centrifugal force F, thus $2A = Kn^2$. When the unit operates under the condition of no-load and no-magnetic current, the oscillation of the shaft or

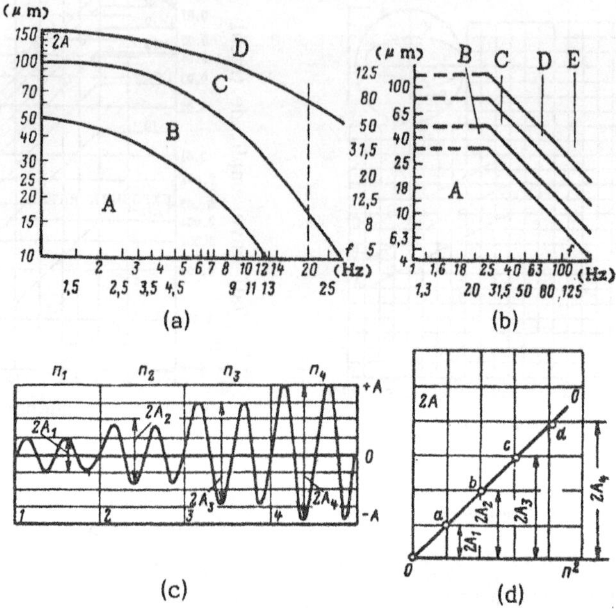

Figure 3.8 Vibration standard of generator and turbine

vibration of the guide bearing can be measured at different speed as shown in Figure 3.8 (c). In Figure 3.8 (d) the line $2A = f(n^2)$ is plotted from Figure 3.8 (c) and the relation between $2A$ and n^2 represents a straight line. But an unbalanced rotor rarely appears alone, even if there is no magnetic current and no hydraulic excitation.

There is still another excited force acting on the rotor. For example, the connective condition of the supporting elements of the unit and unevenness of thrust bearing pads will affect the oscillograms of shaft vibration. In addition, shaft vibration may also be caused by the internal hysteresis of the shaft, dry friction or impact between rotating and fixed elements. In some cases, if the excited frequency is equal to the natural frequency of the shaft or to its multiple, resonance or resonance at doubling frequency may be produced. Therefore an analysis of the excitation source must be based on the character of testing frequency and amplitude in an operating range. In most cases shaft vibration cannot be eliminated by adding balance mass to rotor or runner only.

3.2 Other Excitation Source

3.2.1 Piping Reaction

The general wave equation of a pipe of a hydropower station is

$$\frac{\partial^2 S}{\partial t^2} = c\,\frac{\partial^2 S}{\partial x^2} \tag{3.2}$$

where S is instantaneous amplitude of fluctuation at any point x in the pipeline. The above is a one-dimensional general wave equation — one-dimensional in the sense that waves are travelling parallel to the pipe x-axis, and only this dimension of space appears in the equation. It is a second order partial differential equation. By substitution, a solution of equation (3.2) may be obtained as follows

$$\begin{aligned} S &= a_1 \sin[2\pi(t/T - x/\lambda)] + a_2 \sin[2\pi(t/T + x/\lambda)] \\ &\quad + b_1 \cos[2\pi(t/T - x/\lambda)] + b_2 \cos[2\pi(t/T + x/\lambda)] \end{aligned} \tag{3.3}$$

where a_1, a_2, b_1 and b_2 are arbitrary constants. Each of the terms represents a harmonic wave travelling in the positive or negative x-direction. T is the period of the wave, λ is the distance the wave travels during time T and c is wave velocity, so that $c = \lambda/T$.

Since there is no pressure fluctuation during the operation of hydropower station at the inlet of pipe ($x = 0$), we obtain

$$0 = (a_1 + a_2)\sin 2\pi t/T + (b_1 + b_2)\cos 2\pi t/T$$

which gives

$$a_1 + a_2 = 0 \quad \text{and} \quad b_1 + b_2 = 0$$

or

$$a_2 = -a_1 \quad \text{and} \quad b_2 = -b_1$$

Thus, we now have

$$\begin{aligned} S &= a_1\{\sin[2\pi(t/T - x/\lambda)] - \sin[2\pi(t/T + x/\lambda)]\} \\ &\quad + b_1\{\cos[2\pi(t/T - x/\lambda)] - \cos[2\pi(t/T + x/\lambda)]\} \end{aligned}$$

If there is also no pressure fluctuation at the outlet of pipe ($x = \ell$) at any time

$$\sin\frac{2\pi\ell}{\lambda} = 0 \quad \text{or} \quad \frac{2\pi\ell}{\lambda_m} = m\pi$$

where m is a positive integer. We obtain

$$\ell = \frac{m\lambda_m}{2} \tag{3.4}$$

or

$$f_m = \frac{mc}{2\ell} \tag{3.5}$$

The modes of vibration are thus defined and all frequencies specified by equation (3.5) will be present in the composite vibration. The amplitudes of the various modal vibration depend on the initial conditions. While there is no simple method of determining amplitudes, one of the possible methods may be the graphic method.

Figure 3.9 shows the first few modes of vibration of a water column in a pipeline. (i) is the first mode, for which $\ell = \lambda_1/2$. For the second mode (ii), $\ell = \lambda_2$ and for the third (iii), $\ell = 3\lambda_3/2$. For a general n^{th} mode, $\ell = n\lambda_n/2$, which agrees with equation (3.4).

It is well known that the draft tube pressure pulsations become severe between 30% and 50% load. At partial loads the core makes precession motion around the runner axis at about $1/3 \sim 1/5$ of the turbine speed. This pulsation is called low-frequency pulsation in the draft tube. Besides low-frequency pulsation there is also medium- frequency pulsation in the draft tube. In reference [1] model tests were carried out to determine the characteristics of medium-frequency pulsation in the draft tube. The result is shown in Figure 3.10. When the unit speed $n_1' = nD_1/\sqrt{H}$ is constant, where n is speed, D_1 is runner diameter and H is test head, the medium-frequency of pulsation is also constant.

A field test was carried out at the hydroelectric station Liu JiaXia and Sizitan and the results of medium-frequency of pulsation are shown in Table 3.1.

Table 3.1 Field test data

Name of hydroelectric station	Rotational speed n (Hz)	Medium- frequency of pulsation f (Hz)	f/n
LiuJiaXia unit No.1	2.08	2.41	1.16
LiuJiaXia unit No.5	2.08	2.48	1.19
Sizitan unit No.2	4.50	5.00	1.10

A field test of the penstock of No.5 unit LiuJiaXia (Gong, 1983 [3.1]) was carried out. When the guide vane opening is $a_0 = 25 \sim 30\%$ or $a_0 = 95\%$, there are two vibration regions in the water column of the penstock. The two

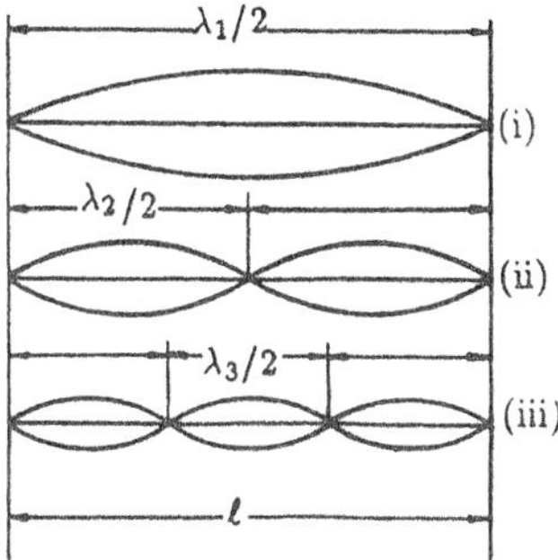

Figure 3.9 Modes of vibration of a water column in the pipeline

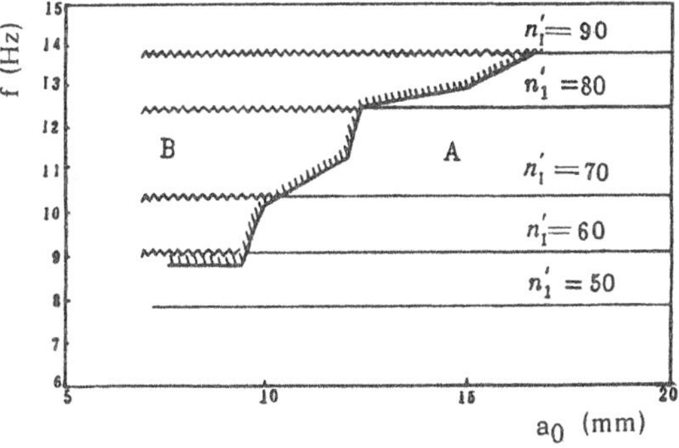

Figure 3.10 Characteristics of medium-frequency pulsation in draft tube
A – Medium-frequency of pulating zone
B – Irregular medium-frequency of pulsating zone

vibration regions have the same measured frequency, that is, 5 Hz. When a_0 = 25 ~ 30% the double-amplitude of vibration is 20 ~ 30 m of water column and when a_0=95% it is 10 ~ 23 m of water column. The same vibration frequency is also measured on the shaft system, on the supporting members of shaft and on the reinforced concrete supporting structure.

The vibration source is analyzed as follows. The length of the penstock of unit No.5 is ℓ=128 m. For $c \approx 1300$ m/s and $m = 1$, the calculated natural frequency of vibration of water column in the penstock is $f = 5.1$ Hz. The excited medium-frequency of pulsation in the draft tube is 2.48 Hz. Therefore, the high vibration amplitude of the water column in the penstock arises from resonance. The reason for inducing resonance is that the natural frequency f of the water column in the penstock is equal to twice the medium-frequency of pulsation in the draft tube.

Therefore, when designing a hydroelectric station, attention must be paid to the relation between the length of penstock and the shaft speed. Otherwise, not only high vibration amplitude would be produced in the penstock, but also this excited force of high vibration amplitude would act on the unit.

3.2.2 Internal Hysteresis of Rotating Shaft as a Cause of Instability

An interesting case of self-excited vibration is caused by the internal hysteresis of the shaft metal. Hysteresis is a deviation from Hooke's stress-strain law. In Figure 3.11 (a) Hooke's law is represented by a straight line P_1 P_3. In reality the fibres of shaft experience alternately tension and compression and the stress-strain relation is represented by a long narrow elliptic loop which is always in the clockwise direction.

Consider a vertical rotating shaft supported by two guide bearings with a central rotor as shown in Figure 3.11 (b). During the whirling motion the centre of the shaft S describes a circle about point B on the bearing centre line. Figure 3.11 (c) shows a cross section at the middle of the shaft. $P_1Q_1P_2P_3Q_2P_4$ is the outline of the shaft and the dotted circle is the path of S during the whirl.

It is assumed that the rotation of the shaft and the whirl are both clockwise. When the shaft is bent, the line AA divides it into two parts so that the fibres of the shaft above AA are elongated while those below AA are shortened. The line AA may be described as the neutral line of strain, which on account of the deviation from Hooke's law does not coincide with the neutral line of stress.

In order to understand the above statement, let us consider the point P_1 in Figure 3.11 (c) which may be assumed as a red mark on the shaft. In the course of the shaft rotation the red mark travels to $Q_1P_2P_3$, etc.

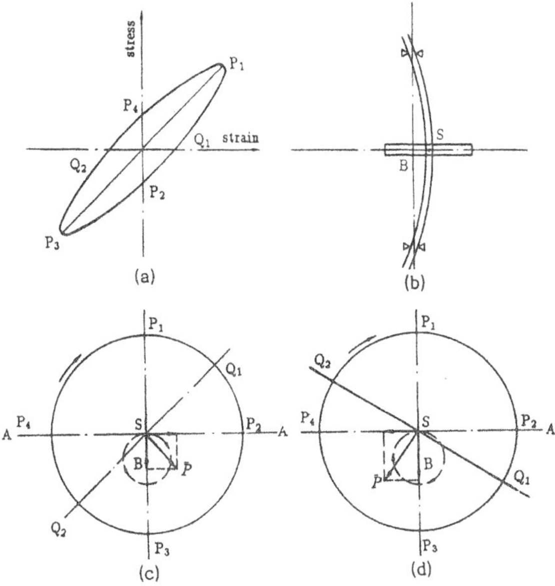

Figure 3.11 Shaft whirl caused by internal hysteresis

When the shaft whirls, both S and line AA revolve around the dotted circle. The speed of rotation and the speed of whirl are independent of each other. In case the speed of rotation is equal to the speed of whirl, the red mark will always be on the extension of the line BS. In case the rotation is faster than the whirl, P_1 will consecutively reach the position P_2, P_3, etc. In case the rotation is slower than the whirl, P_1 will go the other way through the sequence $P_1P_4P_3P_2$, etc.

First consider the case of rotation faster than the whirl. The points P_1, P_2, P_3, P_4 in Figure 3.11 (c) are indicated by the same letters in Figure 3.11 (a). In Figure 3.11 (c), the line Q_1Q_2 is the neutral line of stress. All fibres above Q_1Q_2 have tensile stresses and all fibers below Q_1Q_2 have compressive stresses. The stress system described sets up an elastic force $\bar{P}$. This elastic force $\bar{P}$ has not only a component toward B (the usual elastic force) but also a small component to the right, tending to drive the shaft around in its path of whirl Thus there is a self-excited whirl.

If the rotation is slower than the whirl, the inclination of Q_1Q_2 becomes reversed and the elastic force has a damping instead of an excitation as shown in Figure 3.11 (d). The whirling motion is determined primarily by the elastic force of the shaft toward the centre B combined with the inertia force of the rotor. The very small driving component of the elastic force merely overcomes the damping. When the whirl speed is below the critical speed, the internal hysteresis of the shaft acts as damping, whereas above that speed a self-excited whirl may be built up at the critical frequency.

References

3.1 Gong Shou-zhi (1983). 'Analysis and improvement measures of the hydraulic resonance in penstock of hydropower station LiuJiaXia', *Journal of Hydroelectric Engineering*, No. 1, pp. 21- 30.

3.2 Smilnov, A.M. (1979). Utilization and adjustment of the hydroelectric unit.

3.3 Den Hartog, J.P. (1956). *Mechanical vibration*, McGraw-Hill.

3.4 Sharman, R.V. (1963). *Vibrations and waves*, Butterworths, London.

3.5 Dong Yu-xin, (1986). 'Vibration of vertical hydrogenerator unit', *Large Electric Machine and Hydraulic Turbine*, No.1, pp. 49-54.

3.6 Dong Yu-xin (1989). *Vibration of hydroelectric unit*, The Press of Dalian University of Technology, May.

3.7 Japan Electrical Association (1983). 'Shaft vibration of turbine and generator and investigation on vibration', Part II, No. 142.

3.8 VDI 2059 Part 5 (1982). 'Shaft vibrations of hydraulic machine sets, Measurement and evaluation'.

3.9 Draft proposal ISO/DP 7919/4 (1986). Reference number ISO/TC 108/SC 2N84.

3.10 IEC (1984). TC/4, 'Hydraulic turbines, Working group WG 5'.

3.11 IEC (1963). TC/4, (USSR), 'The level of allowable vibration of hydro-units'.

3.12 Rund, F.O. (1980), 'Vibration criteria for transient operation of hydroelectric unit', IAHR, Symposium 1980, Tokyo, pp. 517.

Chapter 4

Structural Vibration due to Hydraulic Excitation

H. Netsch

4.1 Introduction To Flow Phenomena In Hydraulic Machinery

Hydraulic turbomachines change the energy level of the fluid. In pumps the energy level is increased, in turbines it is decreased. The flow conditions in hydraulic machines are much too complicated to be solved uniquely by calculation, particularly if all possible parameters are considered. The classical hydrodynamic formulae can only be applied to very simplified quasi-stationary steady-state flows when carrying out the analysis of a hydraulic machine.

4.1.1 Idealized Steady Flow

To begin with, the analysis will be restricted to steady flow conditions. The absolute flow through the stationary channels of the turbine and the relative flow through the runner at any given point, will not change with time. Transient phenomena resulting from a regulation sequence are therefore excluded. When studying the absolute flow, an observer must be part of the fixed frame of the turbomachine, whereas for relative flow investigations the observer will move with the rotating system.

Despite the difficulties of rigorous mathematical flow analysis, a sound knowledge of potential flow conditions is necessary. A first approximation based on potential flow theory, which is valid only for irrotational frictionless flow, allows the determination of velocity potentials in stationary guide vane channels. The runner is characterized by rotational flow, a necessary condition for energy exchange between the fluid and the turbine.

Basic Relations

The basic relations characterizing hydraulic turbines will be briefly reviewed. This task is greatly simplified by assuming uniform flow velocities in the runner channel cross sections and neglecting, where possible, the influence of viscous flow losses. Following this, the complicated actual flow conditions of the real fluid will then be discussed. In all cases, results are based on stationary flow.

The specific energy produced by a turbine is expressed by Euler's equation as

$$gH\eta_h = E\eta_h = u_1 v_{u1} - u_{\bar{1}} v_{u\bar{1}} \ , \tag{4.1}$$

where H and $E = gH$ are the effective head (m) and specific hydraulic energy (J/kg) of the turbine, g the gravitational acceleration, u the circumferential runner velocities, v_u the peripheral components of the absolute velocities and η_h the hydraulic efficiency of the turbine representing the effectiveness of energy conversion between the spiral case entrance and the draft tube exit. The suffix 1 and $\bar{1}$ correspond to the inlet and exit of the runner, respectively.

The hydraulic efficiency η_h can only be estimated by experience or an educated guess and can be obtained precisely through model or prototype tests. For a Francis turbine, the velocity triangles at their respective runner entrance and exit edges are illustrated in Figure 4.1.

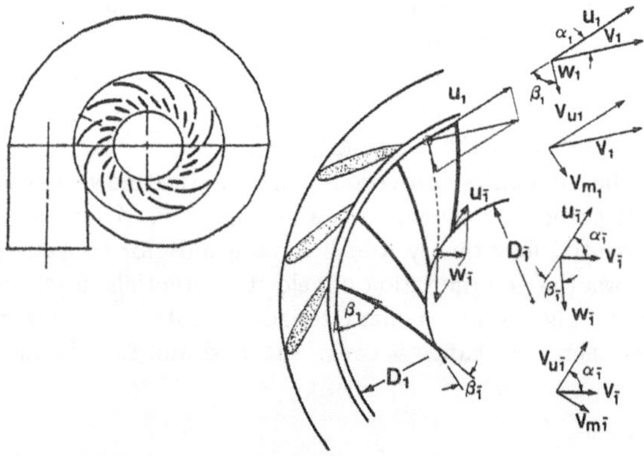

Figure 4.1 Guide vane cascade, bladed runner and velocity triangles of a Francis turbine

Although the moment of momentum theory can be applied with certain restrictions to real fluids with friction losses, equation (4.1) is based on a uniform velocity over the runner entrance and exit cross sections. Generally, the effects of variations of flow and head cannot yet be evaluated purely theoretically in a satisfactory manner. Hope of calculating flow conditions by purely theoretical means becomes even more illusory for unsteady flows, typical of cavitation. For turbines with strongly differing values of u along the entrance and exit runner edges, equation (4.1) must then be applied to axi-symmetric flow surfaces characterizing predetermined fractions of the total water flow.

The power P (W) delivered to the turbine shaft coupling is

$$P = \rho Q g H \eta = \rho Q \eta_t (u_1 v_{u1} - u_{\bar{1}} v_{u\bar{1}}) \,, \tag{4.2}$$

where ρ is the density of the fluid (kg/m^3), Q the volume flowrate (m^3/s), η the overall efficiency and η_t the total mechanical efficiency including disc friction loss and mechanical losses in bearings and seals.

Since the specific energy gH remains constant, any change in power delivered at the turbine shaft coupling can only be achieved by varying the mass flow ρQ. Consequently, all velocity triangles must then be different. However, the circumferential runner velocities u and the direction of the relative velocities w at the runner entrance and exit edge, which are dictated by the blade angles β_1 and $\beta_{\bar{1}}$, remain unchanged. When the turbine is operated at off-rated load, the boundary layers will be different and stall zones are likely to develop.

4.1.2 Idealized Flow Conditions

Once again it is assumed that the absolute velocities at the runner entrance and exit have equal amplitudes and flow directions.

Rated Load Operation

For this mode of operation, the volume flow rate is Q_{opt}. The velocity triangles, formed by v, u and w are shown in Figure 4.2 (a) and the turbine operates with the best efficiency $\eta_{h\,max}$. Since the circumferential component $v_{u\bar{1}} = 0$ at rated load operation, equation (4.1) simplifies to

$$g H \eta_{h\,max} = u_1 v_{u1} \,. \tag{4.3}$$

Often, a very small value of $v_{u\bar{1}}$ is desirable, since a slight flow rotation avoids flow separation in the draft tube. For draft tubes having an unusually large flare angle, a small value of $v_{u\bar{1}}$ may even be necessary.

The blade entrance angle β_1 is obtained from the velocity triangle. The blade exit angle $\beta_{\bar{1}}$ ensures that the fluid either leaves the runner with some small required rotation or no rotation at all.

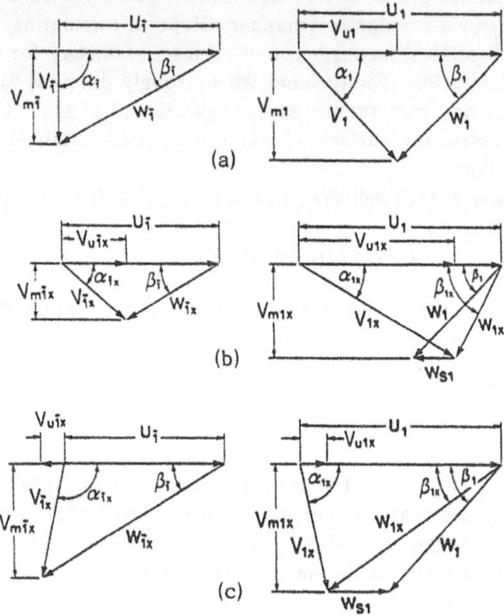

Figure 4.2 Velocity triangles at the runner exit and entrance edge for a high head Francis turbine: (a) rated load (b) partial load (c) overload

Partial Load

Since the specific energy gH available at the turbine entrance remains constant, a reduction in power P_x can only be achieved by a reduction of flowrate Q_{opt} to Q_x. The flow conditions at the runner entrance edge are determined by the flow at the runner exit edge. The various velocities at the runner are shown in Figure 4.2 (b).

Flow conditions at the runner exit edge: For operation under partial load, all turbine cross sections are now oversized and the power P_x is produced at a lower efficiency η_x. The lower efficiency η_x is the direct result of the decrease in hydraulic efficiency η_h and the specific energy is now

$$gH\,\eta_{hx} = (u_1 v_{u1x} - u_{\bar{1}} v_{u\bar{1}x})\,. \tag{4.4}$$

Since the runner exit angle $\beta_{\bar{1}}$ dictates the exit flow direction $w_{\bar{1}x}$, the value of $v_{u\bar{1}x}$ is directly related to $v_{m\bar{1}x}$. The contribution $u_{\bar{1}}v_{u\bar{1}x}$ to the specific energy can now be determined, as indicated in Figure 4.2 (b).

Flow conditions at the runner entrance edge: The value of v_{u1x} depends on Q_x and η_{hx} and the theoretical value of w_{1x} obtained from velocity triangles would require a flow angle β_{1x}, as shown in Figure 4.2 (b). This represents a flow discontinuity at the runner entrance edge, which is physically impossible. The relative flow vector w_{1x} must be deflected and accelerated by w_{s1} to obtain the vector w_1 with β_1 and the meridional velocities $v_{m1x} = w_{m1x}$ are related to Q_x by the runner entrance dimensions.

Since this phenomenon is gradual and irreversible, a kinetic energy loss occurs expressed by

$$E_{ks1} = \frac{1}{2}\phi_1 w_{s1}^2 \tag{4.5}$$

where the factor ϕ_1 corrects for the energy actually lost.

The value of Q_x required to deliver the power P_x depends on η_{hx} and is determined by the guide vane position, obtained from model tests and the final governor adjustment in the power station.

Overload

For this condition, the velocity triangles are illustrated in Figure 4.2.c. The velocity $v_{m\bar{1}x}$ related to Q_x and the runner exit angle $\beta_{\bar{1}}$, determines $v_{u\bar{1}x}$ which is negative, pointing in the opposite direction to u. The specific energy now becomes

$$gH\eta_{hx} = (u_1 v_{u1x} + u_{\bar{1}} v_{u\bar{1}x}) \tag{4.6}$$

and v_{u1x} must therefore decrease. Consequently the guide vane opening must be increased, resulting in a larger angle α_{1x}, bringing the guide vane exit edges closer to the runner. Again, the presence of the deflection velocity w_{s1x} causes an energy loss E_{ks1}.

Whereas for partial load, the deflection velocity impinges on the blade driving face, for overload this phenomenon occurs at its trailing surface. The flow deflection phenomenon is gradual and the following data should be interpreted as a guide. For partial load ϕ_1 is around 0.6 and decreases with Q_x as w_{1x} increases to w_1. For overload ϕ_1 is approximately 0.8 and w_{1x} is larger than w_1.

Values of Hydraulic Efficiency at Off-Design Operation

Model tests provide the efficiency $\eta_x(P_x)$ hill diagram and have permitted the amassing of a great amount of data. However operational turbines are the prime source of the necessary supplementary information for off-design operation. Detailed measurements on such machines are however very difficult to make and results cannot always be generalized. Very often, measurements have only been instigated following operational difficulties. Systematic studies are very expensive and therefore seldom undertaken.

4.1.3 Rotating Stall and Cork Screw Vortex

In the case of rated load operation, the water leaves the runner on surfaces of co-axial cylinders or axi-symmetric cones. For partial load, a cork screw vortex at the draft tube entrance cross section can be observed, turning in the direction of the runner motion, whereas for overload, the vortex rotates in the opposite direction. This is illustrated by the velocity triangles of Figure 4.2 (a), (b) and (c).

Pressure fluctuations due to the vortex at partial load can be propagated through the turbine runner provoking resonance in the penstock. Undesired power fluctuations can result, requiring swift corrective action. In extreme cases, the penstock and consequently the station could be endangered. At first it should be attempted to dampen these oscillations by supplying air to the draft tube entrance cross section.

Since the frequency f_3 of these vortices is smaller than that of the turbine rotational frequency n, with observed data f_3 in the range of

$$f_3 = (\frac{1}{2} \text{ to } \frac{1}{5})\, n \,, \tag{4.7}$$

secondary flow phenomena must be present. The explanation is the presence of a stall, resulting from the interaction between the flow and the boundary layer in the channel. Fanelli, 1988 [4.3], deserves credit for a precise mathematical model of the vortex rope in the draft tube of Francis turbines. Due to many intervening parameters, the following presentation will be much simplified.

In converging channels the pressure decreases in the direction of the flow and the fluid follows the wall geometry. The boundary layer is very thin and the flow pattern shows good agreement with potential frictionless flow. On the other hand, flow in divergent channels is very different, with the velocity gradient strongly dependent on the Reynolds number and angle of divergence. If the flare is sufficiently large, the flow pattern even ceases to be axi-symmetric. Negative velocities, the result of backflow, can develop and the thick boundary layer can separate the flow from either wall of the flared channel.

Both flow conditions are present in hydraulic turbines. The guide vanes and rotating runner channels form converging ducts to extract energy from the water, whilst the the flared draft tube allows energy recuperation behind the runner. For rated operating conditions, the runner blade entrance angles and the angles of the oncoming fluid are matched so that the turbine achieves its highest efficiency.

For partial load or overload, and head or speed variations, the flow incidence will differ, resulting in flow separation from either the trailing or driving blade wall. The flow ceases to be regular, stall will develop and flow losses will increase disproportionally. A partial, even small blockage of one runner channel will result in an increase in the flow through the others, thus modifying their angle of incidence. A cork screw vortex rotating with a frequency f_3 much lower than that of the turbine can then be observed. This phenomenon is difficult to analyze precisely, since the runner construction and blade profile, particularly its maximum thickness/length ratio, and the Reynolds number are important.

Stall can develop in the draft tube entrance zone if the flare is excessive, even under rated operational conditions. This stall is detrimental to the turbine operation and reduces its efficiency. Limiting the flare and providing for a rotational or peripheral component $v_{u\bar{1}}$ can prevent flow separation resulting from the stall. The centrifugal force will offset an unfavourable boundary layer pressure gradient and flow separation is avoided. Despite the fact that centrifugal forces due to v_u cannot provoke energy losses, the value of $v_{u\bar{1}}$ must be kept small to keep friction losses to a minimum.

4.1.4 Actual Flow Conditions

In the preceding sections, it was assumed that the time averaged flow velocity v_2 leaves the guide vane cascade with uniform magnitude and direction. The energy extracted by the runner determines the through-flow in the various sections of the turbine. The absolute velocity direction of the fluid leaving the runner was also assumed to be uniform. Actual flow conditions were then corrected by the hydraulic efficiency η_{hx} which reflects the actual flow pattern. These simplifying assumptions allow the development of the velocity triangles and to determine the various vane angles α and β.

Actual flow conditions, which we will now discuss, are much more complicated. The following discussions are based on the flow in a low specific speed Francis turbine at rated load. The corresponding velocity triangles are presented in Figure 4.2 (a).

Flow Through the Guide Vane Cascade

For the flow through the spiral case and the guide vane cascade, small friction losses are assumed. Smooth entry of the fluid into the guide vane channels is anticipated. The flow, in the absence of shear stresses, thus remains essentially irrotational and absolute velocities can therefore be calculated by potential theory. Corroboration of this assumption would require that virtually no specific energy reduction takes place.

In order to carry out detailed studies of the conditions at the exit section, the cylinder with the diameter D_2 of the guide vane tips has been developed

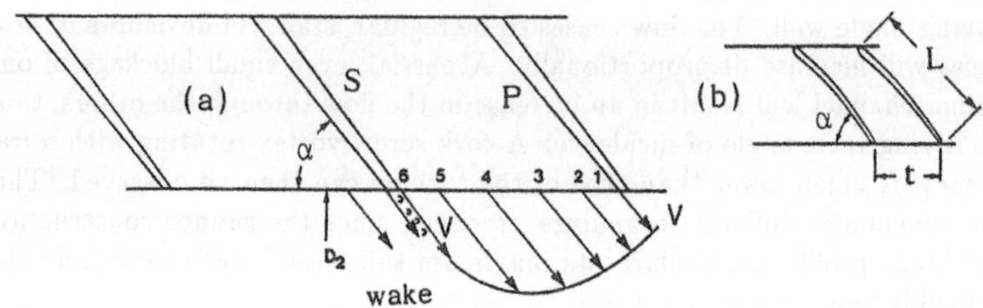

Figure 4.3 Velocity distribution at the guide vane trailing edges with the vane cascade developed into a plane:
(a) Velocity distribution (b) Part of the runner cascade

into a plane and is represented in Figure 4.3 (a). Due to the change in direction along the curved cascade path, mass effects lead to the velocity distribution indicated in Figure 4.3 (a). This velocity profile depends on the guide vane shape, the angle α and particularly the pitch/chord ratio t/L.

The subtraction of circumferential runner velocity vector u from absolute velocity vector v results in relative velocity vector w. For the observer moving with the rotating system, the steady flow v at the guide vane exit leads therefore to varying relative velocities w_1, with different amplitudes at the runner entrance section and consequently also in the runner. In addition, the runner entrance edges pass periodically through wakes, shed by the guide vanes, resulting from the differing velocities v_1 to v_6 over the pressure and suction sides of the guide vanes as shown in Figure 4.3. The influence of the respective boundary layers must also be mentioned.

The difference between v_1 and v_6 is progressively reduced downstream by energy dissipation but this requires an appreciable mixing length. A more uniform relative profile can also be obtained from an accelerated flow. Details of the influence of these wakes will be presented later. The flow thus enters the runner with pulsating relative velocities w_1 and is unsteady from the point of view of the rotating observer.

Flow in the Rotor Vane Channels

To simplify the rather complicated analysis of flow conditions in the runner channels, constant time averaged relative entrance velocities w_1 are assumed. If one considers a closed rotating runner channel, the water in it will rotate with a angular velocity $(-\omega)$ in the opposite sense of the turbine rotation (ω) as shown in Figure 4.4 (a).

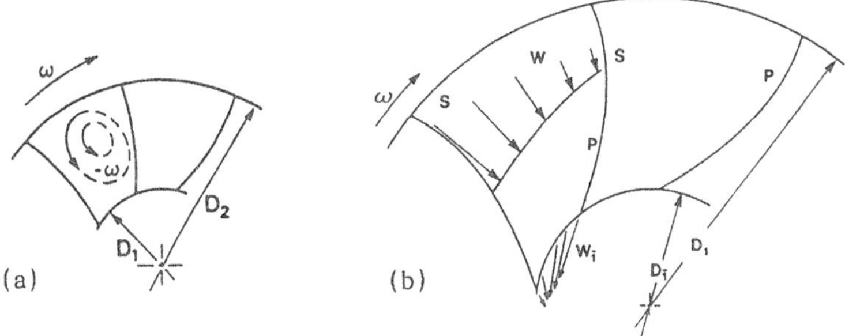

Figure 4.4 Flow conditions in a turbine runner: (a) Relative rotation in the runner for ideal flow conditions (b) Velocity distribution at various runner diameters

Since the purpose of the turbine runner is to reduce the specific energy of the fluid, axi-symmetric irrotational flow must be excluded, since the pressure distribution over the runner blade surfaces P and S in Figure 4.4 (b) would then be equal. The rotational flow is obtained by a finite number of runner blades. In the presence of shear stresses, the flow conditions correspond then to real fluids. The flow in the guide vane cascades and in the runner will now differ distinctly from the case of an ideal fluid.

Contrary to ideal irrotational potential flow, the circulation within the closed runner channel

$$\Gamma = \oint w_s \, ds(4.8) \tag{4.8}$$

depends now on the closed path chosen, where ds is a small segment of the closed path and w_s the component of w parallel to ds.

The application of the momentum and the moment of momentum laws to the runner is theoretically possible. In practice however, the unsteady flow in the runner channels make their solution impossible.

4.1.5 The Flow Acceleration in the Runner

<u>Acceleration of Absolute and Relative Flow</u>

Since the mass flow and the absolute velocities determine the power delivered progressively to the runner between entrance and exit, it is of interest to relate absolute and relative flow accelerations. This gives a valuable insight into the flow mechanism in the runner and therefore also to its optimization possibilities.

The mathematical description of the state of a moving fluid is most conveniently expressed by vector relations. This allows further generalizations, as will be seen later in this chapter. Additional information for specialized vector based treatments on fluid mechanics is given in the book by Vavra, 1960 [4.21].

A few elementary relations will be recapitulated and are indicated in Figure 4.5.

$$\boldsymbol{a} \cdot \boldsymbol{b} = |\boldsymbol{a}||\boldsymbol{b}| \cos(\boldsymbol{a}, \boldsymbol{b}) , \tag{4.9}$$

$$\boldsymbol{a} \times \boldsymbol{b} = \boldsymbol{c} = |\boldsymbol{a}||\boldsymbol{b}| \sin(\boldsymbol{a}, \boldsymbol{b})\, \boldsymbol{e} , \quad |\boldsymbol{e}| = 1 , \tag{4.10}$$

$$\text{grad} = \nabla = \boldsymbol{i}\frac{\partial}{\partial x} + \boldsymbol{j}\frac{\partial}{\partial y} + \boldsymbol{k}\frac{\partial}{\partial z} , \tag{4.11}$$

where $(\boldsymbol{a}, \boldsymbol{b})$ denotes the angle between the two vectors $\boldsymbol{a}$ and $\boldsymbol{b}$.

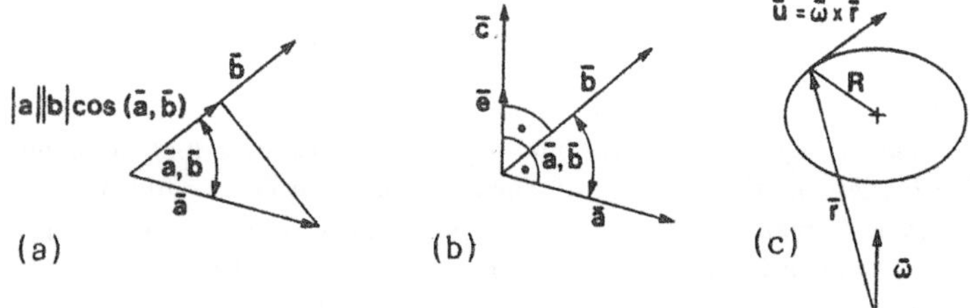

Figure 4.5 Vector relations: (a) Scalar product (b) Vector product (c) Velocities of a body, revolving about a fixed axis

The following sections are devoted to the solution of the flow within a runner. At a certain instant of time, $\boldsymbol{v}$ denotes the velocity vector of a fluid particle seen by an observer who is part of the absolute, therefore, motionless turbine frame. In a coordinate system fixed to the runner, the relative velocity vectors $\boldsymbol{w}$ in the runner channels can be defined. Finally, $\boldsymbol{u} = \boldsymbol{\omega} \times \boldsymbol{r}$ is the rotational runner velocity normal to the plane formed by $\boldsymbol{\omega}$ and $\boldsymbol{r}$.

From the vector representation of the velocity triangles in Figure 4.5 (c) and 4.6, it can be seen that at any instant,

$$\boldsymbol{v} = \boldsymbol{w} + \boldsymbol{\omega} \times \boldsymbol{r} . \tag{4.12}$$

A relative observer, leaving point 1 on the blade at time t as shown in Figure 4.7, moves through the runner channels and arrives at time $t + dt$ at point $\bar{2}$,

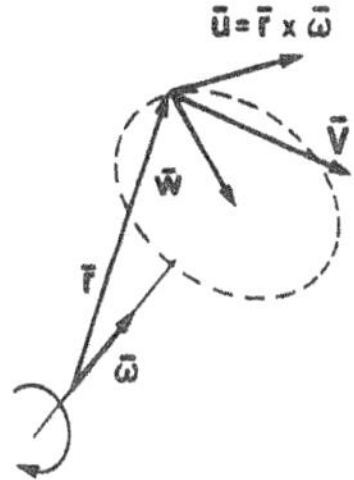

Figure 4.6 The absolute velocity in a runner

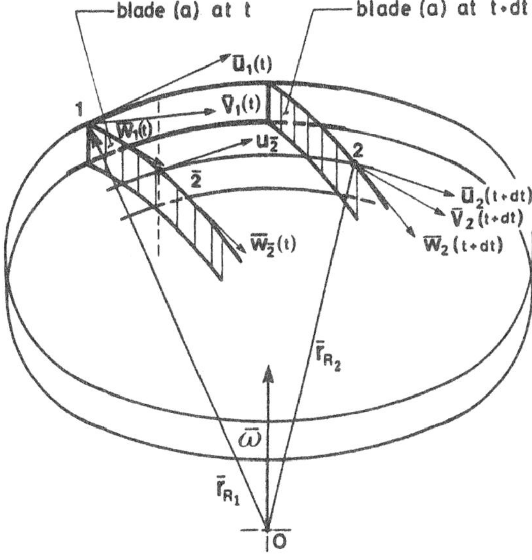

Figure 4.7 Velocities in the runner channel

which is also on the blade surface. Since the runner is subjected to a rotation, the relative observer is then at point 2 for another stationary observer. It is therefore possible to link all observed vector quantities. Conditions at point 1 are denoted by the subscript 1. Subscripts $\bar{2}$ and 2 distinguish velocity vectors at points $\bar{2}$ and 2.

To simplify the following presentations, subscript $_{1-2}$ denotes the variation from 1 to 2.

(a) Stationary Observer

$$\text{absolute velocity:} \qquad \boldsymbol{v}_2 = \boldsymbol{v}_1 + d\boldsymbol{v}_{1-2}$$

The circumferential velocity $\boldsymbol{u}_2$ can also be presented by

$$\boldsymbol{u}_2 = \omega \times \boldsymbol{r}_{R2} \, .$$

(b) Relative Observer

$$\text{relative velocity:} \qquad \boldsymbol{w}_{\bar{2}} = \boldsymbol{w}_1 + d\boldsymbol{w}_{1-\bar{2}}$$

It is also possible to present the circumferential velocity at $\bar{2}$ by the relation

$$\boldsymbol{u}_{\bar{2}} = \omega \times (\boldsymbol{r}_{R1} + \boldsymbol{w}_1 \, dt) \, . \tag{4.13}$$

For the stationary observer, the basic vector relation at 2 between $\boldsymbol{c}$, $\boldsymbol{w}$ and $\boldsymbol{u}$ can also be written as

$$\boldsymbol{v}_2 = \boldsymbol{w}_2 + \omega \times \boldsymbol{r}_{R2} \, . \tag{4.14}$$

When relating this data to all initial values at point 1, the result must then be expressed by

$$\boldsymbol{v}_2 = \boldsymbol{v}_1 + d\boldsymbol{v}_{1-2} = (\boldsymbol{w}_1 + d\boldsymbol{w}_{1-2}) + \omega \times (\boldsymbol{r}_{R1} + \boldsymbol{w}_{1-2} \, dt) \, . \tag{4.15}$$

When linking the data to vector quantities, seen by the relative observer in the runner channels, it is necessary to introduce also $\omega \times dt$, representing the angle of rotation which the runner has undergone in the time increment dt. The vector $\boldsymbol{w}_2$ at 2 in the absolute frame is linked to $\boldsymbol{w}_{\bar{2}}$ at $\bar{2}$ in the runner channel by

$$\boldsymbol{w}_2 = (\boldsymbol{w}_1 + d\boldsymbol{w}_{1-\bar{2}}) + \omega \times (\boldsymbol{w}_1 + d\boldsymbol{w}_{1-\bar{2}}) \, dt \, , \tag{4.16}$$

and consequently also

$$\boldsymbol{u}_2 = \omega \times (\boldsymbol{r}_{R1} + \boldsymbol{w}_{1-\bar{2}} dt) + \omega \times [\omega \times (\boldsymbol{r}_{R1} + \boldsymbol{w}_{1-\bar{2}} dt)] \, dt \, . \tag{4.17}$$

Introducing equations (4.16) and (4.17) into equation (4.14) leads to

$$\begin{aligned} \boldsymbol{v}_1 + d\boldsymbol{v}_{1-2} \; = \; & \boldsymbol{w}_1 + d\boldsymbol{w}_{1-\bar{2}} + \omega \times (\boldsymbol{w}_1 + d\boldsymbol{w}_{1-\bar{2}}) \, dt + \omega \times (\boldsymbol{r}_{R1} + \boldsymbol{w}_{1-\bar{2}} dt) \\ & + \omega \times [\omega \times (\boldsymbol{r}_{R1} + \boldsymbol{w}_{1-\bar{2}} dt)] \, dt \, . \end{aligned} \tag{4.18}$$

Neglecting small velocity differences leads finally to the general expression of acceleration $a = dv/dt$,

$$\underset{\text{absolute value}}{a} = \underset{\text{observed in the runner channel}}{dw/dt + 2\omega \times w + \omega \times (\omega \times r)} \tag{4.19}$$

The vector a on the left side is seen by the stationary observer, all others are in the runner channels. The terms $\omega \times (\omega \times r)$ represents the centripedal acceleration with the magnitude $r\omega^2$ and $2\omega \times w$ is the Coriolis acceleration. These accelerations act on a flow element and lead to body forces.

The above relationship allows a valuable insight into the flow conditions within the runner channels. Since no restrictions were stipulated in its derivation, it can be applied to almost any flow conditions and therefore to arbitrary runner channel shapes. One of the most general cases that can be treated is represented by a Francis turbine runner, where the flow direction changes from radial inflow to axial outflow.

For the runner, revolving around its axis, the local and centripedal forces are self-explanatory. The Coriolis force has a great influence on the flow pattern within the runner and also on the runner layout. The term $\omega \times w$ represents a vector which is perpendicular to both ω and w. The force on a fluid particle arising from this acceleration component is therefore perpendicular to the blade surface P, pointing from the direction of S.

Furthermore, as illustrated in Figure 4.4 (b), the Coriolis force increases with w the closer one approaches the suction surface S. Because of the small velocities in the boundary layer on the runner hub and shroud, and on the pressure and suction surfaces of the blade channel, the boundary layer fluid will be displaced by the Coriolis force from the suction surface S to the pressure surface P. Consequently, the relative velocity on P will increase and the whole flow pattern will be profoundly modified and secondary flows will develop, therefore reducing the hydraulic turbine efficiency η_h.

It can thus be concluded, that the turbine runner extracts most of its power in the channel passages close to the entrance edges and has the tendency to develop pump characteristics towards the exit edge. Large energy changes will lead to large Coriolis forces.

When rigorously applying the vector laws, one sees that for flows where w is parallel to ω, no fluid energy can be extracted by the rotor, since the blade channels are in axi-symmetric planes. In addition, another interpretation of equation (4.19) allows one to evaluate the influence of the runner form on the hydraulic energy losses in the runner channels. The centripedal acceleration $\omega \times (\omega \times r)$ reduces the pressure p through the runner, which is a necessary condition for the conversion of high head hydraulic energy to mechanical power. The other two similar quantities, such as $(\omega \times w) + (\omega \times w)$ represent the time related rotations of vectors $\boldsymbol{w}_{\bar{2}}$ an $\boldsymbol{u}_{\bar{2}}$ into positions $\boldsymbol{w}_2$ and $\boldsymbol{u}_2$ respectively.

Relations for Axial Flow Turbomachines

Again the basic flow relations will be derived for a turbine operating at design conditions, where the stream surfaces can be approximated reasonably well by co-axial cylinders. The relative velocities $\boldsymbol{w}$ at the inlet and exit of the runner thus possess no component normal to the turbine axis. The very general law, relating the absolute fluid particle acceleration at any station in the runner channel to the corresponding relative value of velocity and the rotation, is again expressed by equation (4.19) and the centripedal force acting on a fluid particle as given by

$$\boldsymbol{\omega} \times (\boldsymbol{\omega} \times \boldsymbol{r}) \tag{4.20}$$

is not directed against the through flow and remains constant.

Since the energy production is only due to Coriolis forces, high velocities in the runner channels necessarily makes these machines very sensitive to cavitation. This by the way provides an elegant proof why axial flow turbines are limited in their application to small heads only.

4.2 Pressure Fluctuations Due To Hydraulic Excitation

Due to the sudden flow deflection w_{s1} shown in Figure 4.2 (b) and (c), energy losses E_{ks1} will result which can be calculated by equation (4.5). For rated load operation, $E_{ks1} = 0$. On the other hand, the magnitude of the observed pressure fluctuations and related structural vibrations achieve their highest value for rated conditions where $w_{s1} = 0$. For this reason, these mechanical vibrations cannot originate in the flow deflections.

4.2.1 Interaction of Guide Vanes and Runner Channels

Figure 4.3 shows that the runner entrance edges pass periodically through wakes shed by the guide vane exit edges. The pressure fluctuations to be expected thus possess a frequency f_1 given by

$$f_1 = z_b\, z_g\, n\ , \tag{4.21}$$

where n is the rotational frequency of turbine, z_b the number of runner blades and z_g the number of stationary guide vanes.

The conditions which might lead to a smooth flow pattern at the guide vane exit are twofold: extremely thin exit blade edges and careful equalization of wake velocities and pressures on both sides of the guide vane surface, so that the energy levels over P and S do not differ appreciably. Unfortunately, both these conditions are impossible to achieve in practice.

The action of the perturbing wake can be described as follows: the flow velocities v_1 to v_6 have the same flow angle α but vary in magnitude. Consequently the kinetic energy levels at stations 1 to 6 must differ. These energy variations will gradually disappear over a sufficient long downstream mixing length. Generally, the amplitudes of the pressure fluctuations f_1 will be small and disappear completely for part-load operation.

Any asymmetry in the position or shape of a guide vane will strongly disturb the flow in the channels. According to the Kutta- Joukowsky circulation theory, the vane trailing edge will no longer be a stagnation point. The size of the wake will be increased by boundary layer separation and strong turbulent eddies formed. When a runner blade moves through such a wake, considerable pressure peaks will occur at the runner entrance plane.

If large pressure amplitudes signals of frequency f_1 are detected during operation, the turbine must be immediately taken out of service to prevent severe damage. Very strong disturbances will arise from the combined effect of guide vane interaction reaching an equal number of runner blades. This can be prevented by ensuring that z_b and z_g have no common denominator so that the shock pressure pulses are staggered in successive flow channels of the runner.

Pressure Fluctuations in High Head, Low Specific Speed Francis Turbines

For high head Francis turbines with a relatively low specific speed, the amplitude of the pressure fluctuations f_1 can be relatively large. Due to the combined effect of the very short distance between the trailing edge of the guide vanes and the runner inlet, and the very high flow velocity, the time is too short for an effective energy smoothing between the kinetic energies $v^2/2$ before the impact of the disturbing wake with the runner blades. This situation may be very pronounced in the case of fully opened guide vanes, but is less critical for small openings of the guide vanes, since the distance between the guide vanes and runner inlet is then substantially increased and the pressure fluctuations f_1 tend to disappear.

Pressure Fluctuations in Low Head, High Specific Speed Francis Turbines

For low head, high specific speed Francis turbines, the distance between the guide vane trailing edges and the runner inlet edge at the hub is increased. An optimization of the runner blade entrance edge shape and the runner channels depends on the blade entrance angles β_1, which are very large at the hub and must decrease towards the shroud. Consequently, the runner entrance edge must therefore be inclined and its distance to the guide vane exit edges must decrease along its span. In the runner hub zone, the mixing length of the upstream wakes is appreciable, as the flow changes its direction from radial inwards to axially downwards. This tends to decrease vibration amplitudes.

4.2.2 Interaction of Spiral Case Tongue and Runner Blades

The interaction between the spiral case tongue and the runner blades produces corresponding pressure fluctuations of frequency f_2 defined as

$$f_2 = z_b\, n \,. \tag{4.22}$$

The spiral case tongue, which by necessity is thick on a large turbine, acts as a divider between the incoming flow from the penstock and the residual flow at the end of its travel inside the spiral case.

As a result of the friction losses suffered in the spiral by the residual flow, these two flows are at a different energy level, and their turbulent junction at the trailing end of the spiral case tongue produces a strong wake as shown in Figure 4.8. It takes only a fraction of a second for this wave to pass through the guide vane grid and impinge on the leading edge of a runner blade. The pressure fluctuations thus created can be of rather large amplitude.

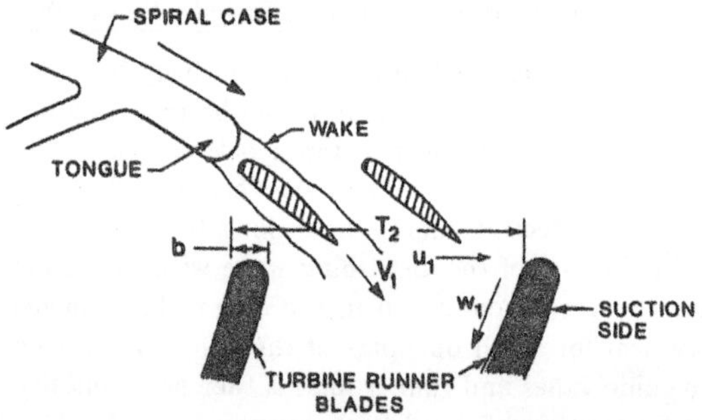

Figure 4.8 Wakes shed by the spiral case tongue

4.2.3 Pressure Fluctuations due to Unstable Cork Screw Vortex

Pressure fluctuations at the exit section of a turbine runner are observed for part-load operation. Fluid particles of different energy level are present in the runner channels and their mixing length is too short. Consequently, the blade exit section can no longer control the flow. The origin of pressure fluctuations with frequencies f_3 will now be discussed. As has been pointed out in section 4.1.3, flow conditions within the runner channels at partial load and overload will result in the formation of an unstable vortex, characterized by the frequency $f_3 = (1/2 \text{ to } 1/5)\, n$ as given by equation (4.7).

Resonance in the penstock can occur, with power fluctuations which could endanger the penstock, the machines and therefore also the power station. This problem is discussed in detail in section 6.3 of this volume.

4.2.4 Periodic Pressure Fluctuations due to Vortex Street

Turbine runner blades are submerged plates. Occasionally periodic vibrations of a turbine and also of the penstock can be observed. These vibrations can be attributed to periodic pressure fluctuations, resulting from a Karman vortex street formed at the runner blade exit edges. Although these pulsations are generally limited to a narrow turbine operating range, corrective action must be performed immediately. Very exhaustive tests to study this phenomenon have been performed on cylinders. The findings apply also to submerged plates having slightly different Reynolds numbers R_e and Strouhal parameters S_t.

In a flow around a solid submerged body, when fluid particles are accelerated they have the tendency to follow the body shape. If the fluid is decelerated, fluid particles close to the wall have their kinetic energy reduced by friction and are consequently slowed down. The external pressure can lead to flow separation from the wall. Since viscous forces are present, the Reynolds similarity law allows a general representation of the flow pattern. For the flow along a plate, see Figure 4.9, R_e is defined by

$$R_e = (w_{\bar{1}} s_{\bar{1}})/\nu \tag{4.23}$$

whereas for a cylinder, see Figure 4.10, the Reynolds number R_e is

$$R_e = (w\, d)/\nu \tag{4.24}$$

If the Reynolds number is below 50 for a cylinder, a laminar boundary layer is formed, enclosing a stable eddy, as shown in Figure 4.10. A similar flow condition at the end of the turbine blade is not possible. At Reynolds numbers between $R_e = 40$ to 5×10^3 oscillations in this laminar layer increase in amplitude, with the boundary layer separating alternatively on the upper and lower surface, see Figure 4.11 (a). Vortices are formed periodically behind the cylinder and shed into the fluid. One eddy will increase and be swept into the fluid with another counter-rotating vortex in the earlier stage

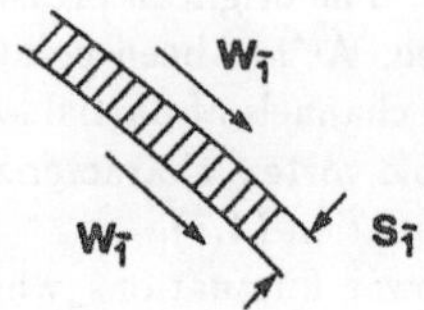

Figure 4.9 Exit section of a Francis turbine runner blade

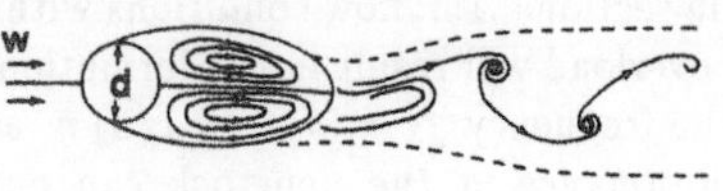

Figure 4.10 Flow field behind a cylinder at very low Reynolds number

of formation. These vortices become stable to small pressure fluctuations in the fluid with a ratio of $h/\ell = 0.286$ for cylinders and $h/\ell = 0.306$ for plates, with ℓ the distance between two vortices Γ of the same sense of rotation. The vortices are swept into the fluid with a velocity v lower than the upstream free velocity w. At some distance from the submerged body, the value of ℓ will also increase. The shedding frequency f_4 can be expressed by

$$f_4 = (w - v)/\ell\,, \tag{4.25}$$

or by the Strouhal number S_t for a cylinder;

$$S_t = f_4 d/w\,, \tag{4.26}$$

whereas for a plate, or a turbine runner blade;

$$S_t = f_4 s_{\bar{1}}/w_{\bar{1}}\,. \tag{4.27}$$

The frequency of the shed vortices increase until the Reynolds number attains $R_e = 10^3$. Pertinent data can be seen in Figure 4.11 (b).

According to Kelvin's theorem, the circulation around a closed contour does not vary with time. Vortices must keep this characteristic, if only mass forces are present resulting from a potential. This indicates, as illustrated in Figure 4.11 (a), that a circulation will be created around the cylinder in the opposite sense to the eddy just previously formed. This fluctuating circulation varying from, $-\frac{1}{2}\Gamma$ to $+\frac{1}{2}\Gamma$, gives rise to periodically varying forces. Since the flow pattern behind the blade trailing edge fluctuates up and down, transversal forces are created, acting on the blade surface close to the trailing edge.

If, in a narrow turbine operating range, vibrations of the turbine or the penstock occur, one should calculate the Reynolds number at the runner blade exit edge, to see if the corresponding Strouhal frequency suggests a Karman vortex street. An eigenfrequency of the blade exit zone close to or

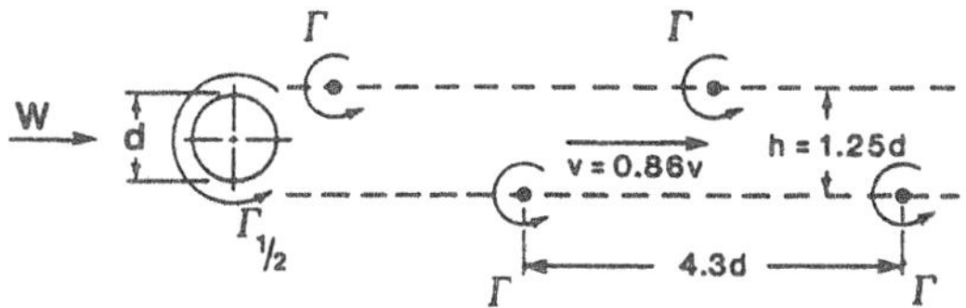

(a) Cylinder with large aspect ratio

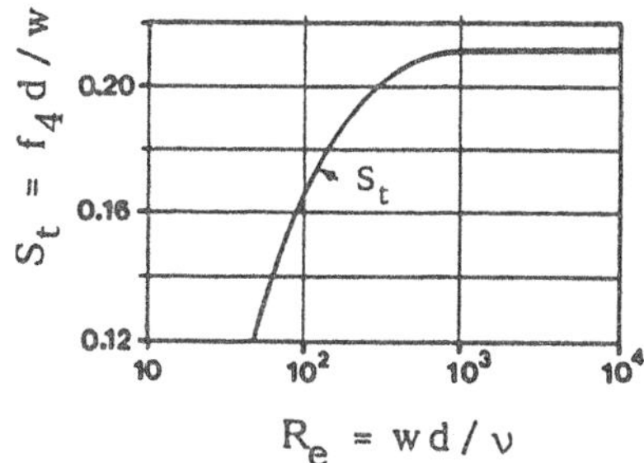

(b) Strouhal number (after Reif and Simmons)

Figure 4.11 Karman vortex street behind a cylinder for high Reynolds numbers

equal to that of the vortex street forces the blade into resonance and cracks will develop rather rapidly in the runner blade exit zone.

Modifying the blade thickness by either reducing or increasing $s_{\bar{1}}$ by welding layers on the blade surface will slightly alter the Reynolds number, but the new geometric form of the blade exit section will also change its eigenfrequency. Rounding off the blade exit edge can also be very helpful, but this could provoke cavitation if the unit is already operating close to the onset of this phenomenon.

Another possibility of eliminating blade vibrations consists in welding profiled struts from one blade to another around half the runner span. The resulting ring stiffens the runner blades and the free exit edge is reduced to half its original value. This possibility is restricted to very small turbines. Since the runner exit edge section must be considered as a beam, clamped at the runner hub and shroud, the danger of vibration due to a Karman vortex street increases with the exit edge length and therefore for Francis turbines having a high specific speed.

Blade vibrations resulting from a Karman vortex street are only possible in Francis turbines, since the water in the runner channels is accelerated between entrance and exit cross sections. In the runner of centrifugal pumps, the fluid is decelerated at the entrance edge and dead water zones will be formed, leading to flow separation at the entrance of the runner channels.

4.3 Level of Hydraulic Excitation

Due to the complex nature of the system, composed of the turbine, its water-filled passages and the generator, its mathematical model cannot be established. When developing a turbine, knowledge of the relative amplitudes of possible vibrations is important. For units in operation subjected to vibration, the diagnostic of their origin allows to apply remedies.

4.3.1 Amplitudes of Pressure Fluctuations

The energy level of pressure variations having the basic frequency f_1 will generally be low. On the other hand pressure shocks with the fundamental frequency f_2 can assume high values. The vortex behind the runner, characterized by the frequency f_3 can lead to resonance in the penstock and provoke power fluctuations. Such vibrations and resonance occur at a particular turbine load and remedies are immediately required. Frequencies f_4 are seldom observed and are limited to a narrow turbine operating range. If they occur, corrective actions must be taken to eliminate resulting vibrations. In the following discussions our attention will be focused on pressure fluctuations with the frequency f_2 and its harmonics.

4.3.2 Periodic Shocks and Fourier Frequency Spectrum

Since a shock represents a transmission of kinetic energy to a system within a very short time interval, the ultimate goal of any investigation is not always the waveform itself, but rather the effect on the mechanical system.

Periodic shocks produce complex effects and are far more common than single frequency perturbations. Any periodic signal, no matter how complex it is, may be looked upon as the source of a range of harmonic vibrations and its Fourier expansion will contain terms of harmonic frequencies above the fundamental. If one of these frequencies happens to be close to the natural or resonant frequency of a structural element in the system, the resulting large amplitude may overshadow the smaller amplitudes of other driving harmonics and make it the dominant part. It is possible to replace impulses of differing forms by rectangular ones of equal area, provided the pulse width

b is small compared to the value of the period T_2. Pertinent parameters are represented in Figure 4.12.

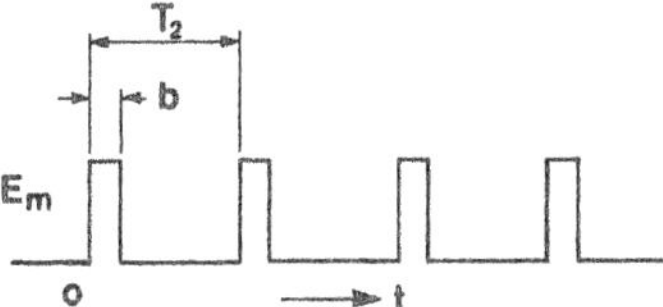

Figure 4.12 Periodic pressure pulses due to the runner leading edges passing through wakes

The complex amplitude spectrum $F(n)$ of such a rectangular pulse is then given by

$$F(n) = E_m \frac{b}{T_2} \frac{\sin(\pi nb/T_2)}{\pi nb/T_2} e^{-\pi jnb/T_2} , \tag{4.28}$$

and makes it possible to estimate actual conditions.

Forced harmonics $3f$, $5f$, $7f$ and higher will be present and in the case of a pulse of a short duration b, additional even harmonics $2f$, $4f$ and higher ones can result. A low level of mixing energy in a wake keeps E_m low but a small relation (b/T) tends to offset this tendency.

4.4 Determination Of Vibration Transmission Path

It has already been indicated in sections 4.2.1 and 4.2.2 that the runner entrance edge passes periodically through wakes shed by either the guide vane exit edges or the spiral case tongue, causing vibrations in the turbine structure or even resonance of structural elements. This phenomenon is illustrated in the following example involving head cover vibrations of the turbine shown in Figure 4.13. To measure all frequencies with their related amplitudes, three accelerometers were installed and two hydrophones placed in the water flow (refer Netsch, (1982) [4.13], for further detail).

4.4.1 Resonance of Turbine Structural Elements

Of all the vibration amplitudes registered, those with frequency f_2 were particularly strong. At the onset of the measurements, the turbine was operated at a slower speed, which resulted in a proportional reduction of

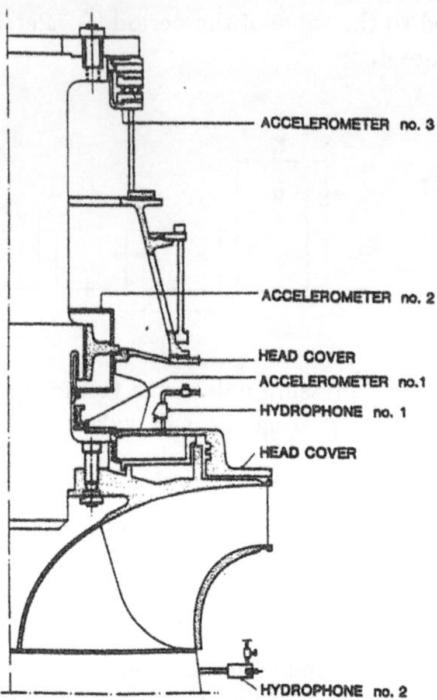

Figure 4.13 Vertical cross section of a turbine indicating the position of the various transducers

all frequencies present. This indicated that the origin of all vibrations was in the turbine runner entrance section. Particularly high amplitudes with frequency $2f_2$ were detected in one area of the head cover surrounding the turbine shaft.

If a signal is indicated by one transducer at time t and by a second transducer at time $t + \tau$, the length of the transmission path is then

$$\ell = c\tau \tag{4.29}$$

where c is the velocity of sound or celerity of the disturbance wave in the turbine structure.

There were only two ways for the vibrations to be transmitted from the runner area to the central part of the head cover:

(i) Up the turbine shaft to the thrust bearing and then down through the thrust pads and the thrust bearing support. The signal will then arrive at the head cover with a time delay τ_2. Along this path, the thrust bearing supports could act as critical elements amplifying the vibrations at a specific frequency.

(ii) The other way was through the water pressure fluctuations above the runner crown with time delay τ_1, causing the head cover to vibrate, in spite of the heavy load exerted on the head cover by the thrust bearing support. This is indicated in Figure 4.14. The signals emitted by the transducers placed at various points in the turbine structure, and the turbine frequency, which was provided by a deflected light beam, were recorded on a magnetic tape.

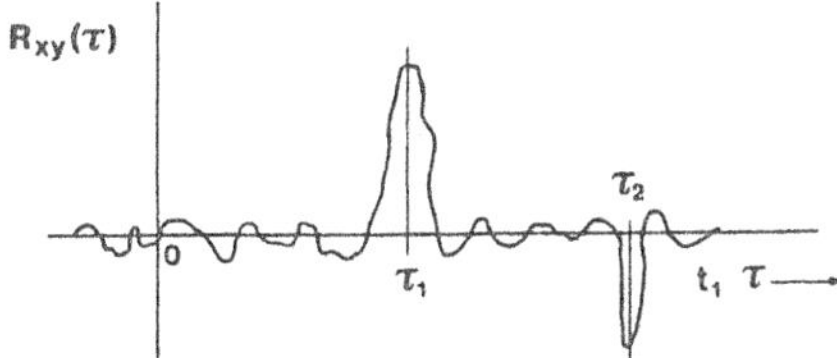

Figure 4.14 Cross-correlation plot of a signal originating at x and arriving at y via two different paths, characterized by time delays t_1 and t_2

4.4.2 Experimental Determination of Vibration Transmission Path

Filtering Technique

The signals, emitted by accelerometers No.1, 2 and 3 and the hydrophone No.1 were registered on magnetic tape and played back. A series of accurate narrow pass band operational filters tuned to $2f_2$ were inserted between the play-back signals and a multichannel recorder. The small damping of these filters resulted in sizable signal pikes.

The recordings must be inspected very carefully. If the signal frequency remains identical to the filter frequency, the signal transmission path can then be determined. If, on the other hand, short periods exist where the signal and filter frequencies differ, this interesting technique must then be abandoned. These instabilities must be attributed to the low damping level of the filters.

Cross-Correlation Techniques

A signal $x(t)$ is sampled at time t during the time interval $-T$ to $+T$ and stored in the computer memory. The correlation between $x(t)$ and its version with time delay τ is indicated by the autocorrelation function $R_{xx}(\tau)$;

$$R_{xx}(\tau) = \lim_{T\to\infty} \frac{1}{2T} \int_{-T}^{+T} x(t)\, x(t+\tau)\, dt\,. \tag{4.30}$$

For $\tau = 0$, $R_{xx}(0)$ reaches a maximum value.

When it is desired to find the relationship, if any, between two events in a physical system, cross-correlation methods may be used. The cross-correlation function $R_{xy}(\tau)$ of two signals $x(t)$ and $y(t)$ is defined in an analogous way as

$$R_{xy}(\tau) = \lim_{T\to\infty} \frac{1}{2T} \int_{-T}^{+T} x(t)\, y(t+\tau)\, dt\,, \tag{4.31}$$

and describes the general dependence of one signal on the other. The signal $y(t)$ has been sampled at an arbitrary time t during the same time interval $-T$ to $+T$ and is also stored in the computer memory.

If the signal $y(t)$ is shifted gradually for various values of the time parameter τ towards the stored signal $x(t)$, the value of τ which corresponds to the signal propagation time from x to y, will result in the maximum value of the correlation function $R_{xy}(\tau)$, since only then will the stored functions match each other. On the other hand, $R_{xy}(\tau)$ will always remain zero, if the two signals $x(t)$ and $y(t)$ are completely independent of each other.

Since each signal transmission path through a structure will generally be associated with a different delay time τ, a separate peak will occur in the cross-correlogram for each path which contributes significantly to the output, as is shown in Figure 4.14. Using equation (4.29), the length of various transmission paths can be calculated and the transmission path be found.

The actual signal $x(t)$ acting on the system is composed of harmonic components $s(t)$ and random noise $r(t)$ created by the turbulence in the water,

$$x(t) = s(t) + r(t) \tag{4.32}$$

and

$$y(t+\tau) = s(t+\tau) + r(t+\tau)\,. \tag{4.33}$$

Following the previously outlined methods, the combined signal $y(t+\tau)$ of equation (4.33) is again τ-shifted towards $x(t)$ expressed by equation (4.32).

For a stochastic signal with a large frequency band, $R_{rr}(\tau)$ is a sharp spike, see Figure 4.15, indicating that the turbulent signal has little conservation tendency. For turbulence limited in frequency, as can be in the case of water flowing through the turbine, the peak value $R_{rr}(\tau)$ will decrease and reaches zero at a sufficiently large value of τ. This is indicated in Figure

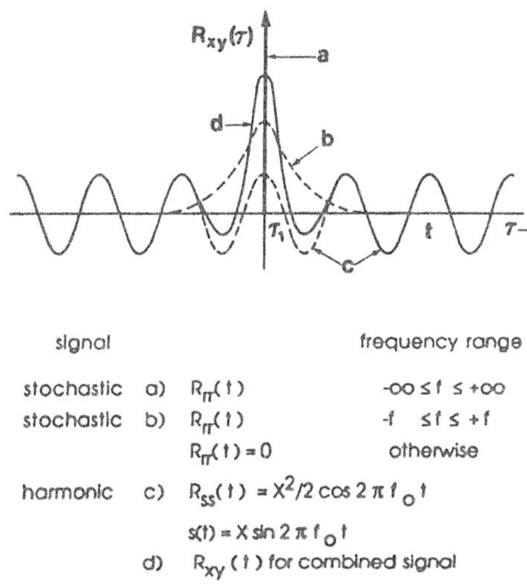

Figure 4.15 Auto-correlation function $R_{xx}(t)$ of periodic and stochastic signals

4.15. Limiting the frequencies further produces a $R_{rr}(\tau)$ close to a decaying cosine-function.

On the other hand, harmonic functions are represented by a periodic cosine-type $R_{ss}(\tau)$, since harmonic functions have a strong conservation tendency. It must be mentioned however that phase relations of harmonic waves travelling from x to y are lost in the $R_{xy}(\tau)$ representation. A typical plot of $R_{xy}(\tau)$ for a combined signal is shown in Figure 4.15.

Actual measurements reproduced in Figure 4.16 do not always indicate random noise spikes and the signal transmission path cannot then be determined. The reason for the absence of random noise spikes could be due to the fact that:

(i) The selected storage time $-T \leq t \geq +T$ was too short.

(ii) The $R_{xy}(\tau)$ is not very peaked due to a limited number of frequencies contained in the stochastic signal. In the case of a hydraulic turbine, the degree of turbulence of the water in the turbine is not sufficiently high. Generally, the frequency band of a stochastic signal must be far greater than the transmission band of the structure.

(iii) The lobes of a decaying type $R_{xy}(\tau)$ function are large in comparison

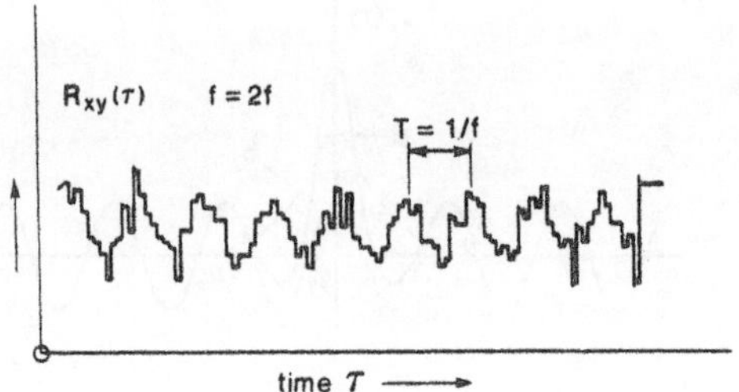

Figure 4.16 Cross-correlation function $R_{xy}(\tau)$ resulting from actual measurements. (The signal at x recorded by hydrophone No.1 was cross-correlated with the signals at y emitted by accelerometers No.1, 2 and 3)

with the signal transmission path and therefore the time of the signal through the structure.

Cross-Power Spectral Density Techniques

The last remaining method to find the signal transmission path is provided by the cross-power spectral density technique. Again, the excitation signal $x(t)$, the input, is captured by the hydrophone No.1 and its response, the output, measured at different points y is provided by the accelerometers No.1, 2 and 3.

If the signal $x(t)$ is harmonic, the appropriate frequency response has the same frequency for linear acting elements, but will generally have a different amplitude and must also display a phase shift. To describe these relations, the power spectral technique is used. The input power spectral-density function $G_{xx}(f_0)$ obtained at x is related to the history record $x(t) = X \sin 2\pi f_0 t$ and its autocorrelation function $R_{xx}(\tau)$ of equation (4.30) by the Fourier transformation;

$$G_{xx}(f_0) = \int_{-\infty}^{+\infty} R_{xx}(\tau)\, e^{-2\pi j f_0}\, d\tau \,. \tag{4.34}$$

The cross-power spectrum density function is defined by

$$G_{xy}(f_0) = \int_{-\infty}^{+\infty} R_{xy}(\tau)\, e^{-2\pi j f_0}\, d\tau \,. \tag{4.35}$$

Both functions contain the same amount of information and the application of either $G_{xx}(f)$ or $R_{xx}(\tau)$ is only dictated by the physical conditions of the system as we have seen in the preceding relations. As already noted in equation (4.32) the signal will contain harmonics and a stochastic component. Harmonic frequencies f may have large amplitudes whereas from energy considerations of the stochastic signal, its continuous frequency spectrum must have small amplitudes.

Considering that $G_{xx}(f)$ and $G_{xy}(f)$ were calculated from the harmonic signals at x and y, the displacement output and phase shift can be expressed by

$$G_{xy}(f) = G_{xx}(f)\, H(f)\,, \tag{4.36}$$

with $H(f)$ the complex transfer characteristics given by

$$H(f) = G_{xy}(f)/G_{xx}(f) = C_{xy}(f) - jQ_{xy}(f)\,. \tag{4.37}$$

The amplitude gain factor between x and y is then

$$|H(f)| = \sqrt{C_{xy}(f)^2 + Q_{xy}(f)^2}\,. \tag{4.38}$$

The phase delay angle $\Phi_{xy}(f)$ is calculated by

$$\Phi_{xy}(f) = \arctan \frac{Q_{xy}(f)}{C_{xy}(f)}\,. \tag{4.39}$$

For a signal having harmonic components and stochastic fluctuations, the form $G(f)$ must contain functions of both. The power- spectrum must contain sharp spikes, corresponding to the various frequencies. The stochastic part is a continuous curve, limited by its frequency band range. This is illustrated in Figure 4.17.

The phase delay angle $\Phi_{xy}(f)$ now allows the calculation of the phase shift time t_1 by

$$t_1 = \frac{\Phi_{xy}(f)}{2\pi} T \tag{4.40}$$

as illustrated in Figure 4.18. According to equation (4.29) the length of the signal transmission is then

$$\ell = \frac{c}{f} \frac{t_1}{T} \tag{4.41}$$

with c being the celerity (speed of sound) of the signal through the structure.

Measurement results for $2f_2$, the multiple harmonics of f_2 introduced in section 4.2.2, are shown in Figure 4.19. A critical inspection of these results shows that the values of $\Phi_{xy}(f)$ change with the turbine output power, since stress changes in the structural material represent a feedback. This is of rather academic interest and will not influence the determination of the signal transmission path.

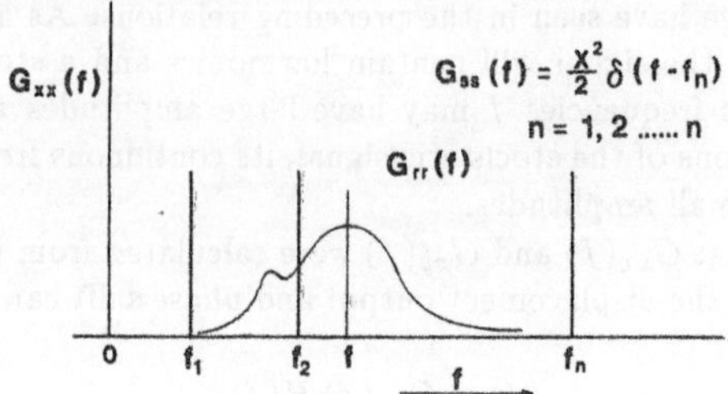

Figure 4.17 Power-spectrum of a stochastic signal with periodic components

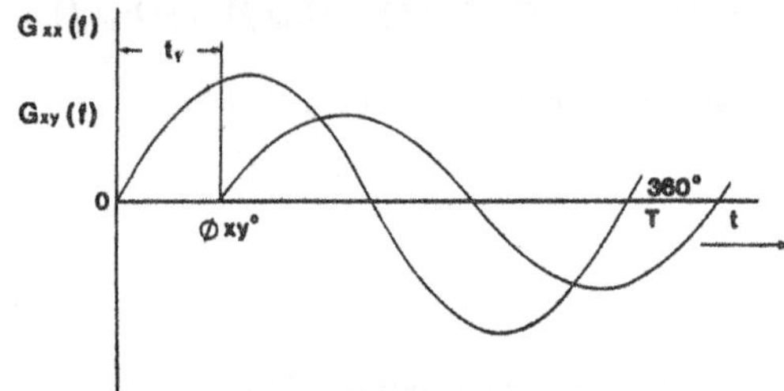

Figure 4.18 Time delayed harmonic signals in x and y of frequency f

4.4.3 Interpretation of Test Results

The interpretation of Figure 4.19 proved conclusively that, in the present case, the particularly strong head cover vibrations indicated by accelerometer No.2 with $f = 2f_2$ were caused by the interaction of the spiral case tongue and the runner channels. The resulting water pressure fluctuations of frequency f_2 were measured by hydrophone No.1, considered as point x, in the leakage flow between the runner and the case. For some heads, the shift angles Φ registered by the accelerometer No.1 relative to hydrophone No.1 were positive. This contradiction is attributed to the fact that hydrophone No.1 was connected by a short soft copper tube to the turbine head cover which caused the apparent time delay. Calculations confirmed this hypothesis.

Recorded signals played back through accurately tuned narrow band pass filters can indicate the vibration transmission path if the filters are stable. If the previous method proves unsatisfactory, correlation methods must then be used with a preference for spectral power density techniques.

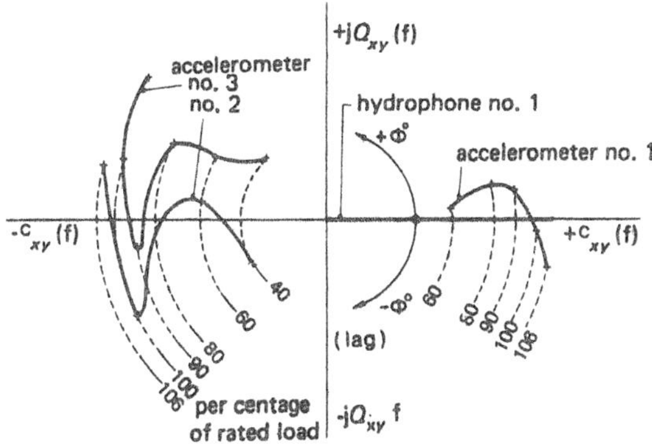

Figure 4.19 Experimental results showing the phase angle $\Phi_{xy}(2f_2)$ between hydrophone No.1 and accelerometer No.1

4.5 Reduction of Pressure Fluctuations

4.5.1 Prediction

During the development stage of a turbine model on a test stand, pressure pulsations and vibrations should be monitored by installing accelerometers and hydrophones at various points of the turbine. The additional expenses involved in such tests are usually justified, since theoretical predictions of vibration amplitudes are virtually impossible.

During the commissioning of large units, similar tests of various elements of the structure of a turbine and measurements of pressure pulsations are recommended. Since the instrumentation and test protocols are the same, the latter tests can be largely justified by the resulting extra detailed knowledge of actual operating conditions. Based on model test results, the required modifications of structural elements subjected to resonance with a forcing frequency are generally not too expensive. This is certainly not the case when emergency repairs on the prototype are required. Such tests could prove both valuable and necessary in the development of highly optimized turbines.

4.5.2 Methods for Reducing Pressure Pulsations

Various solutions are possible to reduce pressure variations at the runner entrance edge arising from the wakes shed by the guide vane edges or the spiral case tongue. The most important ones are summarized as follows:

For large, high head Francis turbines, the energy level of the water flows on both sides of the tongue wall can be equalized by providing a number of connecting holes.

If the leading edges of the runner blades are made oblique in relation to the trailing edges of the guide vanes and the spiral case tongue, this results in a more gradual sweep of the wakes by the runner blades. The total energy disturbance of amplitude E_m is then spread over a certain time period which is a function of the blade inclination. For a given total wake energy, the effective pulse energy decreases in the same proportion as b is increased, where b is the total time taken for the oblique leading edge of the blade to sweep through the vertical wake. This is illustrated in Figure 4.20. Only a fraction of the pulse is active at a given time, thus considerably reducing the disturbance and the related amplitudes at frequencies f_1 and f_2 with their harmonics, so that resonance is practically avoided.

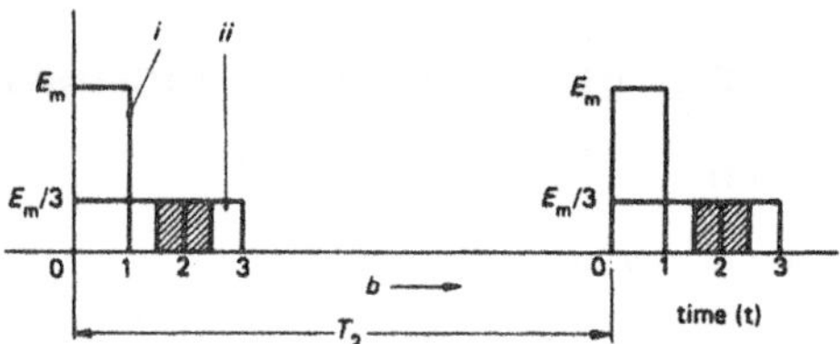

Figure 4.20 Shock pulses of equal energy, where i = the vertical leading edge of a runner blade and ii = inclined edge (here taking three times longer to sweep the wake). (The shaded sections show where the runner blade section crosses the wake)

Obviously, the maximum inclination of the blade leading edges is limited by efficiency requirements to avoid dead water zones in the inclined blade-hub region, and by manufacturing considerations when stainless steel layers are to be welded on these oblique surfaces of the blades.

A slight improvement could be obtained from the injection of compressed air at selected locations in the water filled area just underneath the head cover, but one must ensure that the rotor labyrinths are always operating in water. Vibrations, if not completely eliminated, could break this external piping and lead to a dangerous situation. In units where the disc-friction has been reduced by injecting air between the head cover and the runner, this suggestion has already been applied.

Finally it is also possible to eliminate resonance of certain elements in the turbine structure by welding on stiffeners, thus moving the resonant frequency to a less critical value.

References

4.1 Bendat, J.S. and Piersol, A.G. (1966). *Measurement and analysis of random data*, John Wiley & Sons, New York, London, Sydney.

4.2 Chen, Y.N. (1961). *Wasserdruckschwingungen in Spiralgehäusen von Speicher-pumpen*, Forschungsheft Technische Rundschau, Sulzer Rev. 534, pp. 1-14.

4.3 Fanelli, M. (1988). The Vortex rope in the draft tube of Francis turbines operating at partial load, A proposal for a mathematical model, ENEL Centro Ricerca Idraulica e Strutturale, 20162, Milano.

4.4 Gerich, R. (1974). 'Untersuchungen über die instationäle Strömung in einer schnel-läufigen Francis-Modellturbine mit besonderer Berücksichtigung des Teillastverhaltens', Thesis, Faculty of Machinery and Electrotechnology, Technical University Munich.

4.5 Grein, H. and Bachmann, P. (1976). 'Hydraulic torque on misaligned guide vanes', *Water Power and Dam Construction*, February.

4.6 Kemp, N.H. and Sears, W.R. (1953). 'Aerodynamic interference between moving blade rows', *Journal of Aeronautic Sciences*, September, pp. 580-589.

4.7 Lange, F.H. (1967). *Correlation techniques* (translation of *Korrelationselektronik*, VEB Verlag 1959, Berlin), Van Nostrand, Princeton.

4.8 Lee, Y.W. (1960). *Statistical theory of communication*, John Wiley & Sons, New York, London, Sydney.

4.9 Mataix, C. (1989). *Turbomaquinas hidraulicas*, Editiones ICAI, Madrid.

4.10 Mayer, R.X. (1955). 'Interference due to viscous wakes between stationary and rotating blades in turbo-machines', Thesis, The Johns Hopkins University, Baltimore.

4.11 Netsch, H. (1956). 'Strömungsvorgänge in Turbinen und Pumpen und deren Auswirk-ung auf die Druckrohrleitung von Kraftwerken und Pumpstationen', *Die Wasserwirtschaft*, February, pp. 1-9.

4.12 Netsch, H. (1958). 'Die Saugrohrausbildung von Francisturbinen grosser Fallhöhe', *Die Wasserwirtschaft*, September, pp. 223-228.

4.13 Netsch, H. and Giacometti, A. (1982). 'Axial flow induced vibrations in large high-head machines', *Waterpower and Dam Construction*, August, pp. 21-27.

4.14 Pfleiderer, C. (1955). *Die Kreiselpumpen*, Springer Verlag, Berlin, Göttingen, Heidelberg.

4.15 Pfleiderer, C. and Petermann, H. (1964). *Strömungsmaschinen*, Springer Verlag, Berlin, Göttingen, Heidelberg.

4.16 Raabe, J. (1985). *Hydro power* (translation of *Hydraulische Maschinen*), VDI-Verlag, Düsseldorf.

4.17 Schlichting, H. (1960) *Boundary layer theory* (translation of *Grenzschichttheorie*, Verlag G. Braun, 1958, Karlsruhe), McGraw-Hill, New York, Toronto.

4.18 Solodovnikov, V.V. (1965). *Dynamique statistique des systèmes lineaires de commande automatique*, Dunod, Paris.

4.19 Vivier, J. (1966). *Turbines hydrauliques et leur régulation*, Editions Albin Michel, Paris.

4.20 Stern, J., Barbeyrac, J. and Poggi, R. (1967). *Méthodes pratiques des fonctions aléatoires*, Dunod, Paris.

4.21 Vavra, M.H. (1960). *Aero- thermodynamics and flow in turbomachines*, John Wiley & Sons, New York, London.

Chapter 5

Rotordynamics

R. Nordmann and W. Diewald

5.1 Introduction

The vibration characteristics of rotating machinery can be subdivided into lateral vibrations and torsional vibrations of the rotating parts. The latter usually occur in large power trains and are excited for example by generator and motor moments due to network disturbances or short circuits. This chapter deals with the lateral vibrations only.

Lateral bending vibrations can be caused by different reasons. Unbalance forces which are always present in a rotor lead to synchronous vibrations with running speed. Other whirl frequencies can come from self excitation mechanisms e.g. in journal bearings or seals. Finally transient motions can occur due to start up or shut down, blade losses or earthquake excitations.

The aim of this chapter is to give a brief overview for these vibration features. In particular, the theoretical background is explained and characteristic vibration results are shown.

The basis of calculations are models which have to include the most important effects. Starting from the rotating machinery shafts with stiffness and mass distributions the vibrations of these parts are strongly influenced by fluid-forces acting in journal bearings, seals and so on. In some cases also the bearing pedestals, the housing and the piping system have to be included in a model.

In order to understand the basic dynamic mechanisms in rotating machinery it is however more helpful to start with a very simple rotor model, and to include further effects later on. It will be seen that typical vibrations present in a real machine can qualitatively also be observed with simple models like the Jeffcott-rotor.

5.2 Basic Equations For Lateral Vibrations

5.2.1 Simple Rotor Model (Jeffcott-Rotor)

A basic model for rotor dynamics is a simple rotor with one mass at midspan as shown in Figure 5.1. The shaft is flexible with negligible mass and the bearings are supposed to be rigid. In general the mass is not balanced, so that the centre of rotation (W) is not identical with the centre of gravity (S).

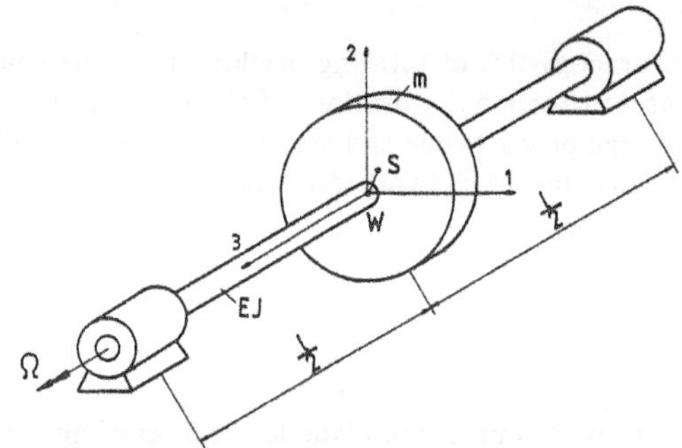

m=mass; EJ=stiffness; Ω=running speed;
W=centre of rotation; S=centre of gravity; ℓ=length

Figure 5.1 Model of a Jeffcott-rotor

The pure radial motion of such a simple rotor model can be described with the two coordinates q_{1s}, q_{2s} for S or q_1, q_2 for W and with the angle ψ for the torsion as shown in Figure 5.2. The coordinates q_{1s}, q_{2s} can be expressed in terms of q_1, q_2, ψ and eccentricity e as

$$\begin{aligned} q_{1s} &= q_1 + e\cos\psi \\ q_{2s} &= q_2 + e\sin\psi \end{aligned} \tag{5.1}$$

The time derivatives are

$$\begin{aligned} \dot{q}_{1s} &= \dot{q}_1 - e\dot{\psi}\sin\psi \\ \ddot{q}_{1s} &= \ddot{q}_1 - e\dot{\psi}^2\cos\psi - e\ddot{\psi}\sin\psi \\ \dot{q}_{2s} &= \dot{q}_2 + e\dot{\psi}\cos\psi \\ \ddot{q}_{2s} &= \ddot{q}_2 - e\dot{\psi}^2\sin\psi + e\ddot{\psi}\cos\psi \end{aligned} \tag{5.2}$$

It can be seen that the accelerations of the centre of gravity consist of the accelerations of the centre of rotation plus a part with $\dot{\psi}^2$, which is a centrifugal acceleration, and a part with $\ddot{\psi}$, a tangential acceleration.

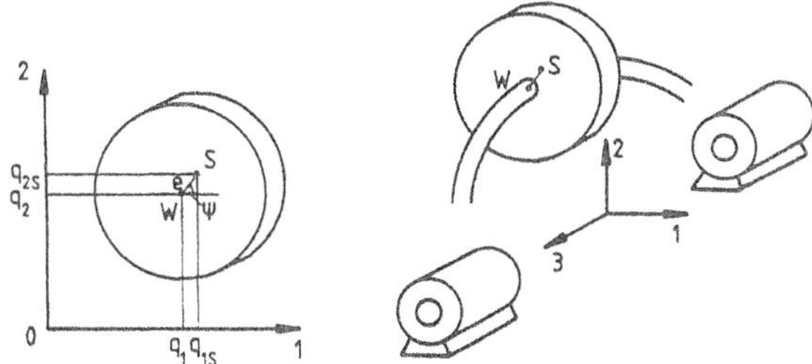

Figure 5.2 Coordinates for the motion of a Jeffcott-rotor

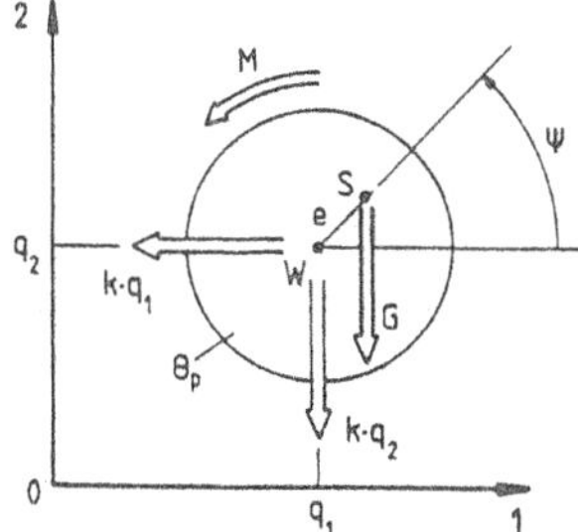

k=stiffness of the shaft; m=mass; G=weight force;
M=external moment; θ_p=polar moment of inertia

Figure 5.3 Force equilibrium for a rotating mass

Equation of Motion for the Model

Using these kinematic relations, one can derive the equations of motion for the centre of gravity of the rotating mass, based on equilibrium conditions of restoring forces, inertia forces, the weight G and the external moment M shown in Figure 5.3.

$$m\ddot{q}_{1s} + kq_1 = 0 \tag{5.3}$$

$$m\ddot{q}_{2s} + kq_2 = -G \tag{5.4}$$

$$\theta_p\ddot{\psi} = ke\,(q_2 \cos\psi - q_1 \sin\psi) + M \tag{5.5}$$

Combining with equation (5.2) yields

$$m\ddot{q}_1 + kq_1 = me\dot{\psi}^2 \cos\psi + me\ddot{\psi}\; sin\psi \tag{5.6}$$
$$m\ddot{q}_2 + kq_2 = me\dot{\psi}^2 \sin\psi - me\ddot{\psi} \cos\psi - G \tag{5.7}$$
$$\theta_p\ddot{\psi} = ke\,(q_2 \cos\psi - q_1 \sin\psi) + M \tag{5.8}$$

These equations of motion can be simplified using assumptions such as:
(a) The rotor operates in a steady state condition. Then the external moment vanishes, $M = 0$.
(b) The polar radius of inertia ${i_p}^2 = \theta_p/m$ is much larger compared to the eccentricity e and the displacements q_1 and q_2. Then the tangential acceleration can be set to zero.

$$\ddot{\psi} = 0\,, \quad \dot{\psi} = \Omega = \text{const.}, \quad \psi = \Omega t + \beta \tag{5.9}$$

This yields the well known equations for an unbalanced Jeffcott-rotor:

$$\begin{aligned} m\ddot{q}_1 + kq_1 &= me\,\Omega^2 \cos(\Omega t + \beta) \\ m\ddot{q}_2 + kq_2 &= me\,\Omega^2 \sin(\Omega t + \beta) - G \end{aligned} \tag{5.10}$$

Here Ω is the rotational frequency and β an phase angle. The equations show the equilibrium of inertia forces, restoring forces and centrifugal forces. They are linear inhomogeneous differential equations of second order with constant coefficients. Because they are not coupled the equations can be solved separately.

Solutions of Equations of Motion

The solutions for the equation (5.10) consist of a homogeneous and an inhomogeneous part

$$\begin{aligned} q_1 &= q_{1hom} + q_{1part} \\ q_2 &= q_{2hom} + q_{2part} \end{aligned} \tag{5.11}$$

The homogeneous parts describe the free vibrations of the shaft. Forced vibrations by unbalance forces and static deflections due to the weight go into the inhomogeneous solution.

In order to solve the free vibrations, the right hand sides of equations (5.10) are set to zero

$$m\ddot{q}_i + kq_i = 0 \qquad (i = 1, 2) \tag{5.12}$$

The solution of this equation has the form

$$q_{ihom} = A_i \cos\omega t + B_i \sin\omega t \qquad (i = 1, 2) \tag{5.13}$$

where $\omega = \sqrt{k/m}$ is the eigenfrequency of the rotor. A_i and B_i have to be calculated using the initial conditions.

The inhomogeneous solution of equations (5.10) can be achieved with a statement of the form of the right hand side

$$\begin{aligned} q_{1part} &= \hat{q}_1 \cos(\Omega t + \beta) \\ q_{2part} &= \hat{q}_2 \sin(\Omega t + \beta) + q_{2stat} \end{aligned} \tag{5.14}$$

with

$$\hat{q}_1 = \hat{q}_2 = \frac{me\,\Omega^2}{k - \Omega^2 m} = e\frac{w^2}{1 - w^2}\,; \qquad w = \frac{\Omega}{\omega} \qquad \text{and} \qquad q_{2stat} = -\frac{G}{k}$$

So the complete solution can be written as:

$$\begin{aligned} q_1 &= A_1 \cos\omega t + B_1 \sin\omega t + e\frac{w^2}{1 - w^2}\cos(\Omega t + \beta) \\ q_2 &= A_2 \cos\omega t + B_2 \sin\omega t + e\frac{w^2}{1 - w^2}\sin(\Omega t + \beta) - \frac{G}{k} \end{aligned} \tag{5.15}$$

where A_1, A_2, B_1 and B_2 can be calculated from the initial conditions. Looking at the inhomogeneous part of the solution only, it is interesting to draw the vibration amplitudes versus running speed as shown in Figure 5.4. The amplitudes for the centre of rotation W are

$$r_w = e\left|\frac{w^2}{1 - w^2}\right| \tag{5.16}$$

and for the centre of gravity S

$$r_s = r_w + e = e\left|\frac{1}{1 - w^2}\right| \tag{5.17}$$

The distance between the two curves is always equal to the eccentricity e. When the running speed is equal to the eigenfrequency $\Omega = \omega$, the vibration level goes to infinity.

Therefore this frequency is called critical speed. At running speeds lower than the critical, the centre of gravity rotates outside the centre of rotation and vice versa for overcritical speeds. At very high running frequencies the shaft rotates around the centre of gravity. This effect is called self balancing (self centring).

5.2.2 Jeffcott-Rotor in Flexible Supports

Up to now rigid supports have been assumed, but in the field there is always a flexibility in the bearings and the pedestals. Due to the design of the supports they have also often a different stiffness in two directions.

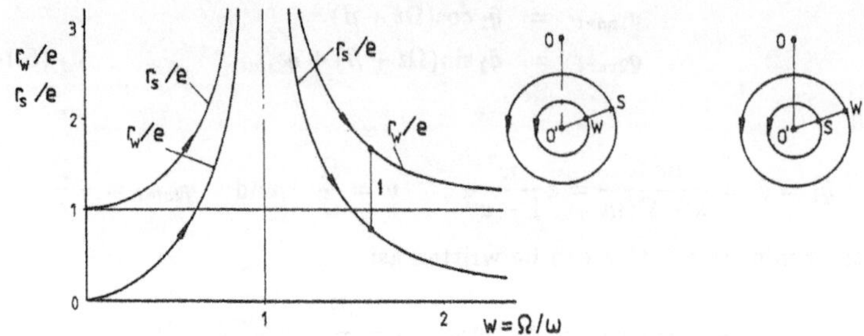

Figure 5.4 Vibration amplitude for the centre of rotation and the centre of gravity

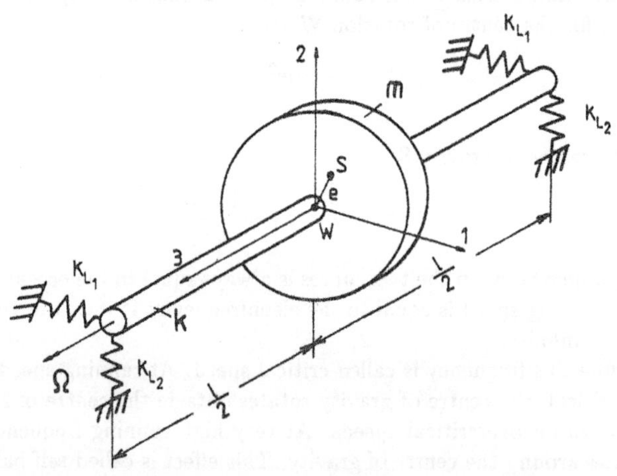

Figure 5.5 Jeffcott-rotor in flexible supports

The model for a Jeffcott-rotor in flexible supports is shown in Figure 5.5.

Equations of Motion

The equations of motion of the Jeffcott-rotor in rigid supports have to be modified with respect to the stiffness parameters. The additional flexibility of the bearings acts as a spring in series arrangement.

The rest of equation (5.10) is unchanged

$$\begin{aligned} m\ddot{q}_1 + k_1 q_1 &= me\,\Omega^2 \cos(\Omega t + \beta) \\ m\ddot{q}_2 + k_2 q_2 &= me\,\Omega^2 \sin(\Omega t + \beta) - G \end{aligned} \tag{5.18}$$

with

$$k_i = \frac{2k_{Li}\,k}{2k_{Li} + k} \qquad (i = 1, 2)\,; \qquad k_{Li}\text{: bearing stiffness}$$

Solutions of Equations of Motion

The solutions of these equations are analog to those of equations (5.10). Due to the different stiffness parameters there are now two eigenfrequencies in the homogeneous solutions

$$\begin{aligned} q_{1hom} &= A_1 \cos\omega_1 t + B_1 \sin\omega_1 t \\ q_{2hom} &= A_2 \cos\omega_2 t + B_2 \sin\omega_2 t \end{aligned} \tag{5.19}$$

with

$$\omega_i = \sqrt{\frac{k_i}{m}} \qquad (i = 1, 2)$$

The inhomogeneous solutions are different in amplitude compared to the rigid case. This leads also to two different resonances for the two directions

$$\begin{aligned} q_{1part} &= \hat{q}_1 \cos(\Omega t + \beta) \\ q_{2part} &= \hat{q}_2 \sin(\Omega t + \beta) + q_{2stat} \end{aligned} \tag{5.20}$$

with

$$\hat{q}_i = \frac{me\,\Omega^2}{k_i - \Omega^2 m} = e\frac{w^2}{\bar{\omega}_i^2 - w^2} \quad ;$$

$$\bar{\omega}_i = \frac{\omega_i}{\omega} \quad (i = 1, 2)\,, \quad w = \frac{\Omega}{\omega} \quad \text{and} \quad \omega = \sqrt{\frac{k}{m}}$$

The two resonances are the two critical speeds of this model.

$$\begin{aligned} w_1 &= \bar{\omega}_1 \\ w_2 &= \bar{\omega}_2 \end{aligned} \tag{5.21}$$

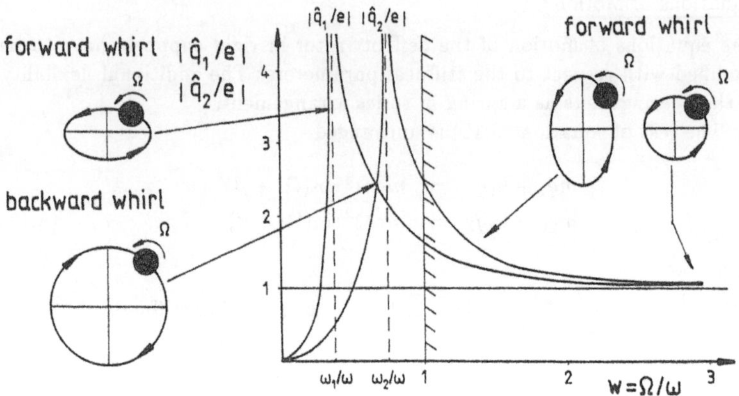

Figure 5.6 Vibration amplitude of the centre of rotation for a Jeffcott-rotor in flexible supports

They are lower than the critical speed with rigid supports due to the additional flexibility of the bearings as shown in Figure 5.6. The orbits in general have an elliptical shape which changes to straight lines in the resonances and to a circle between the criticals or at high speeds.

As one can see from the solutions of equation (5.20), the direction of the vibration orbit changes between the resonances. Below the first and above the second critical the direction of the orbit is the same as the shaft rotation, as it was for the whole range with rigid supports. This is called forward whirl. Between the critical speeds it is vice versa. This is called backward whirl.

5.2.3 External Damping

The last two models of a rotor did not include any damping. In a real machine there are always damping forces present either coming from the material itself (internal damping) or from contact to the surrounding. External damping has always a positive effect on the dynamic behaviour of a machine, because it lowers the vibration amplitudes at the resonances. Internal damping can also destabilize a system if it operates above the critical speed.

In the following, only the influence of external damping is shown. A simple model shown in Figure 5.7 for the damping forces consists of restoring forces which are proportional to the velocities of the rotor mass.

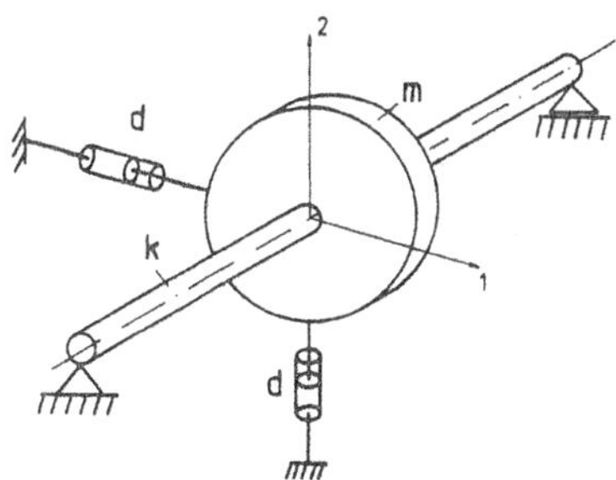

Figure 5.7 Model for external damping at a Jeffcott-rotor

The corresponding equations of motion for the Jeffcott-rotor with rigid bearings can be written as:

$$\begin{aligned} m\ddot{q}_1 + d\dot{q}_1 + kq_1 &= me\,\Omega^2 \cos(\Omega t + \beta) \\ m\ddot{q}_2 + d\dot{q}_2 + kq_2 &= me\,\Omega^2 \sin(\Omega t + \beta) - G \end{aligned} \tag{5.22}$$

These again are inhomogeneous, linear differential equations. They are uncoupled, and so it is possible to solve them separately.

The homogeneous parts of the solutions are now different from equation (5.13), due to the damping coefficients. Using an exponential statement leads to the eigenvalue problem and the characteristic equation for the eigenvalues.

$$\lambda_i^2 m + \lambda_i d + k = 0 \qquad (i = 1, 2) \tag{5.23}$$

The resulting eigenvalues are conjugate complex

$$\lambda_i = -\alpha \pm j\sqrt{\omega^2 - \alpha^2} \tag{5.24}$$

with

$$\alpha = \frac{d}{2m} \quad \text{and} \quad \omega^2 = \frac{k}{m}$$

So the homogeneous solutions have the form

$$q_{ihom} = e^{-\alpha t}\left(A_i \cos\sqrt{\omega^2 - \alpha^2}\, t + B_i \sin\sqrt{\omega^2 - \alpha^2}\, t\right) \tag{5.25}$$

For large times t the displacements go to zero due to the real part of the eigenvalue α, which is proportional to the damping coefficient d.

For the oscillatory inhomogeneous solution one can work with a complex notation for the unbalance excitation. The right hand sides of equations (5.22) can be represented as real parts of a complex function

$$\begin{aligned} m\ddot{\bar{q}}_1 + d\dot{\bar{q}}_1 + k\bar{q}_1 &= me\,\Omega^2\,\Re\{e^{i(\Omega t+\beta)}\} \\ m\ddot{\bar{q}}_2 + d\dot{\bar{q}}_2 + k\bar{q}_2 &= me\,\Omega^2\,\Re\{e^{i(\Omega t+\beta+\frac{\pi}{2})}\} \end{aligned} \tag{5.26}$$

Then the real parts of the solutions of these equations give the unbalance forced vibrations of a damped rotor

$$\begin{aligned} q_1(t) &= \hat{q}\cos(\Omega t - \epsilon) \\ q_2(t) &= \hat{q}\sin(\Omega t - \epsilon) \end{aligned} \tag{5.27}$$

with

$$\hat{q} = \frac{ew^2}{\sqrt{(1-w^2)^2 + (2Dw)^2}}\,, \qquad \epsilon = \arctan(\frac{2Dw}{1-w^2});$$

$$\beta = 0\,, \quad w = \frac{\Omega}{\omega}\,, \quad \omega = \sqrt{\frac{k}{m}} \quad \text{and} \quad D = \frac{d}{2m\omega}$$

In Figure 5.8 the results of equation (5.27) are plotted versus the related running speed for different damping values. For higher damping the resonance amplitudes decrease clearly.

5.2.4 Transient Vibrations

As mentioned before, the vibrations of a rotor consist of a homogeneous and an inhomogeneous part. For unbalance excitation it is easy to determine the inhomogeneous part, because the excitation force is a harmonic function of the running speed.

In practice the unbalance is the most important excitation, but besides that there are also other forces acting on a rotor. This can be loads from surrounding fluids, basement accelerations coming from other machines or from earthquakes. Such oscillations, which are caused by these non-harmonic forces, are called transient vibrations.

For the vibrations caused by arbitrary forces it is convenient to start with the forced vibration due to an unit momentum. The force signal $f(t)$ has the form:

$$\lim_{t\to 0}\int_0^t f(t)\,dt = 1 \tag{5.28}$$

This means, that the right hand side of the equation of motion of a single degree of freedom system is zero for $t > 0$ and so the solution consists only of the homogeneous part. The initial conditions are given from the momentum conservation with:

$$q_i(t=0) = 0 \quad \text{and} \quad \dot{q}_i(t=0) = \int_0^t \frac{f(t)\,dt}{m} = \frac{1}{m} \tag{5.29}$$

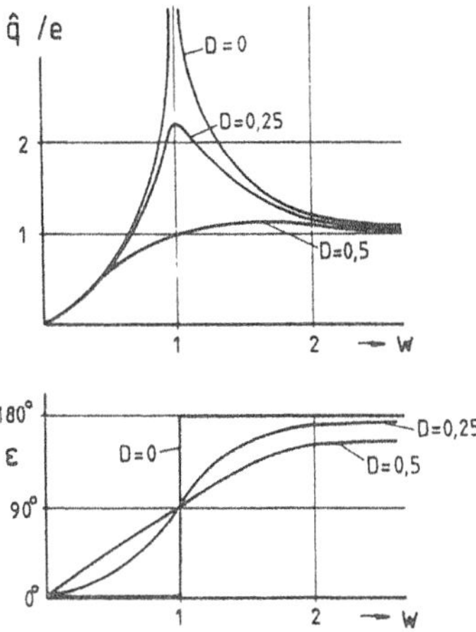

Figure 5.8 Unbalance responses of a damped rotor for different damping values

This leads to a solution, which is the impulse response function

$$q_i(t) = \frac{1}{m\omega\sqrt{1-D^2}}\, e^{-D\omega t} \sin \omega\sqrt{1-D^2}\, t \tag{5.30}$$

with

$$\omega = \sqrt{\frac{k}{m}} \quad \text{and} \quad D = \frac{d}{2m\omega}$$

An arbitrary force $f(t)$ can now be considered as a sum of very short momentums, starting at different times as shown in Figure 5.9. So it is possible to write down an integral equation for the forced vibration due to any excitation.

$$q_{ipart}(t) = \frac{1}{m\omega\sqrt{1-D^2}} \int_0^t f(t')\, e^{-D\omega(t-t')} \sin \omega\sqrt{1-D^2}\,(t-t')\, dt' \tag{5.31}$$

This formula is called Duhamel-integral or convolution integral.

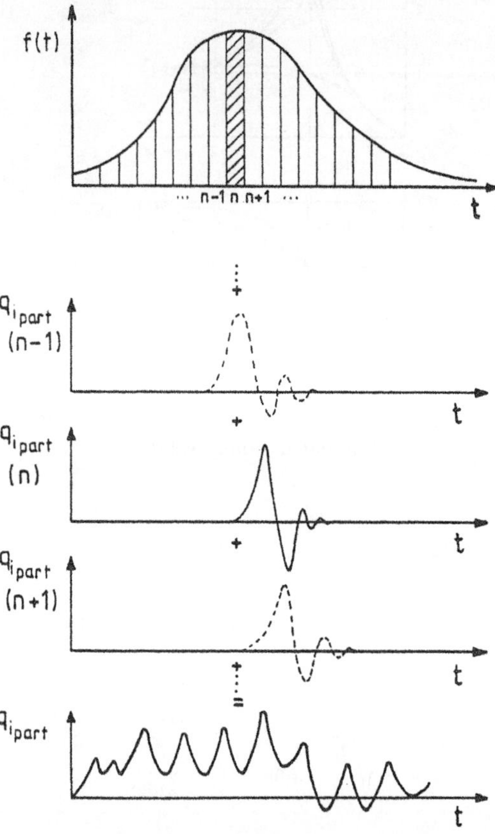

Figure 5.9 Forced vibrations as a sum of momentum excited motions

5.3 Influence of Fluid Forces in Journal Bearings and Seals

In the last chapter it is shown that there are two kinds of vibration in rotating machinery. Synchronous motions are caused by unbalance forces. Oscillations with the eigenfrequencies can be evoked by initial conditions, transient loads and by self-exciting mechanisms as will be seen in this chapter.

Especially in high performance turbomachinery there may be strong forces in journal bearings, seals, balance pistons etc. acting on the moving shaft. These elements can transfer energy from rotation to the bending of the rotor via the shear forces in the fluids. According to the shaft characteristics, they act not only as a spring and a damper but also have the capability to excite vibrations. This makes the fluid-elements very important factors for the stability of rotating machinery.

In the following, the dynamic behaviour of journal bearings and seals is considered more exactly. In the case of small motions, it is possible to linearize the force displacement relations. The dynamic characteristics of bearings and seals can than be expressed by stiffness and damping coefficients.

5.3.1 Journal Bearings

Journal bearings are widely used to support rotating shafts. Between the rotor and the stator there is a small clearance filled with oil where a pressure distribution is built up due to the rotation of the shaft as shown in Figure 5.10.

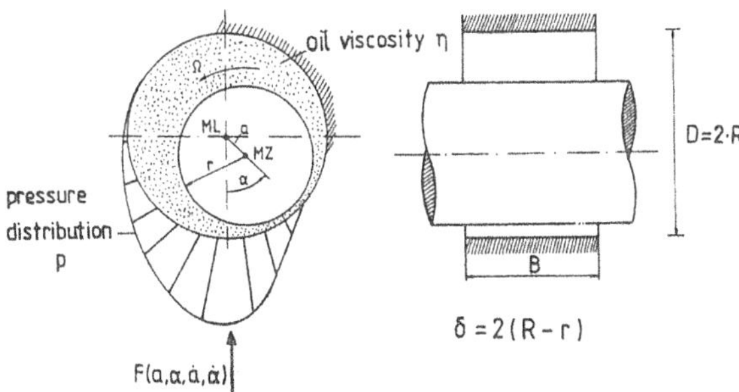

Figure 5.10 Journal bearings

For every operational point there exists an equilibrium position given by the external force and the resulting force of the pressure field. The description of the fluid properties is possible using the Reynolds differential equation which is based on the assumptions of laminar flow and an ideal fluid behaviour. Integration of the resulting pressure distribution over the shaft surface leads to the resulting oil film force acting on the rotor.

The theory shows that the dynamic properties of similar journal bearings are a function of the dimensionless Sommerfeld number

$$So = \frac{F_{stat}\,\psi^2}{B D\,\eta\,\Omega} \tag{5.32}$$

with

F_{stat}	:	static external load
B	:	width of the bearing
D	:	bearing diameter
ψ	:	relative clearance
η	:	oil viscosity
Ω	:	rotational frequency of the shaft

This number is a function of the position of the shaft MZ_0 in the bearing as shown in Figure 5.11. The shaft is centered for $So = 0$ and sink at the bottom for $So = \infty$.

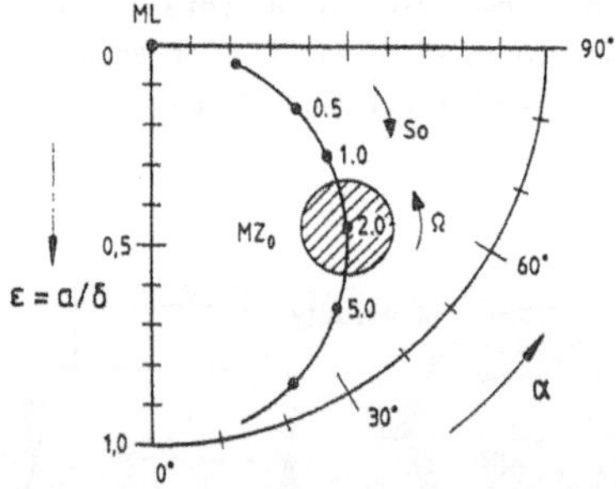

Figure 5.11 Displacement of a shaft in journal bearings

For small motions of the shaft around the equilibrium position, the oil film produces forces which can be expressed as functions of the displacements and the velocities as shown in Figure 5.12.

$$\begin{aligned} Z_1 &= k_{11}q_1 + k_{12}q_2 + d_{11}\dot{q}_1 + d_{12}\dot{q}_2 \\ Z_2 &= k_{21}q_1 + k_{22}q_2 + d_{21}\dot{q}_1 + d_{22}\dot{q}_2 \end{aligned} \tag{5.33}$$

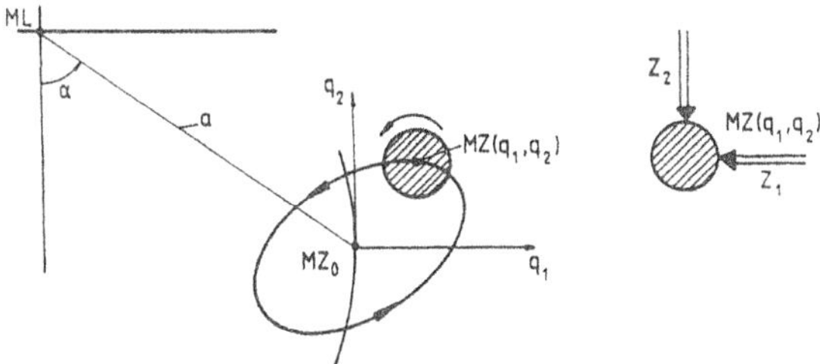

Figure 5.12 Small motions of a shaft in a journal bearing around an equilibrium position

This is a linear formulation for the forces in a journal bearing. The coefficients for stiffness k_{ij} and damping d_{ij} are usually given in a dimensionless form in dependence of the Sommerfeld number as shown in Figure 5.13. Values for the coefficients are available from measurements or calculations where the basis is the Reynolds differential equation. The stiffness and damping properties of a journal bearing are not symmetric in general, which means $k_{11} \neq k_{22}$ and $d_{11} \neq d_{22}$ and the stiffness coupling parameters are different $k_{12} \neq k_{21}$. The latter can cause instability for which the difference between k_{12} and k_{21} is responsible.

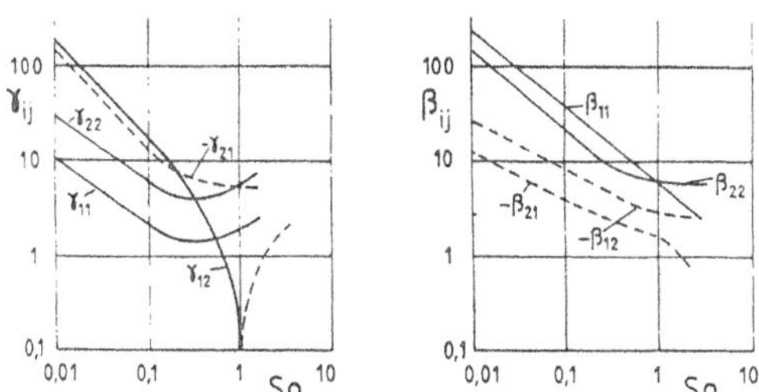

Figure 5.13 Dimensionless coefficients for stiffness and damping of a journal bearing with $B/D = 0.8$ ($\gamma_{ij} = k_{ij}\,\delta/F_{stat}$, $\beta_{ij} = d_{ij}\,\delta\Omega/F_{stat}$; $i,j = 1,2$)

As mentioned before, these linear expressions are only valid for small motions around an equilibrium position. In practice this means that there has to be a major external load which is constant in order to give a equilibrium condition. Especially for vertical shaft machines this is not always fulfilled, and so in these cases a linear theory is normally not applicable.

Equations of Motion for a Jeffcott-Rotor in Journal Bearings

Figure 5.14 shows the model for a Jeffcott- rotor in journal bearings. For symmetry reasons one can reduce the problem to four degrees of freedom.

$$\begin{aligned} m\ddot{q}_1 + kq_1 - kq_{1L} &= me\,\Omega^2 \cos\Omega t \\ m\ddot{q}_2 + kq_2 - kq_{2L} &= me\,\Omega^2 \sin\Omega t \\ -\frac{k}{2}q_1 + (k_{11} + \frac{k}{2})q_{1L} &+ k_{12}q_{2L} + d_{11}\dot{q}_{1L} + d_{12}\dot{q}_{2L} = 0 \\ -\frac{k}{2}q_2 + k_{21}q_{1L} &+ (k_{22} + \frac{k}{2})q_{2L} + d_{21}\dot{q}_{1L} + d_{22}\dot{q}_{2L} = 0 \end{aligned} \tag{5.34}$$

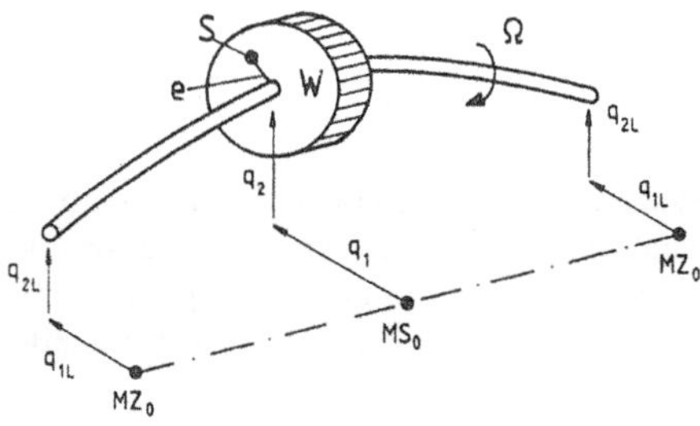

Figure 5.14 Model for a Jeffcott-rotor in journal bearings

These equations can be represented more clearly using dimensionless values and arranging them in matrix form

$$\boldsymbol{M}\boldsymbol{q}'' + \boldsymbol{D}\boldsymbol{q}' + \boldsymbol{K}\boldsymbol{q} = \boldsymbol{f}_c \cos w\tau + \boldsymbol{f}_s \sin w\tau \tag{5.35}$$

with

$$\boldsymbol{M} = \left[\begin{array}{cc|cc} 1 & & & \\ & 1 & & \\ \hline & & & \\ & & & \end{array}\right], \quad \boldsymbol{D} = \frac{a}{w}\left[\begin{array}{cc|cc} & & & \\ \hline & & \beta_{11} & \beta_{12} \\ & & \beta_{21} & \beta_{22} \end{array}\right],$$

$$\boldsymbol{K} = \left[\begin{array}{cccc} 1 & & -1 & \\ & 1 & & -1 \\ -\frac{1}{2} & & \frac{1}{2} + a\gamma_{11} & a\gamma_{12} \\ & -\frac{1}{2} & a\gamma_{21} & \frac{1}{2} + a\gamma_{22} \end{array}\right],$$

$$q = \begin{Bmatrix} q_1 \\ q_2 \\ q_{1L} \\ q_{2L} \end{Bmatrix}, \quad f_c = ew^2 \begin{Bmatrix} 1 \\ 0 \\ 0 \\ 0 \end{Bmatrix}, \quad f_s = ew^2 \begin{Bmatrix} 0 \\ 1 \\ 0 \\ 0 \end{Bmatrix}$$

and

ω^2	=	k/m	: eigenfrequency with rigid supports
τ	=	ωt	: dimensionless time
$(\,)'$	=	$d(\,)/d\tau$	: derivative with respect to τ
w	=	Ω/ω	: related running speed
a	=	$F_{stat}/\delta k$	: related shaft flexibility, δ: bearing clearance

Homogeneous Solution

The free vibrations of the system are calculated via the eigenvalue problem. With the exponential statement $q = \hat{q}\, e^{\mu\tau}$ we obtain

$$(\mu^2 M + \mu D + K)\,\hat{q} = 0\,; \qquad \mu_{i_{1,2}} = \frac{\lambda_{i_{1,2}}}{\omega}\,, \qquad \lambda_{i_{1,2}} = \alpha_i \pm j\omega_i \tag{5.36}$$

with the related eigenvalues μ_i.

The solutions are 6 eigenvalues which in general are conjugate complex. It is not possible to give an analytical expression for the eigenvalues due to the complex way of determining the journal bearing coefficients. Figure 5.15 shows a typical result for eigenvalues versus rotating frequency. The bearing coefficients are speed dependent and so the eigenvalues change as well.

An important point is the speed where the first real part of an eigenvalue changes to a positive sign. This is the stability limit, above which the rotor is unstable. This operation limit is influenced by the bearing properties and so it is interesting to investigate different types of bearings with respect to their stability behaviour. Figure 5.16 shows stability limits versus Sommerfeld number for different bearing designs. Multilobe bearings show better performance compared to cylindrical bearings. Tilting pad bearings are even better, but they are more expensive to produce.

Unbalance Forced Vibrations

At stable running speeds, the free vibrations are damped and only the unbalance forced vibrations remain. The solution of equation (5.35) for the inhomogeneous part leads to a complex set of equations with conjugate complex solutions for the vibrations.

$$(K - w^2 M + jwD)\hat{\hat{q}} = \hat{\hat{F}} \tag{5.37}$$

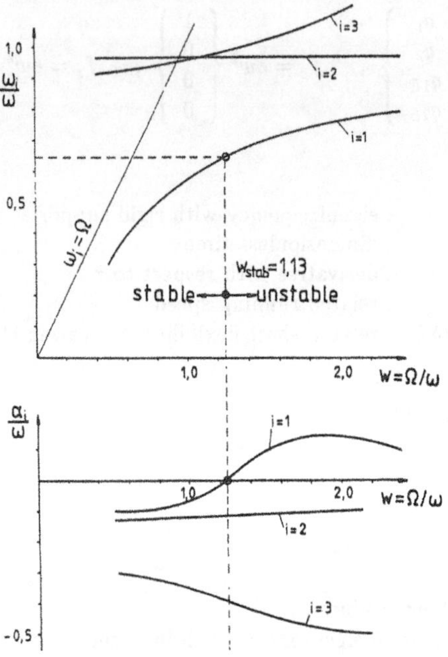

Figure 5.15 Eigenvalues versus rotating frequency for a Jeffcott-rotor in journal bearings

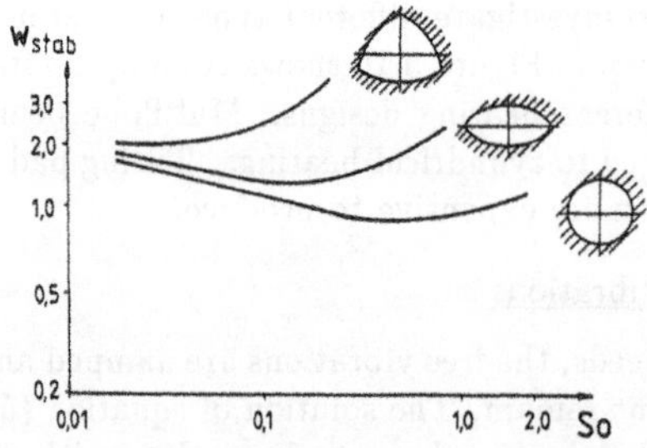

Figure 5.16 Stability limits for different bearing designs

where $\hat{\bar{q}}$ is the complex response amplitude corresponding to the excitation vector

$$\bar{F} = \Re\{(\boldsymbol{f}_c - j\boldsymbol{f}_s)(\cos w\tau + j\sin w\tau)\} = \Re(\hat{\bar{F}}e^{jw\tau})$$

The solution of equation (5.37) leads to elliptical orbits for the rotor showing resonances at the eigenfrequencies as shown in Figure 5.17. The resonance speeds are lower than ω due to the flexibility of the bearings and are split in two because the bearing properties are not symmetric.

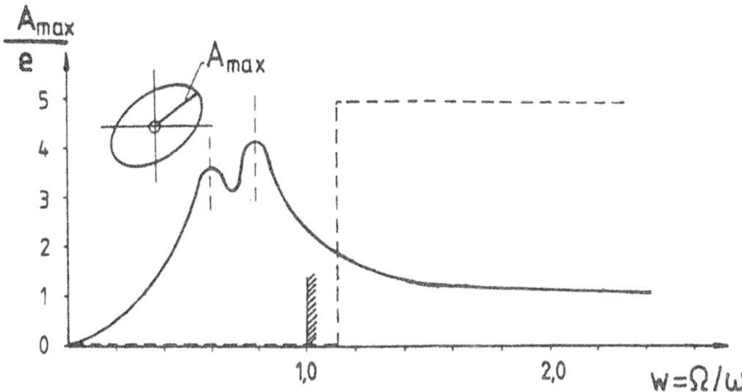

Figure 5.17 Unbalanced response versus running speed

5.3.2 Seals

If contactless seal elements are used in rotating machinery (pumps, turbines etc.) the fluid mechanical interactions have to be considered for the prediction of the vibration behaviour. However, there is often an uncertainty concerning the data for the dynamic coefficients. There is a need for additional research in this area, particularly in the case for grooved seals, which are very common in practice. Different research projects have been started to investigate the dynamics of seals by theoretical models as well as experimental procedures. In the following a possible model is presented, based on a bulk flow theory.

<u>Modelling of Annular Seals with Turbulent Flow Conditions</u>

To explain the seal model, a very simple geometrical form will be considered, consisting of a cylindrical shaft with circular cross section, surrounded by a cylindrical housing as shown in Figure 5.18. This annular seal separates the two chambers with pressure p_1 and pressure p_2, respectively. The pressure difference is $\Delta p = p_1 - p_2$ and it produces a leakage flow in axial direction, which is almost always a turbulent flow with average velocity V. A velocity in circumferential direction is superimposed due to the rotation of the shaft with angular of velocity Ω. In order to obtain the governing equations for the presented seal, translational movements of the shaft in radial direction are assumed.

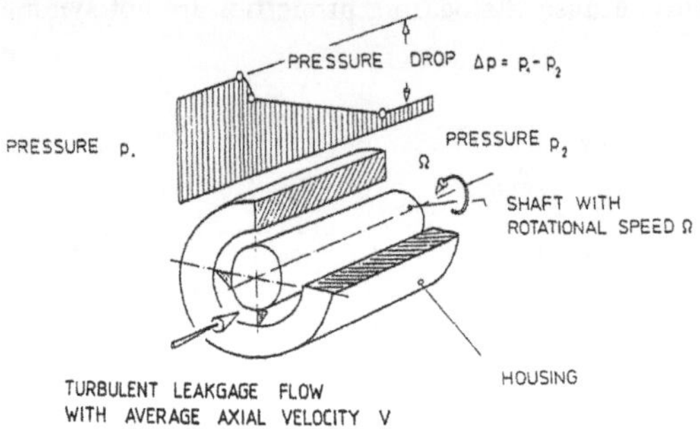

Figure 5.18 Modelling of an annular seal with turbulent flow

A bulk flow model is used to derive the pressure around the shaft and then the force-motion relationships for the vibrating rotor. The first basic idea of this theory is, that the fluid velocity distributions in radial direction are substituted by average velocities. The second basic assumption is the empirical finding, that the relationship between the shear stress at the wall and the mean velocity of the bulk flow – relative to the wall – can be expressed by simple formulas.

One can derive equations for the momentums and the continuity, which lead to the pressure distribution on the rotor in motion. Integrating this pressure yields to the dynamic forces which can be expressed in a linear differential equation, if small motions around the seal centre are assumed.

$$\begin{bmatrix} M_D & 0 \\ 0 & M_D \end{bmatrix} \begin{Bmatrix} \ddot{q}_1 \\ \ddot{q}_2 \end{Bmatrix} + \begin{bmatrix} D_D & d_D \\ -d_D & D_D \end{bmatrix} \begin{Bmatrix} \dot{q}_1 \\ \dot{q}_2 \end{Bmatrix} +$$

$$\begin{bmatrix} K_D & k_D \\ -k_D & K_D \end{bmatrix} \begin{Bmatrix} q_1 \\ q_2 \end{Bmatrix} = \begin{Bmatrix} f_1 \\ f_2 \end{Bmatrix} \tag{5.38}$$

The main diagonal elements in each of the matrices are equal and the cross-coupled terms are opposite in sign. The coefficients are mainly dependent on the pressure drop, the average axial velocity V, the rotational speed Ω of the shaft, the seal geometry (seal length L and Radius R) and on some quantities characterizing the friction in a seal. It is important to note, that the cross-coupled stiffness k_D is strongly influenced by the rotational speed Ω and the fluid entry swirl, which is the circumferential velocity of the fluid at the seal entrance. This effect may cause serious instability problems in high speed rotating machinery, if the cross-coupled stiffness terms become dominant.

Black, 1969 [5.2], has derived stiffness, damping and inertia coefficients for short seals. Figure 5.19 presents this coefficients in dependence of the most important influence parameters Δp, V, Ω, $T = L/V$ and some friction coefficients μ_0, μ_1, μ_2. In Black's derivatives the entry swirl was assumed to be half of the circumferential velocity of the shaft; $R\Omega/2$.

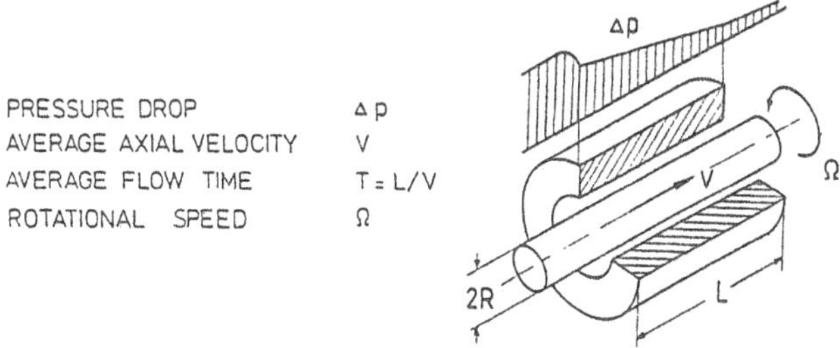

$$M = \frac{\pi R \Delta p}{\lambda} \begin{bmatrix} \mu_2 T^2 & \\ & \mu_2 T^2 \end{bmatrix}$$

$$D = \frac{\pi R \Delta p}{\lambda} \begin{bmatrix} \mu_1 T & \mu_2 \Omega T^2 \\ -\mu_2 \Omega T^2 & \mu_1 T \end{bmatrix}$$

$$K = \frac{\pi R \Delta p}{\lambda} \begin{bmatrix} \mu_0 - \mu_2 \Omega^2 T^2/4 & \mu_1 \Omega T/2 \\ -\mu_1 \Omega T/2 & \mu_0 - \mu_2 \Omega^2 T^2/4 \end{bmatrix}$$

Figure 5.19 Dynamic coefficients of a short seal

Using these expressions it is possible to look at the influence of seals to the dynamics of a simple Jeffcott-rotor. The bearings are assumed to be rigid as shown in Figure 5.20, so that the pure influence of the seal forces can be studied.

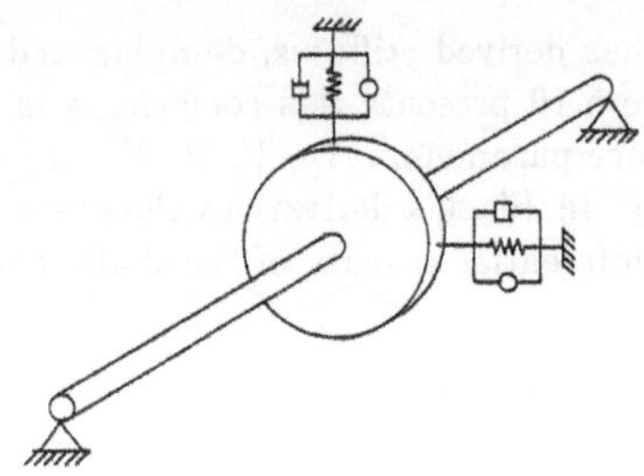

Figure 5.20 Model of a Jeffcott-rotor with seals at the impeller

Equation of Motion and Homogeneous Solution

Due to the rigid supports there are only two degrees of freedom in the model.

$$\begin{bmatrix} M^* & 0 \\ 0 & M^* \end{bmatrix} \begin{Bmatrix} \ddot{q}_1 \\ \ddot{q}_2 \end{Bmatrix} + \begin{bmatrix} D^* & d^* \\ -d^* & D^* \end{bmatrix} \begin{Bmatrix} \dot{q}_1 \\ \dot{q}_2 \end{Bmatrix} +$$

$$\begin{bmatrix} K^* & k^* \\ -k^* & K^* \end{bmatrix} \begin{Bmatrix} q_1 \\ q_2 \end{Bmatrix} = \begin{Bmatrix} f_1 \\ f_2 \end{Bmatrix} \tag{5.39}$$

with

$M^* = m + M_D$: mass of the rotor plus inertia coefficient of the seal

$D^* = D_D$: direct damping of the seal

$K^* = K_D + k_w$: stiffness of the shaft plus direct stiffness of the seal

$k^* = k_D$: cross-coupled stiffness of the seal

The corresponding eigenvalue problem gives four solutions for conjugate complex eigenvalues

$$\begin{aligned} \lambda_{1,2} &= \alpha_1 \pm j\omega_1 \\ \lambda_{3,4} &= \alpha_2 \pm j\omega_2 \end{aligned} \tag{5.40}$$

with

$$\begin{array}{lcll}
\alpha_{1,2} & = & \frac{1}{2}M^*(-D^* \pm \sqrt{a^2+b^2}\cos\frac{\varphi}{2}) & \\
\omega_{1,2} & = & \frac{1}{2}M^*(d^* + \sqrt{a^2+b^2}\sin\frac{\varphi}{2}) & \\
a & = & D^{*2} - d^{*2} - 4M^*K^* & \\
b & = & 4M^*k^* - 2d^*D^* & \\
\varphi & = & \arctan(b/a) & \text{for } a > 0 \\
\varphi & = & \arctan(b/a) + \pi & \text{for } a \leq 0\,,\ b \geq 0 \\
\varphi & = & \arctan(b/a) - \pi & \text{for } a < 0\,,\ b < 0
\end{array}$$

For instability onset speed ($\alpha_i = 0$) two equations can be derived

$$K^*D^{*2} - k^{*2}M^* + k^*d^*D^* = 0\,, \qquad K^* - \omega_i^2 M^* \pm \omega_i d^* = 0 \qquad (5.41)$$

where $\omega_i = \pm k^*/D^*$.

The seal coefficients can be expressed by physical parameters using Black's theory. This gives an impression of the effects these parameters have on the eigenvalues

$$\omega_i = \frac{\Omega_G}{2}\,; \qquad \Omega_G = \sqrt{\frac{8\mu_o \tau R\Delta p}{m\,\lambda^*} + (2\omega)^2} \qquad (5.42)$$

with

Ω_G : onset speed

$\omega = \sqrt{\frac{k}{m}}$: critical speed of the *dry* shaft

λ^* : friction factor

Δp : pressure drop

R : seal radius

μ_o : constant factor

It can be seen that instability occurs at rotational speeds higher than two times the critical speed for the dry shaft. The eigenfrequency is half of the running speed at the onset point of instability.

Influence of Seal Parameters

As shown before the influence of seals to the dynamic behaviour of turbopumps is pretty strong. Therefore it is also important to know more in detail how single physical parameters of the seal influence the characteristics.

In the following the results for eigenvalues and unbalance excitations of a Jeffcott-rotor are shown for several parameter variations.

The tangential velocity of the fluid entering the seal v_0 (normalized with respect to speed of a full developed flow) strongly affects the cross-coupled coefficients and so the stability of the system. The direct stiffness is however not influenced and keeps the eigenfrequencies constant as seen in Figure 5.21.

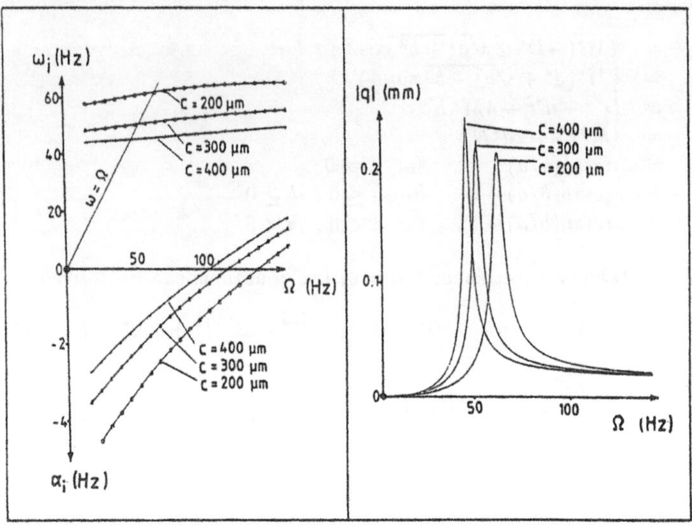

Figure 5.21 Influence of inlet swirl of a seal on a Jeffcott-rotor

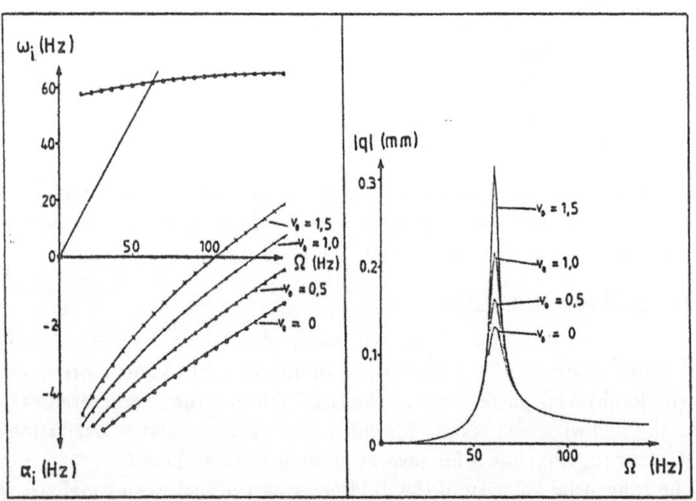

Figure 5.22 Influence of seal clearance on a Jeffcott-rotor

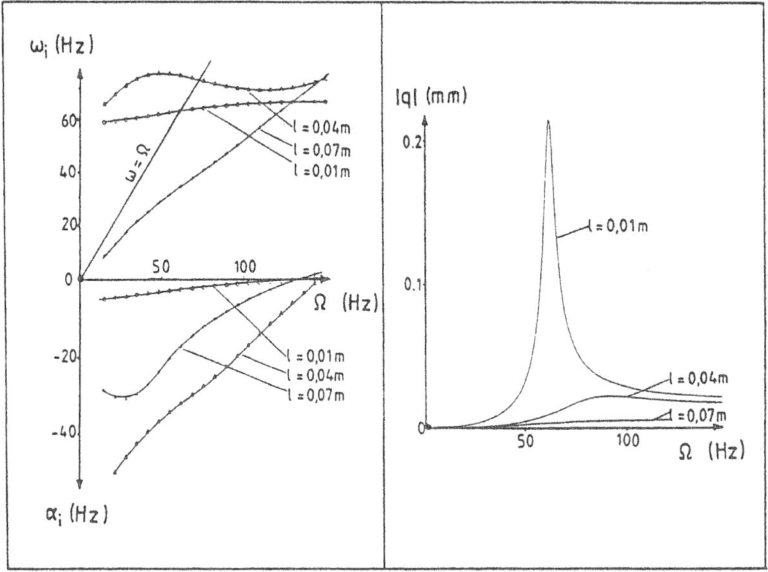

Figure 5.23 Influence of seal length on a Jeffcott-rotor

During the lifetime of a machine the clearance C of the seals can increase due to erosive effects. This lowers the critical speed and the stability limit respectively as shown in Figure 5.22.

Figure 5.23 shows the very interesting influence of the seal length ℓ on the dynamics of a Jeffcott-rotor. As one can expect, a longer seal stiffens the system and introduces more damping to the shaft motions. If the seal length is increased, the critical speed may disappear completely and no resonance for unbalance excitation is present. This is a very positive effect from a practical point of view. It means that unbalance behaviour is no more a critical design problem for machines with long seals or with a large number of seals, for example, as in the case of multistage turbopumps.

5.4 Vibrations of Large Rotating Machinery

In the previous section simple rotor models were treated to describe the basic rotordynamic behaviour. The basic equations could be solved analytically and the vibrations were expressed in simple formulas. But already for the characterization of the fluid effects it was not possible to give analytical formulations for the solutions of the equations of motion. This holds in an even stronger way for real rotating systems in the fields. Here, many degrees-of-freedom have to be considered in the model and it leads to a system of equations of motion where only numerical solutions are possible.

5.4.1 Modelling for Large Rotor Systems

For the modelling of large rotor systems two different approaches are used today. Both start with a discretization of the rotor into single elements of constant properties. The vibrations are not described as a function of the axial coordinate but as the motions of the nodal points where the elements are connected.

Based on these considerations the transfer function method calculates the transfer behaviour from one point to the neighbour point for the displacements and the forces. This method is easy to apply if the structure can be modelled as a chain of elements. It needs only little computer storage availability, because only one element is treated at a time. But the disadvantages are obvious. Separated branches are difficult to handle and rigid supports can cause problems.

A different approach is the finite element procedure which is the most applied method today. The complete dynamic behaviour of a system is described in a matrix equation, where the matrices for inertia, damping and stiffness are usually very large. This requires a huge storage in a computer, which can cause problems even for advanced machines. But in this matrix formulation it is easy to calculate the dynamic behaviour for different excitations.

In the following, only the finite element method is described and used for the calculation of large rotor systems. For each discretized finite element one can derive an equation of motion for the degrees-of-freedom of the element. For the basic beam elements of the shaft there are eight degrees-of-freedom if the lateral vibrations are described, two deflections and two angular displacements in each of the two planes. Figure 5.24 shows a typical model for a double suction centrifugal pump including journal bearing and seals.

Besides the beam elements one can use rigid discs to describe the impellers and dynamic elements which express the fluid forces acting on the moved shaft in seals, journal bearings and so on. All these elements are described with equations of motion which can be superimposed as shown in Figure 5.25 to get the complete equation of motion of a rotor system. The system matrices are band structured and consist of a symmetric and

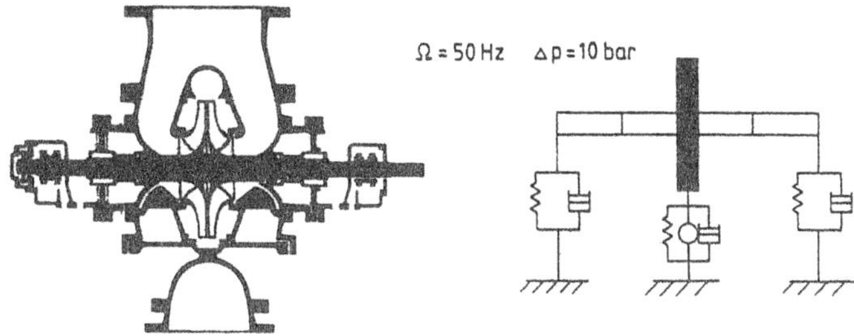

Figure 5.24 Model for a double suction centrifugal pump including journal bearings and seals

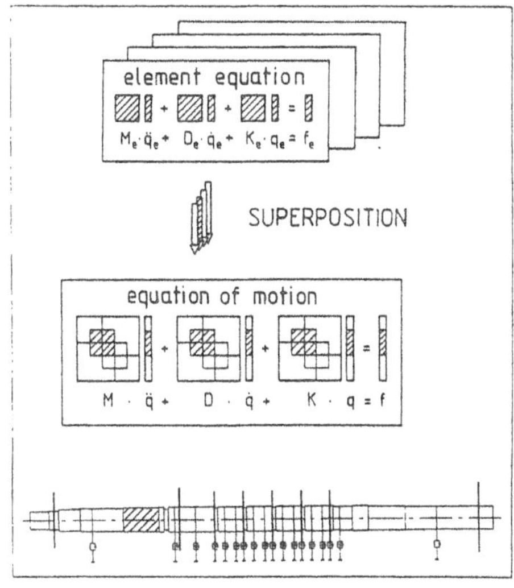

Figure 5.25 Superposition of the element related equations of motion

antimetric part due to the modelled effects

$$\boldsymbol{M}\ddot{\boldsymbol{q}} + \boldsymbol{D}\dot{\boldsymbol{q}} + \boldsymbol{K}\boldsymbol{q} = \boldsymbol{f}(t) \tag{5.43}$$

with

$\boldsymbol{M}$	:	inertia matrix $(n \times n)$
$\boldsymbol{D}$	:	damping matrix $(n \times n)$
$\boldsymbol{K}$	:	stiffness matrix $(n \times n)$
$\boldsymbol{q}, \dot{\boldsymbol{q}}, \ddot{\boldsymbol{q}}$	:	vectors of displacements, velocities and acceleration $(n \times 1)$
$\boldsymbol{f}(t)$	:	vector of external forces $(n \times 1)$
n	:	number of elements

All informations on the dynamic properties of the system are included in this equation. The solutions depend on the external loads, the system matrices and the initial conditions.

5.4.2 Dynamic Calculations

The dynamic calculations can be performed in a similar way as for the basic systems.

Homogeneous Solution

If the external loads in equation (5.43) are zero, the system can vibrate in its natural motions if an impact or non zero initial conditions were present. The exponential statement for the displacements $\boldsymbol{q} = \hat{\boldsymbol{q}}e^{\lambda t}$ leads to the eigenvalue problem.

$$(\lambda^2\boldsymbol{M} + \lambda\boldsymbol{D} + \boldsymbol{K})\,\hat{\boldsymbol{q}} = 0 \tag{5.44}$$

The resulting modal parameters, eigenvalues λ and eigenvectors $\hat{\boldsymbol{q}}$ are in general conjugate complex

$$\lambda_{i_{1,2}} = \alpha_i \pm j\omega_i \tag{5.45}$$

$$\hat{\boldsymbol{q}}_{i_{1,2}} = \boldsymbol{s}_i \pm j\boldsymbol{t}_i \tag{5.46}$$

As one can see from this, again the real part of an eigenvalue α determines whether a system is stable $(\alpha < 0)$ or unstable $(\alpha > 0)$ and the imaginary part gives an eigenfrequency.

The total homogeneous solution is obtained as a sum of the single results.

$$\boldsymbol{q} = \sum_{i=1}^{2n} A_i\,\hat{\boldsymbol{q}}_i\,e^{\lambda_i t} \tag{5.47}$$

The constants A_i have to be calculated from the initial conditions.

Two conjugate complex eigenvectors can be added to give a real representation, where each nodal point describes an elliptical orbit.

Figure 5.26 shows the results of the homogeneous equation for the model of Figure 5.24. The eigenvalues are split in real and imaginary parts and are plotted versus the rotational frequency Ω. Two characteristic running speeds can be observed. If an eigenfrequency is identical to the rotating frequency, crossing with the line $\omega = \Omega$, the eigensolution is excited by rotation and the speed is called critical speed Ω_C. The second point Ω_{stab} is where one of the real parts changes to positive sign. This is the instability onset speed or the stability limit. The corresponding eigenvectors show that the first two eigensolutions are determined by the journal bearings and the third and the fourth can be called bending modes of the shaft. These similar looking vibration modes differ in the direction of circulation. The third vibrates in the opposite direction of the rotation of the shaft and is called backward whirl, whereas the forth mode is a forward whirl.

Unbalance Response

Unbalance forces go into the equation of motion (5.43) as a harmonic right hand side, where the frequency Ω equals running speed. The force vector $f(t)$ is composed as a sum of all single unbalances. With complex notation one gets

$$\boldsymbol{M}\ddot{\bar{q}} + \boldsymbol{D}\dot{\bar{q}} + \boldsymbol{K}\bar{q} = \Re\{\hat{\bar{\boldsymbol{F}}}e^{j\Omega t}\} \tag{5.48}$$

where $\hat{\bar{\boldsymbol{F}}}$ is a function of m_i, e_i, ϵ_i and Ω with

m_i	:	single unbalance mass
e_i	:	single unbalance eccentricity
Ω	:	rotational frequency
ϵ_i	:	orientation of a single unbalance

In general only the inhomogeneous part of the unbalance forced vibrations is calculated by using the same statement as for the external force.

$$\bar{q} = \Re\{\hat{\bar{q}}e^{j\Omega t}\} \tag{5.49}$$

This yields to

$$(\boldsymbol{K} - \Omega^2\boldsymbol{M} + j\Omega\boldsymbol{D})\,\hat{\bar{q}} = \hat{\bar{\boldsymbol{F}}} \tag{5.50}$$

which can be solved with a complex algorithm. The real part of (5.49) gives the unbalance forced vibration.

Figure 5.27 shows for the mentioned test model calculated vibrations for a given speed and the relation between amplitude $|q|$ versus running speed Ω. The orbits look similar to those of the eigenvectors but the amplitudes have physical dimensions. The amplitude response shows clearly the critical

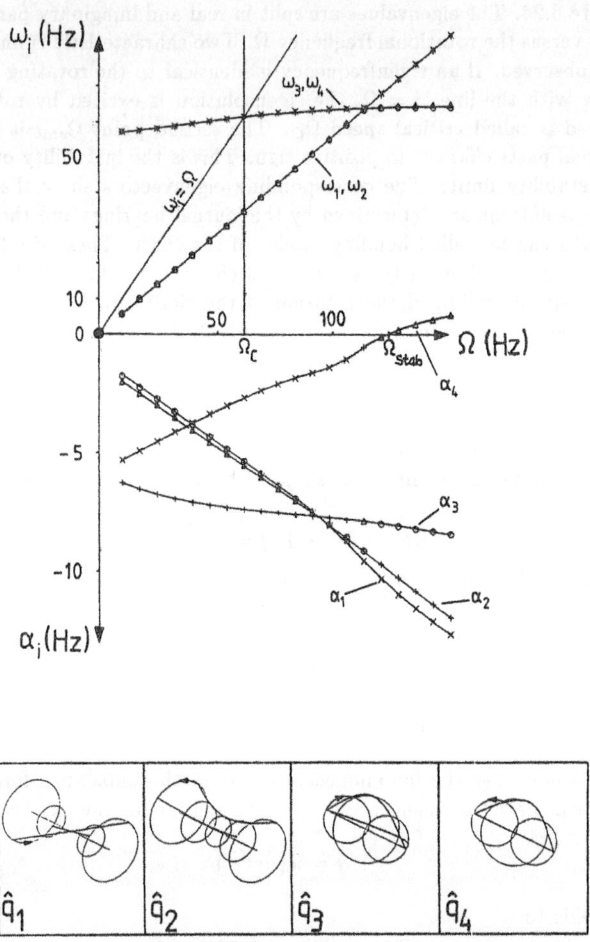

Figure 5.26 Eigenvalues versus running speed and eigenvectors for the rotor model

speed Ω_C where the first bending eigenfrequency is excited by the unbalance forces. The stability limit cannot be seen from the inhomogeneous solution, because it is determined by the homogeneous part.

Transient Vibrations

Transient vibrations can occur in several occasions as start up, shut down, sudden unbalances, earthquake excitations and so on. There are different possibilities to solve the equations of motion for arbitrary external loads.

If the forces can be expressed in an analytically simple way, like sudden unbalances, it is possible to derive the exact solution consisting of a homogeneous and inhomogeneous part. The displacements $\boldsymbol{q}(t)$ can be calculated very efficiently.

If the forces can only be evaluated point by point it is possible to integrate equation 5.43 directly. Here one assumes a constant acceleration for one time interval and calculates the velocities and displacements from this. The computational effort is approximately proportional to the number of time steps, which have to be rather small to achieve a good result.

A third calculation method uses the modal properties to decouple the equations of motion. Therefore equation (5.43) is written in state space formulation.

$$\boldsymbol{A}\eta - \dot{\eta} = \boldsymbol{g}(t) \tag{5.51}$$

with

$$\boldsymbol{A} = \begin{bmatrix} 0 & \boldsymbol{I} \\ -\boldsymbol{M}^{-1}\boldsymbol{K} & -\boldsymbol{M}^{-1}\boldsymbol{D} \end{bmatrix} \quad : \text{system matrix}$$

$$\eta = \begin{Bmatrix} \boldsymbol{q} \\ \dot{\boldsymbol{q}} \end{Bmatrix} \quad : \text{state space vector}$$

$$\boldsymbol{g}(t) = \begin{Bmatrix} 0 \\ -\boldsymbol{M}^{-1}\boldsymbol{f}(t) \end{Bmatrix} \quad : \text{modified external load}$$

$$\boldsymbol{I} \quad : \text{unit matrix}$$

For this equation two eigenvalue problems can be solved which lead to right hand and left hand eigenvectors. For the latter a transposed system matrix is used

$$\begin{aligned} (\boldsymbol{A} - \lambda \boldsymbol{I})\hat{\boldsymbol{r}} &= 0 \\ (\boldsymbol{A}^T - \lambda \boldsymbol{I})\hat{\boldsymbol{l}} &= 0 \end{aligned} \tag{5.52}$$

with

$\hat{\boldsymbol{r}}$: right hand eigenvector
$\hat{\boldsymbol{l}}$: left hand eigenvector

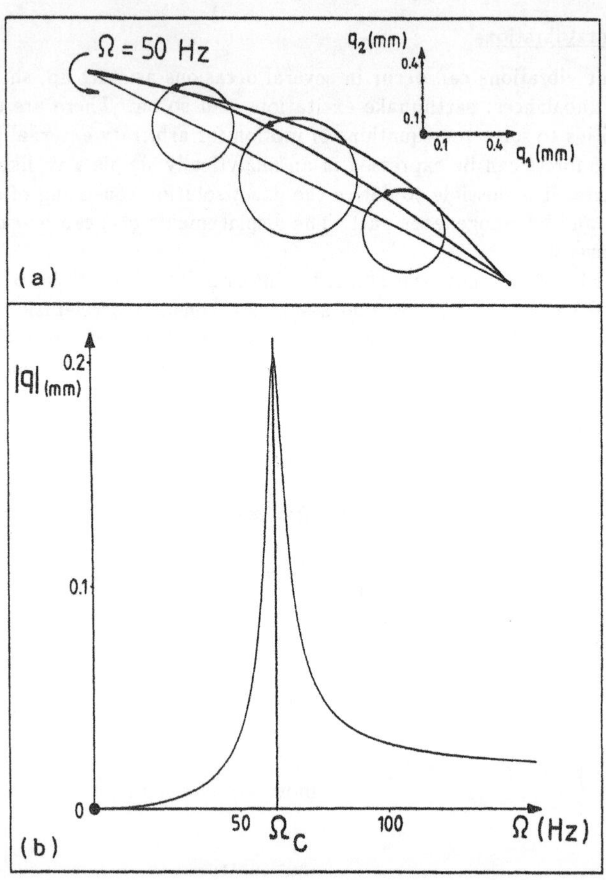

Figure 5.27 Unbalanced forced vibrations of the model:
(a) At operational speed
(b) Amplitude of impeller versus running speed

The eigenvalues λ are in both cases the same and are identical to those of equation (5.44). The eigenvectors are arranged in modal matrices, which are biorthogonal to each other.

$$\boldsymbol{L}^T \boldsymbol{R} = \boldsymbol{I} \tag{5.53}$$

with

$$\boldsymbol{L} = \left[\hat{\boldsymbol{l}}_1 \, \hat{\boldsymbol{l}}_2 \cdots \hat{\boldsymbol{l}}_{2n}\right] \quad \text{and} \quad \boldsymbol{R} = \left[\hat{\boldsymbol{r}}_1 \, \hat{\boldsymbol{r}}_2 \cdots \hat{\boldsymbol{r}}_{2n}\right]$$

Application of these properties to equation (5.51) yields a set of uncoupled differential equations of the first order to describe the dynamic behaviour of a rotor system

$$\boldsymbol{I}\dot{\boldsymbol{\varphi}} - \operatorname{diag}\{\lambda_i\}\,\boldsymbol{\varphi} = \boldsymbol{\gamma}(t) \tag{5.54}$$

with

$$\begin{aligned}
\operatorname{diag}\{\lambda_i\} &= \boldsymbol{L}^T \boldsymbol{A} \boldsymbol{R} &&: \text{diagonal matrix of eigenvalues}\\
\boldsymbol{\varphi} &= \boldsymbol{L}^T \boldsymbol{\eta} &&: \text{vector of modal coordinates}\\
\boldsymbol{\gamma} &= \boldsymbol{L}^T \boldsymbol{g}(t) &&: \text{vector of modal external loads}
\end{aligned}$$

Each modal equation can be solved separately using an iteration scheme or a polygonal approach. The time step size can be adopted efficiently as a function of the eigenfrequencies and it is possible to neglect single equations if the modal load is small. Figure 5.28 shows as an example the transient vibrations of the presented model for a sudden unbalance force.

5.4.3 Realistic Example

Looking to realistic machines one has to take a more complex model to achieve good results in a dynamic calculation. Figure 5.29 shows a multistage centrifugal pump and a section of the mechanical model. The shaft is not uniform and fluid forces act on the rotor at several positions in journal bearings, seals, balance pistons and so on.

To point out the effect of these fluid forces it is interesting to look at the influence of the model to the dynamic results, which is concentrated in Figure 5.30. This comparison can also be seen historically. Up until the sixties it was state of the art to calculate bending vibrations based on the *dry* shaft in rigid supports (Model A). For this example this would mean that the machine operates far beyond the first critical.

If only the stiffening effect of the seals is taken into account in addition (Model B), the critical raises above the operational speed and no resonances would be expected during start up or shut down. This would be similar for Model C where the complete dynamic behaviour of the seals is introduced,

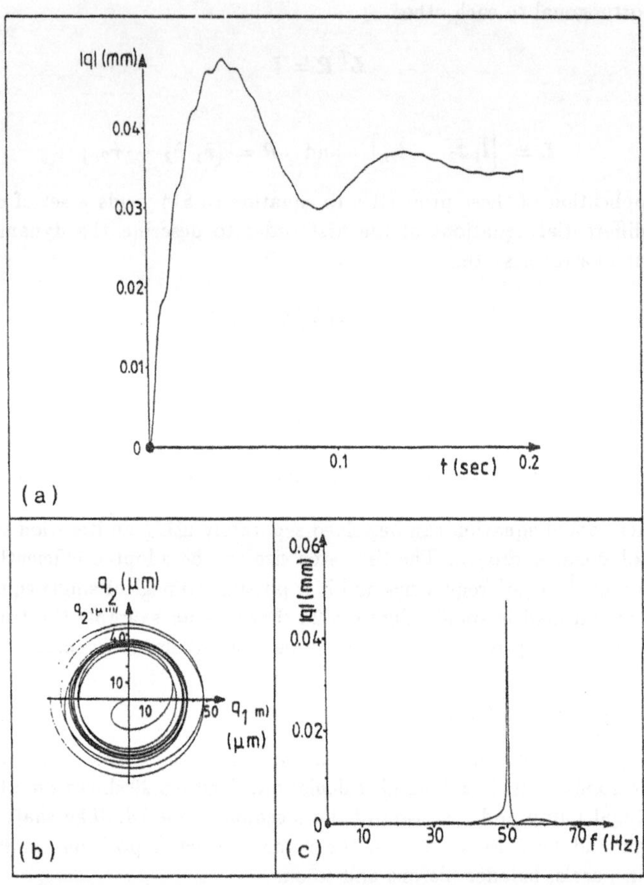

Figure 5.28 Transient response of the model to a sudden unbalance excitation

(a) Amplitude versus time

(b) Orbit of the impeller

(c) Amplitude in the frequency domain

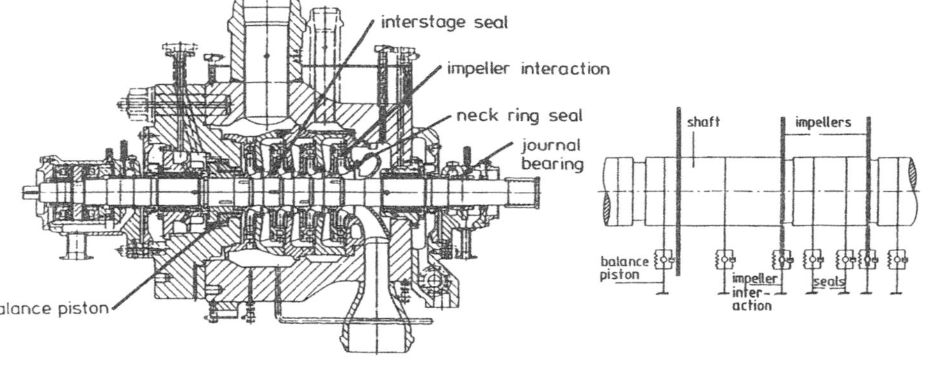

Figure 5.29 Turbopump system with a section of the mechanical model

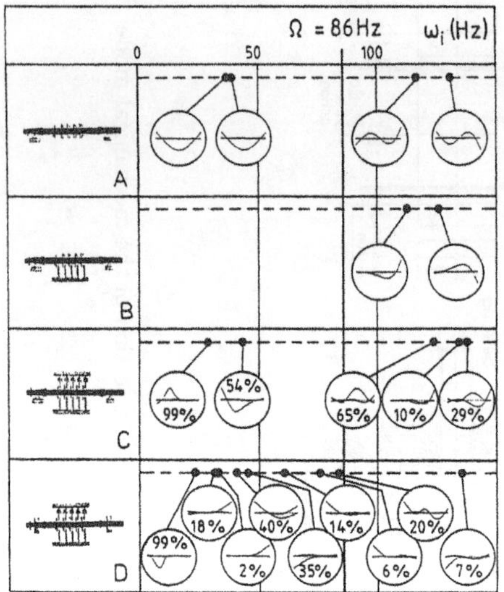

Figure 5.30 Influence of the model to the eigenvalues and mode shapes of a multistage boiler feed pump

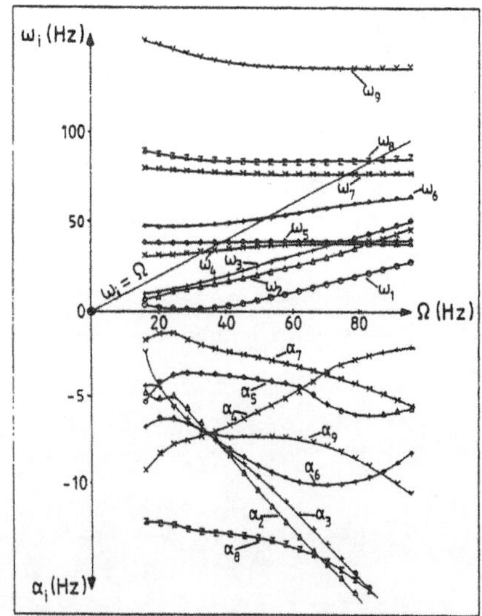

Figure 5.31 Eigenvalues versus running speed for model D

because the two lower eigenfrequencies are very highly damped. The values of modal damping are shown in this figure ($D_i = -\alpha_i/\omega$ in %).

The last case, Model D, is calculated using all available information of the dynamic behaviour of the centrifugal pump. It operates in journal bearings and shows a lot of eigenfrequencies in the operational speed. Some of them are only slightly damped and one is even very close to running speed. This comparison shows that all fluid forces are very important for the dynamics of rotating machinery and the neglect of some does not only slightly change the results but produces a different mechanical model.

The dynamic behaviour of the complete model shall finally be investigated more carefully. Figure 5.31 shows the relation of eigenvalues versus running speed which looks a little confusing at a glance. But one has to concentrate only on some important points. Every cross-over of an eigenfrequency with the diagonal line $\omega = \Omega$ can be a resonance, if the corresponding damping value is small. Here we have five intersections but only ω_5 and ω_7 are dangerous due to low damping.

A stability limit cannot be observed but α_4 approaches the zero line rapidly. In Figure 5.32 the corresponding mode shapes are sampled for different rotational speeds. The modes which can be excited in a resonance show major deflections at the shaft ends, whereas the centre of the machine seems to be well damped and stiffened. This can also be seen at the unbalance responses in Figure 5.33. The shaft ends vibrate in resonances at the mentioned speeds and the vibration levels at the centre of the machine are rather low.

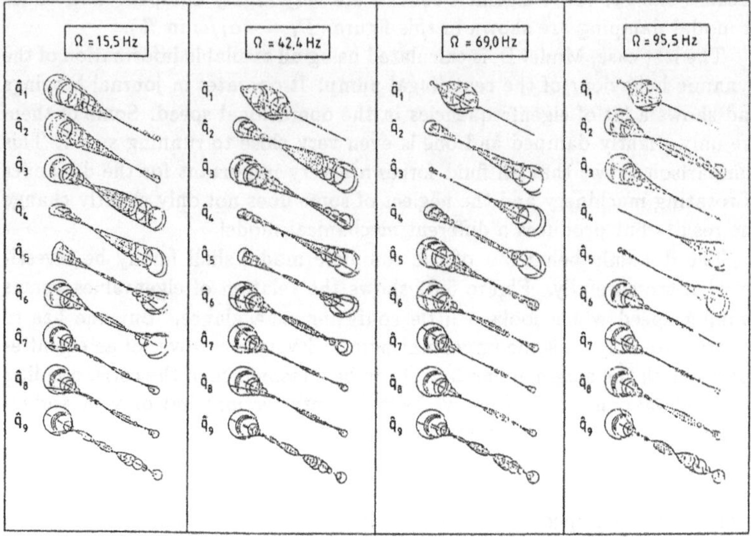

Figure 5.32 Mode shapes of model D for different running speeds

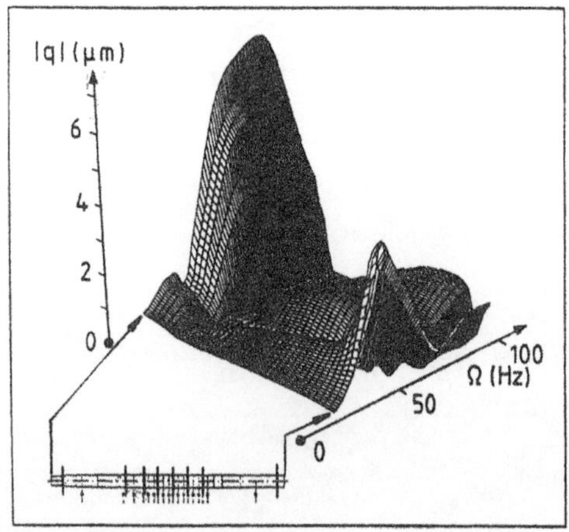

Figure 5.33 Unbalance forced vibrations versus running speed for all nodal points of the model

References

5.1 Bathe, K.J. (1982). *Finite Element Procedures in Engineering Analysis*, Prentice-Hall, New Jersey.

5.2 Black, H.F. (1969). 'Effects of Hydraulic Forces in Annular Pressure Seals on the Vibrations of Centrifugal Pump Rotors', *Journal of Mechanical Engineering Science*, Vol. II, No. 2.

5.3 Diewald, W. and Nordmann, R. (1987). 'Dynamic Analysis of Centrifugal Pump Rotors with Fluid-Mechanical Interactions', 11th Biennial Conference on Mechanical Vibrations and Noise, Boston, September 27-30.

5.4 Gasch, R. and Pfuetzner, H. (1975). *Rotordynamik*, Springer Verlag, Berlin/Heidelberg.

5.5 Nordmann, R. (1976). 'Schwingungsberechnung von nichtkonservativen Rotoren mit Hilfe von Links- und Rechtseigenvektoren', *VDI-Berichte*, Nr. 269.

Chapter 6

System Instability Caused by Hydraulic Machinery

R. Guarga, M.A. Fanelli and M. Nishi

6.1 Instability Due to Performance Characteristics

The problems of instability, whose cause lies in the characteristic curves of hydraulic machines, are dealt with in this section. The characteristic curves are obtained from performance and dynamic tests. The instability originating from the static or dynamic characteristic curves of the machine, differs from the vibration problems due to the internal excitations. The latter type of problems such as those produced by rotating and stationary vane interaction, Karman vortex shedding, etc. are dealt with in Chapter 4.

The type of problems that will be examined in this section is also different from the problems of power swing, or steady oscillations of pressure and discharge that originate from the existence of draft tube swirl flow. These problems are dealt with in sections 6.2 and 6.3. The problems that originate in the existence of a gaseous cavity in the machine due to cavitation, free air, air injection or the pumping of two-phase flow, are also of great interest. These problems are treated in section 6.4.

6.1.1 Concepts of Stability

To deal with the instability problems which stem from the characteristic curves of the machine, two stability concepts must be employed.

One concept is linear stability which is applicable to all cases except for the ones where hysteresis phenomena play a predominant role. Another concept is non-linear stability which is specifically applied to the latter. Both concepts are formulated hereafter in the way which is accepted and widely employed also in other technical areas (see Ogata, 1970 [6.20], chapter 6 and 11).

Linear Stability

A hydraulic system is in an *equilibrium* state if the system remains in the same state in the absence of any perturbation. This state is *unstable* if the variables that define the system are subjected to perturbations of arbitrarily small amplitude and the system does not return to its equilibrium state. The state is *stable* if it tends to return to its equilibrium state and is neutrally stable if the system oscillates in the neighbourhood of the equilibrium point.

In accordance with the given definition, the system will never be able to operate in a state of unstable equilibrium, because even a small perturbation during the operation can bring the system far from the equilibrium. On the other hand, the displacement from the state of equilibrium is materially limited by the dissipation and the resistance of the most vulnerable parts of the system. This indicates that the linear model employed for the study of the stability of a determined state of equilibrium should be substituted by another linear model or a non-linear model, if the behaviour of the system on departing from the said equilibrium is to be described. The study of the stability by means of linear models permits, therefore, the judgement on whether an equilibrium state is unstable, stable or neutrally stable.

Non-Linear Stability

When the behaviour of a system allows for jumps that depend on the previous evolution of the variables of the system (hysteresis phenomena), the linear model is incapable of defining such behaviour.

In a one-dimensional system, the aforementioned behaviour can be represented by Duffing's equation, which formulates the most fundamental case of non-linear forced vibration. At present, a conceptual formulation that is equivalent to the Duffing's equation does not exist for hydraulic systems of industrial interest. Thus, for the presentation of non- linear instability phenomenon in a hydraulic machinery, any conceptual formulation has still not been elaborated due to its complexity.

6.1.2 Formulation of Linear Stability

In order to analyze the behaviour of the hydraulic variables which describe the state of the system, the instantaneous discharge Q and the piezometric head h are referred to:

$$Q = Q_0 + Q^* \quad \text{and} \quad h = h_0 + h^* , \tag{6.1}$$

where Q_0 = mean discharge, h_0 = mean piezometric head, Q^* = instantaneous discharge deviation from the mean, h^* = instantaneous piezometric head deviation from the mean. Q_0 and h_0 are functions of the position in the system and Q^* and h^* are functions of the position and time. For the purpose of the analysis of the stability, Q^* and h^* are expressed by

$$Q^* = \Re(\tilde{Q}e^{\tilde{s}t}) \quad \text{and} \quad h^* = \Re(\tilde{h}e^{\tilde{s}t}) , \tag{6.2}$$

where $\tilde{Q}$ and $\tilde{h}$ are complex variables and function of the position in the system; $\tilde{s} = \sigma + j\omega$ being $j = \sqrt{-1}$, σ damping rate in 1/s, ω frequency in rad/s and $\Re$ the real part of a complex variable.

The so-called transfer matrix method (see Brown, 1969 [6.5] and Chaudry, 1970 [6.7]) has been developed for steady oscillatory phenomena ($\tilde{s} = j\omega$). Based on the linearization that is assumed in the dynamic and mass conservation equations of each hydraulic device with concentrated or distributed parameters, the associated transfer matrix can be obtained by substituting $\tilde{s} = \sigma + j\omega$ in place of $j\omega$ in frequency response functions.

The aforementioned transfer matrices allow the formulation of the relationship between the values of $\tilde{Q}$ and $\tilde{h}$ at the inlet and the outlet of any hydraulic element including hydraulic machines. The relationship is expressed by means of a matrix $\boldsymbol{T}$ with 2 × 2 elements, constituted by four complex functions t_{ij} of the variable $\tilde{s}$. The matrix $\boldsymbol{T}$ describes the variation around an equilibrium state by:

$$\left\{ \begin{array}{c} \tilde{Q} \\ \tilde{h} \end{array} \right\}_O = \overbrace{\left[\begin{array}{cc} t_{11} & t_{12} \\ t_{21} & t_{22} \end{array} \right]}^{\boldsymbol{T}} \left\{ \begin{array}{c} \tilde{Q} \\ \tilde{h} \end{array} \right\}_I , \tag{6.3}$$

where the subscripts I and O mean at inlet and outlet. It must be pointed out that for a turbomachine, the matrix $\boldsymbol{T}$ implies knowing the function between T_e (external torque) and n (rotational speed) of the machine. When the inertia of the rotating masses is significant, it is normal to assume that n is constant.

If the hydraulic system of the concern is a network as shown in Figure 6.1 with b branches, e end points and n nodes, the unknown functions will be $2b$ values of $\tilde{Q}$ at the both ends of the branches and $e + n$ values of $\tilde{h}$ at the end points and the nodes. When a set of hydraulic devices connected in a series is considered, the transfer matrix of the series is obviously obtained as the product of the transfer matrix of each device. The number of available equations is as follows: $2b$ equations for the branches, e equations for the end points and n equations for the continuity of flow quantity at the nodes. These equations have the forms:

$$\left\{ \begin{array}{c} \tilde{Q} \\ \tilde{h} \end{array} \right\}_O = \boldsymbol{T}_i \left\{ \begin{array}{c} \tilde{Q} \\ \tilde{h} \end{array} \right\}_I \quad \cdots \text{ equation for the branch } i, \tag{6.4}$$

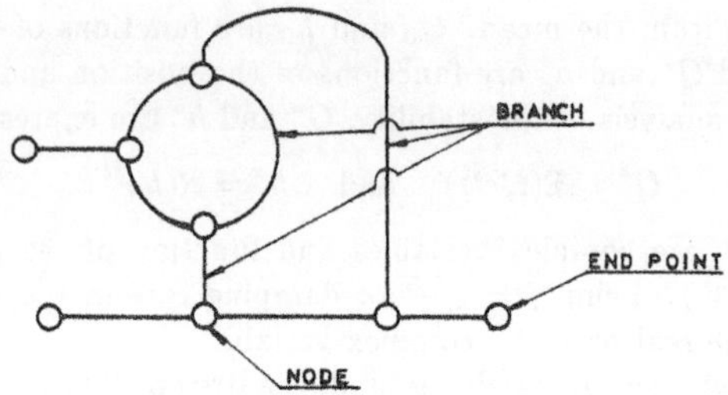

Figure 6.1 Hydraulic system network

$$\tilde{h}_i = Z_i \tilde{Q}_i \qquad \cdots \text{ equation for the end point } i \text{, where } Z_i \text{ is the known impedance at the point,} \tag{6.5}$$

$$\sum \tilde{Q} = 0 \qquad \cdots \text{ equation of continuity at a node.} \tag{6.6}$$

Therefore, the number of equations equals the number of unknown quantities and due to the linearization of the equations, the problem could be formulated as:

$$U \left\{ \begin{array}{c} \tilde{Q}_1 \\ \vdots \\ \tilde{Q}_{2b} \\ \tilde{h}_1 \\ \vdots \\ \tilde{h}_{e+n} \end{array} \right\} = \mathbf{o} \,. \tag{6.7}$$

As $\boldsymbol{U}$ is a square matrix with $(2b + e + n)^2$ elements, the problem is therefore to obtain the eigenvalues and eigenvectors. As the coefficients of $\boldsymbol{U}$ are functions of $\tilde{s}$ and of the physical parameters of the system associated with the analyzed state of equilibrium, the eigenvalues will be obtained by solving

$$\det(\boldsymbol{U}) = 0 \,, \tag{6.8}$$

where the above equation represents two equations each for real and imaginary number. For each eigenvalue $\tilde{s}_i$ obtained as a solution, an eigenvector $(\tilde{Q}_1, \ldots \tilde{Q}_{2b}, \tilde{h}_1, \ldots \tilde{h}_{e+n})$ can be determined by solving equation (6.7). If any of the eigenvalues $\tilde{s}_i$ has a positive real part ($\sigma > 0$), the equilibrium state under study will then be unstable because the $(2b+e+n)$ variables of equation (6.2) tend to depart from the equilibrium state.

The displacement from the equilibrium state in the real installations is limited, as has already been explained. In consequence, the aforementioned linear model will permit the detection of the instability but not the description of the behaviour of the system once this has moved away from the unstable equilibrium state. In real systems the instability is usually accompanied by a phenomenon of self-oscillation that should be described by means of either a non-linear model or by a new linear model whose response in the state of self-oscillation is neutrally stable ($\sigma = 0$). The calculation of the imaginary part of $\tilde{s}_i$ in the said state of auto-oscillation will provide the frequencies ω_i at which the system oscillates.

6.1.3 Stability Analysis of Industrial Installations

First, the role of the transfer matrix of the hydraulic machine in the stability analysis of industrial installations is examined. Figure 6.2 shows a diagram that symbolizes a very common installation of a hydraulic machinery in industrial application. The installation consists of one branch ($b = 1$), two end points ($e = 2$) and no nodes ($n = 0$). In this diagram, the electrical two port representation is used. $\boldsymbol{M}$ indicates the transfer matrix of the hydraulic machine, $\boldsymbol{A}$ the matrix of the part of the installation that is located upstream and $\boldsymbol{B}$ the one for downstream. Z_I and Z_O are the impedances of the inlet and outlet of the system. If the methodology introduced in 6.1.2 is applied, the following relation holds:

$$U \begin{Bmatrix} \tilde{Q}_I \\ \tilde{Q}_O \\ \tilde{h}_I \\ \tilde{h}_O \end{Bmatrix} = \mathbf{o} , \tag{6.9}$$

where $\boldsymbol{U}$ is

$$U = \begin{bmatrix} c_{11} & -1 & c_{12} & 0 \\ c_{21} & 0 & c_{22} & -1 \\ Z_I & 0 & -1 & 0 \\ 0 & Z_O & 0 & -1 \end{bmatrix} , \tag{6.10}$$

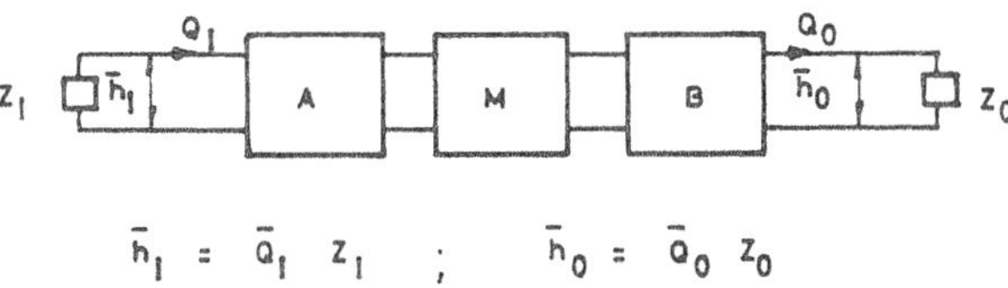

Figure 6.2 Two port representation of an industrial hydraulic installation

being c_{ij} the elements of the matrix $\boldsymbol{C} = \boldsymbol{BMA}$. To calculate the eigenvalues of the matrix equation (6.9), the following equation should be solved in the complex field.

$$\det(\boldsymbol{U}) = 0 \tag{6.11}$$

Calculation of the above determinant leads to

$$c_{11}Z_O + c_{12}Z_I Z_O - c_{21} - c_{22}Z_I = 0 , \tag{6.12}$$

where

$$\begin{aligned} c_{11} &= (m_{11}a_{11} + m_{12}a_{21})\, b_{11} + (m_{21}a_{11} + m_{22}a_{21})\, b_{12} , \\ c_{12} &= (m_{11}a_{12} + m_{12}a_{22})\, b_{11} + (m_{21}a_{12} + m_{22}a_{22})\, b_{12} , \\ c_{21} &= (m_{11}a_{11} + m_{12}a_{21})\, b_{21} + (m_{21}a_{11} + m_{22}a_{21})\, b_{22} , \\ c_{22} &= (m_{11}a_{12} + m_{12}a_{22})\, b_{21} + (m_{21}a_{12} + m_{22}a_{22})\, b_{22} . \end{aligned} \tag{6.13}$$

As m_{ij} are the elements of the matrix $\boldsymbol{M}$, a_{ij} of matrix $\boldsymbol{A}$ and b_{ij} of matrix $\boldsymbol{B}$, they are all complex and functions of $\tilde{s}$. Expressions in (6.12) and (6.13) are of great interest since all the elements m_{ij} of matrix $\boldsymbol{M}$ of the hydraulic machine intervene in equation (6.12) whose unknown quantity is $\tilde{s}$.

If any of the roots $\tilde{s}_i$ of the above equation satisfies the condition $\Re(\tilde{s}_i) > 0$, the system is unstable, according to what has been established in section 6.1.2.

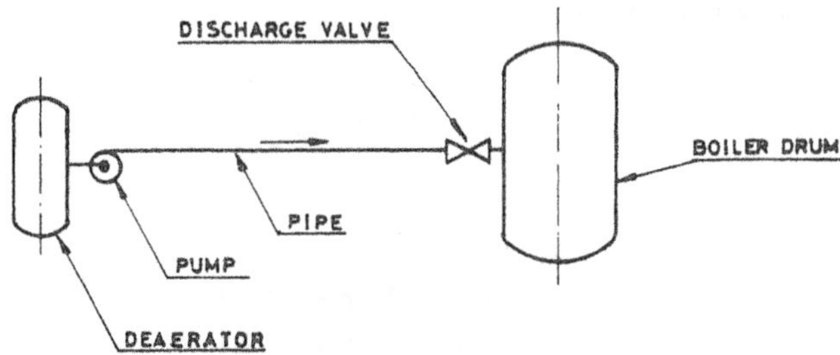

Figure 6.3 Case study scheme

The previous development is applied to a particular case illustrated in Figure 6.3. The system in consideration represents in a simplified way the feed pump system of a power station. The inlet to the system is located in the deaerator, and the outlet in the boiler drum. A discharge valve is installed at the boiler drum inlet and the discharge pipe can be considered as a potential resonator. To simplify the analytical treatment, the friction of the system is assumed to be concentrated at the valve.

Since the pressures at the deaerator and the drum are constant, the values of inlet and outlet impedances are:

$$Z_I = 0 \quad \text{and} \quad Z_O = 0\,. \tag{6.14}$$

By calculating equation (6.11), it can be concluded that:

$$\det(U) = -c_{21} = 0\,. \tag{6.15}$$

The transfer matrices, which intervene in the calculation of c_{21}, are $\boldsymbol{A}$, $\boldsymbol{M}$ and $\boldsymbol{B}$ and their expression is:

$$\boldsymbol{A} = \begin{bmatrix} 1 & 0 \\ 0 & 1 \end{bmatrix}, \quad \boldsymbol{M} = \begin{bmatrix} m_{11} & m_{12} \\ m_{21} & m_{22} \end{bmatrix}, \quad \boldsymbol{B} = \boldsymbol{VP}\,, \tag{6.16}$$

where $\boldsymbol{V}$ and $\boldsymbol{P}$ are the transfer matrices of the discharge valve and the pipe respectively, and are given by:

$$\boldsymbol{V} = \begin{bmatrix} 1 & 0 \\ Z & 1 \end{bmatrix}, \quad \boldsymbol{P} = \begin{bmatrix} \cosh(\tilde{s}\frac{\ell}{c}) & -\frac{1}{Z_c}\sinh(\tilde{s}\frac{\ell}{c}) \\ -Z_c\sinh(\tilde{s}\frac{\ell}{c}) & \cosh(\tilde{s}\frac{\ell}{c}) \end{bmatrix}, \tag{6.17}$$

being ℓ = the length of the pipe, c = the wave velocity, $Z_c = c/gA$ = characteristic impedance, g = gravity acceleration, A = pipe cross-sectional area, $Z = -2Q_0K$ = valve impedance, where K is the valve head loss constant. Applying equation (6.13) yields

$$c_{21} = (m_{11}Z + m_{21})\cosh(\tilde{s}\frac{\ell}{c}) - (m_{11}Z_c + m_{21}\frac{Z}{Z_c})\sinh(\tilde{s}\frac{\ell}{c})\,. \tag{6.18}$$

In order to simplify the analysis of the example, it is assumed that m_{11} and m_{21} are constant, that is to say, they are not functions of $\tilde{s}$. By solving equation (6.15) analytically, the following complex roots are obtained:

$$\tilde{s}_n = \frac{c}{2\ell}\left\{\log\left|\frac{(\frac{m_{21}}{m_{11}Z_c}+1)(Z+Z_c)}{(\frac{m_{21}}{m_{11}Z_c}-1)(Z-Z_c)}\right| + j\arg\left[\frac{(\frac{m_{21}}{m_{11}Z_c}+1)(Z+Z_c)}{(\frac{m_{21}}{m_{11}Z_c}-1)(Z-Z_c)}\right]\right\}, \tag{6.19}$$

being n a natural number. As explained in section 6.1.2 the system will be unstable if $\Re(\tilde{s}_n) > 0$. From equation (6.19) instability occurs if:

$$\left|\frac{\frac{m_{21}}{m_{11}Z_c}+1}{\frac{m_{21}}{m_{11}Z_c}-1}\right| > \left|\frac{Z-Z_c}{Z+Z_c}\right|\,. \tag{6.20}$$

On the other hand, the frequencies ω_n that correspond to the unstable solutions are:

$$\omega_n = \frac{c}{2\ell}\arg\left[\frac{(\frac{m_{21}}{m_{11}Z_c}+1)(Z+Z_c)}{(\frac{m_{21}}{m_{11}Z_c}-1)(Z-Z_c)}\right]\,. \tag{6.21}$$

From the ineqality of (6.20), it can easily be concluded that a necessary condition (even though it is not sufficient) for the instability of the particular case under study is $\Re(m_{21}/m_{11}Z_c) > 0$. Likewise, the frequencies of the oscillatory phenomenon associated with $\sigma > 0$ will correspond to the fundamental frequency ($n = 0$) and harmonics given by equation (6.21). It is to be noted that the presence of non-rational functions in equation (6.18) gives way to the existence of infinite solutions for ω_n with only one damping rate σ. In equation (6.21) it can be similarly observed that not only the machine parameters m_{11} and m_{21} but also the hydraulic system parameters ℓ, c, Z and Z_c intervene in ω_n.

6.1.4 Transfer Matrix Based on Static Characteristics

In the hydraulic discipline, it is common to analyze the dynamic phenomena as a succession of stationary states. As an example of dynamic behaviour of pumps and turbines, Knapp's diagram (Knapp, 1937 [6.17]) can be mentioned, which was obtained from the static experiments of the machine in order to calculate the transitory phenomena; or the use of static characteristics for the study of control problems (Borel, 1960 [6.4]).

It seems, therefore, natural to calculate the coefficients m_{ij} from the characteristic curves obtained from the static experiments of the machine. In actual fact, this is the line that has traditionally been followed and that is still frequently employed without the necessary critical analysis.

If the calculation of the coefficients m_{ij} is performed under the assumption that a behaviour of the machine in dynamic conditions is similar to the static ones, and that the compressibility of the fluid, the deformation of machine casing and the cavitation are negligible, then $\tilde{Q}_I = \tilde{Q}_O$. Since

$$m_{21} = Z_m = \frac{\tilde{h}_O - \tilde{h}_I}{\tilde{Q}_I} , \tag{6.22}$$

the transfer matrix is written as:

$$M = \begin{bmatrix} 1 & 0 \\ Z_m & 1 \end{bmatrix} , \tag{6.23}$$

where Z_m is the denominated *impedance of the machine*, which can be obtained from the static characteristic curve of the machine, $H_0(Q_0)$, H_0 being the mean total head (increase of the mean total head through the machine). The definition of Z_m is then:

$$Z_m = \frac{dH_0}{dQ_0} - \frac{Q_0}{g}\left(\frac{1}{A_I^2} - \frac{1}{A_O^2}\right) , \tag{6.24}$$

being A_I and A_O the inlet and outlet cross-sectional areas, respectively. The evaluation of dH_0/dQ_0 shall be carried out at the point of operation to be studied. In such conditions, Z_m is a real number that only depends on the machine operation point.

In the references there are numerous texts and papers which employ static characteristic curves of the machines for examining the stability of the hydraulic installations. In the most widely used classical texts, this is a commonly adopted point of view. Stepanoff, 1957 [6.25], chapter 14, deals with the stability of pumping by means of static characteristics and gives recommendations for obtaining stable static characteristics. Pfleiderer, 1955 [6.22], section 91, and Sedille, 1967 [6.23], chapter 4, give similar recommendations.

Other articles of interest which successfully employ the static characteristics for the analysis of the stability of diverse hydraulic or ventilation systems, are the works of Katto, 1960 [6.14]; Simpson et al. 1974 [6.24]; Black et al., 1975 [6.3]; Arshenevskii et al., 1981 [6.2], and Greitzer, 1981 [6.12], chapter 1.

6.1.5 Transfer Matrix Based on Dynamic Behaviour

In section 6.1.4 the transfer matrix M was obtained from the static characteristic curve $H_0(Q_0)$ of the turbomachine. This procedure is only valid, as it will be seen, when the changes are *slow* with respect to a time scale which is set by the machine itself. If, on the other hand, the changes are no more *slow*, the matrix M can modify itself considerably. The purpose of this section is to study the matrix M when the dynamic phenomena in the machine are also taken into consideration.

Experimental Determination of Transfer Matrix

The experimental determination of M implies calculating four complex values m_{ij}. In order to do so, the machine is installed in a hydraulic test rig which is capable of generating perturbations to the stationary flow. Let us suppose that the perturbations are sinusoidal with frequency f and that it is possible to measure $\tilde{Q}_I, \tilde{h}_I, \tilde{Q}_O, \tilde{h}_O$ in the circuit as indicated in Figure 6.4. By modifying the circuit without changing the conditions of steady operation, the test for the same frequency f is repeated and another set of measured values, $\tilde{Q}'_I, \tilde{h}'_I, \tilde{Q}'_O, \tilde{h}'_O$, is obtained. It can then be defined that:

$$\begin{bmatrix} \tilde{Q}_O & \tilde{Q}'_O \\ \tilde{h}_O & \tilde{h}'_O \end{bmatrix} = M \begin{bmatrix} \tilde{Q}_I & \tilde{Q}'_I \\ \tilde{h}_I & \tilde{h}'_I \end{bmatrix} . \qquad (6.25)$$

If the sets $\tilde{Q}_I, \tilde{h}_I, \tilde{Q}_O, \tilde{h}_O$ and $\tilde{Q}'_I, \tilde{h}'_I, \tilde{Q}'_O, \tilde{h}'_O$ are linearly independent, which can be achieved by modifying the hydraulic circuit properly, equations (6.25) gives four complex equations by which the required four complex values m_{ij} can be determined.

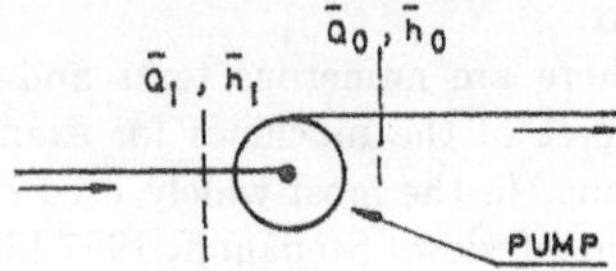

Figure 6.4 Measurement of flow and head fluctuations

In the bibliography it is possible to find various works which describe the hydraulic circuits employed, the methodology of measuring $\tilde{Q}$ and $\tilde{h}$, the way to produce perturbations in the steady flow and the type of signal processing. The following papers deal with this point: Ohashi, 1968 [6.21]; Anderson et al., 1971 [6.1]; Fanelli, 1972, '80, '82 [6.8-11]; Ng, 1976 [6.18]; Ng et al. 1978 [6.19]; Stirnemann et al., 1985 [6.26]; Carolus, 1985 [6.6]; Stirnemann, 1987 [6.27]; Kawata et al., 1985, '87 [6.15,16]; and Jacob et al., 1988 [6.13].

The complementary data regarding the type of machine used for the experiments in the respective papers are presented in Table 6.1. Figure 6.5 illustrates IMHEF PF-4 test rig built in 1986 in the Institut de Machines Hydrauliques et de Mecanique des Fluides de l'Ecole Polytechnique Federale de Lausanne (see Jacob et al., 1988 [6.13] for detail). This facility allows special investigation to be made on the dynamic behaviour of hydraulic machines and other system components. Instrumented piping on either side of the studied model provide a determination of the wave propagation speed and of the oscillatory flow-rate.

Forced hydraulic oscillations are generated by a rotating valve. The idea in this system (sketched in Figure 6.6) is to use the variable speed drive only for the modulation of a steady disturbance. Five measuring sections are equipped on the low pressure side and three on the high pressure side. Two sensors are mounted across a pipe diameter in sections 8 and 9 of Figure 6.5, so that the axial symmetry of the pressure waves can be controlled. Torque is measured using a strain gauges device. The rotational speed is measured with an instrumented DC generator.

Experimental Results

The previously mentioned authors (Ohashi, Anderson, Fanelli, Ng, Stirnemann, Carolus, Kawata and Jacob) present measurements of complex values

Table 6.1 Complementary data on dynamic pump tests

Author (year)	Pump type	Impeller type	Diffuser type	Specific speed (rpm, m^3/s, m)	Rotat. speed (Hz.)
Ohashi (1968)	centrifugal 1 stage	radial, backward vanes	vaneless	17.35	60, 30, 15, 0.
Anderson (1971)	centrifugal 1 stage	radial bladed	vaneless volute, conical diffuser	—	—
Fanelli (1972-82)	centrifugal 1 stage	—	without distributor	—	25
Ng (1976-78)	helical and axial	helical and axial	—	—	150
Kawata (1985)	centrifugal 2 stage	—	—	—	41.66
Kawata (1987)	centrifugal 3 stage	—	—	19.33	33.3
Stirnemann (1985)	centrifugal	—	single volute	50	—

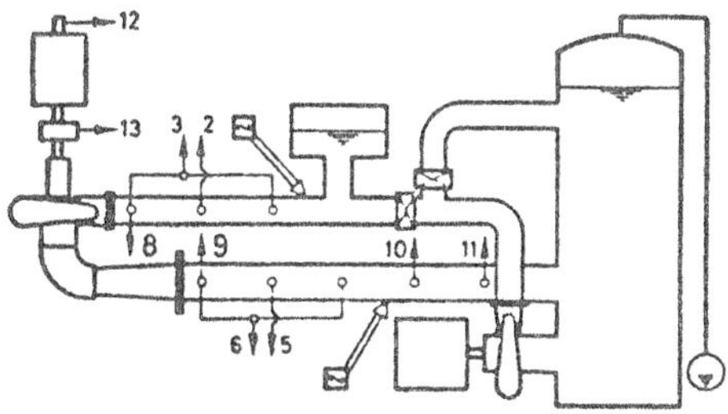

Figure 6.5 IMHEF PF-4 test rig with location of sensors and excitation device

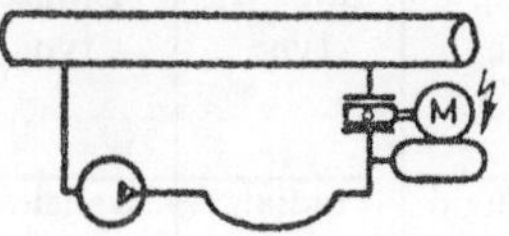

Figure 6.6 Sketch of excitation device

m_{ij}. The diversity in the way in which the results are presented, the variety of machines used for the experiments, and in many cases the absence of a precise description of the machine, makes it difficult to arrive at a synthesis of the results. In general, the following can be summarized from the above mentioned results which were all obtained during cavitation-free operation of the machine.

(a) The elements m_{11}, m_{12} and m_{22} take the following values if the flow is perfectly incompressible and cavitation-free:

$$m_{11} = 1, \quad m_{12} = 0, \quad m_{22} = 1.$$

These results have been verified or assumed as true by the above mentioned authors. However, it is pointed out that the element m_{12} is very sensitive to the presence of very little bubbles of gas or vapour in the machine.

(b) The element m_{21}, which is undoubtedly of the greatest interest when there is no gaseous phase, has been measured by all the above authors. Figure 6.7 shows a summary of the measured results at the design operating point. Due to the diversity in the ways in which the value m_{21} appears in the references mentioned above, the results are presented in Figure 6.7 by using following parameters:

$$X(\delta) = \Re\left(\frac{m_{21}(\delta)}{m_{21}(0)}\right), \quad Y(\delta) = \Im\left(\frac{m_{21}(\delta)}{m_{21}(0)}\right), \tag{6.26}$$

where $\delta = f/n$, n is the rotational speed of the machine and $\Im$ the imaginary part of a complex variable. By employing this expression, it is possible to compare the results of different authors on the same basis. The element m_{21} is also dependent on the rotational speed of the machine (see Ohashi, 1968 [6.21]).

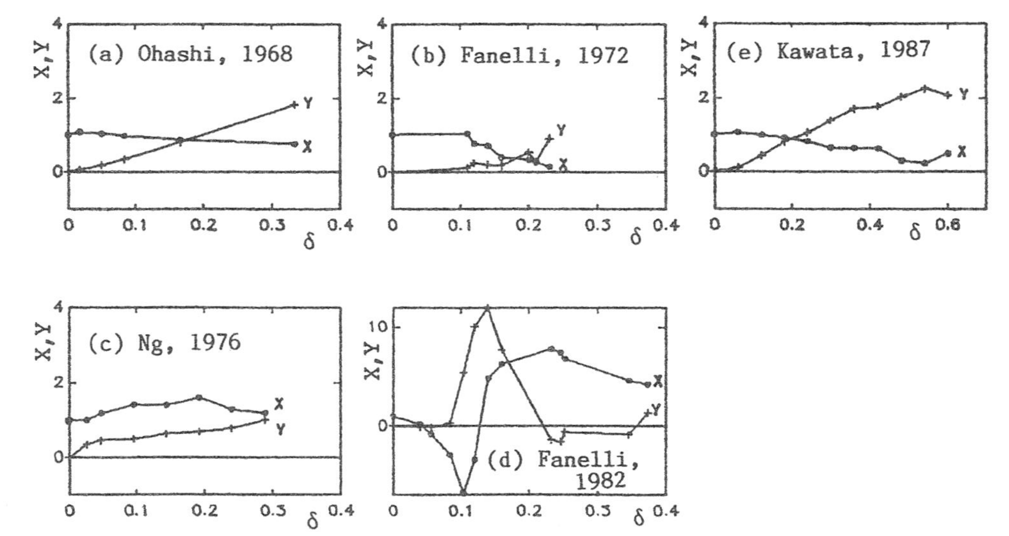

Figure 6.7 X, Y data for different machines at cavitation-free design operating condition

Parametric Synthesis of Experimental Data

An attempt to express the experimental data on the dynamic behaviour of the machine ($f > 0$), has been made by using a reduced number of parameters. For this purpose the scheme shown in Figure 6.8 was proposed by Stirnemann, 1985 [6.26]. Using transfer matrices, this scheme is expressed by:

$$\begin{Bmatrix} \tilde{Q} \\ \tilde{h} \end{Bmatrix}_O = \begin{bmatrix} 1 & -Y_3 \\ 0 & 1 \end{bmatrix} \begin{bmatrix} 1 & 0 \\ -Z_2 & 1 \end{bmatrix} \begin{bmatrix} 1 & -Y_1 \\ 0 & 1 \end{bmatrix} \begin{Bmatrix} \tilde{Q} \\ \tilde{h} \end{Bmatrix}_I . \tag{6.27}$$

After multiplication of the matrices:

$$\begin{Bmatrix} \tilde{Q} \\ \tilde{h} \end{Bmatrix}_O = \begin{bmatrix} 1 + Z_2 Y_3 & -(Y_1 + Y_3 + Y_1 Y_3 Z_2) \\ -Z_2 & 1 + Y_1 Z_2 \end{bmatrix} \begin{Bmatrix} \tilde{Q} \\ \tilde{h} \end{Bmatrix}_I , \tag{6.28}$$

where Y_1 and Y_3 are admittances and Z_2 is an impedance.

With the notation used for the transfer matrix M the proposed parametric synthesis implies:

$$\begin{aligned} m_{11} &= 1 + Z_2 Y_3 , & m_{12} &= -(Y_1 + Y_3 + Y_1 Y_3 Z_2) , \\ m_{21} &= -Z_2 , & m_{22} &= 1 + Y_1 Z_2 . \end{aligned} \tag{6.29}$$

The scheme above implies a parametric concentrated model which allows for compressibility in the inlet (Y_1) and the outlet of the machine (Y_3) and postulated incompressibility and structural rigidity between the inlet and outlet (Z_2). As an example, Figure 6.9 shows the general scheme applied to a machine with cavitation or free gases only in the inlet ($Y_1 = C_1 j\omega$) and with Z_2 consisting of a resistance and a inductance ($Z_2 = R_2 + L_2 j\omega$). The corresponding matrix is then:

$$M = \begin{bmatrix} 1 & -C_1 j\omega \\ -R_2 - L_2 j\omega & 1 + (C_1 j\omega)(R_2 + L_2 j\omega) \end{bmatrix} . \tag{6.30}$$

If the machine does not contain free gases or cavitation, which is the case studied in this section, then the transfer matrix M is written in accordance with the proposed scheme as:

$$M = \begin{bmatrix} 1 & 0 \\ -R_2 - L_2 j\omega & 1 \end{bmatrix} . \tag{6.31}$$

The system represented by this matrix is illustrated in Figure 6.10.

In order to express the experimental data set out in Figure 6.7 with the proposed scheme, X and Y must be calculated in the following manner:

$$X + jY = \frac{m_{21}(\delta > 0)}{m_{21}(\delta = 0)} = X + \frac{L_2}{R_2} \cdot 2\pi\delta n j , \tag{6.32}$$

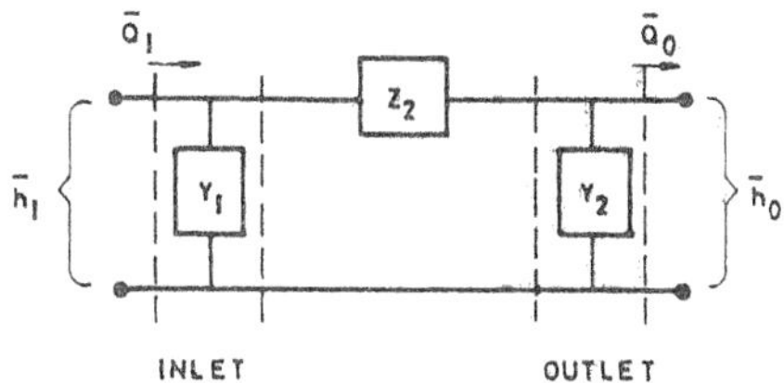

Figure 6.8 Parametric synthesis of experimental data (Stirnemann, 1985 [6.26])

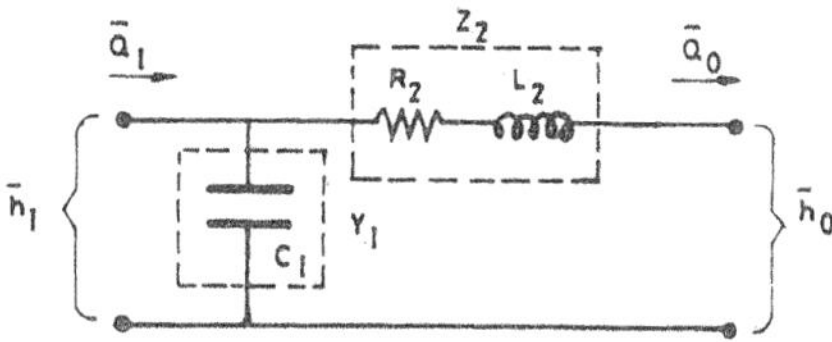

Figure 6.9 Parametric synthesis of a pump with cavitation or free gases at the inlet

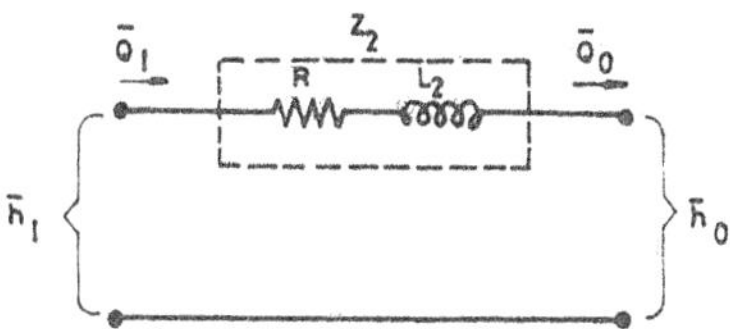

Figure 6.10 Parametric synthesis of a pump without cavitation or free gases

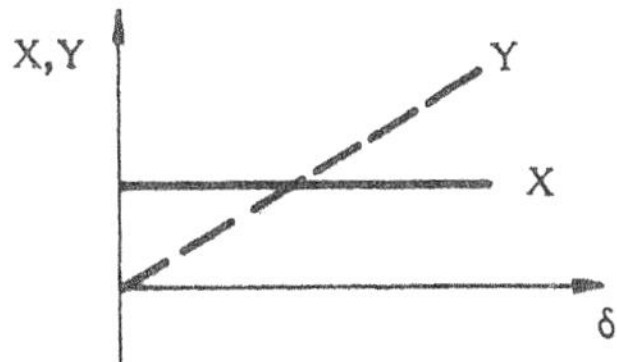

Figure 6.11 Real (X) and imaginary (Y) parts of Z_2

where X is a constant given by $\Re(m_{21}(\delta > 0)/m_{21}(\delta = 0))$. Equation 6.32 is graphically and generically shown in Figure 6.11. When comparing the generic Figure 6.11 with the experimental data of Figure 6.7 presented by different authors, it can be observed that the scheme proposed by Stirnemann represents, in an acceptable way, the results of Ohashi where X is approximately constant and Y is proportional to δ. On the other hand, the proposed scheme seems to be too simple to simulate the experimental data found by Ng, Fanelli and Kawata. In particular, the proposed scheme cannot express a change of sign in $X(\delta)$ as measured by Fanelli, 1982 [6.11] and Kawata, 1985 [6.15].

As a more general alternative to Stirnemann's proposal, it is interesting to examine the following proposal suggested by Ng et al. 1978 [6.19]. The idea is to approximate the elements m_{ij} of the matrix M by means of the developments in power series of $j\omega$. Then the following expression is obtained:

$$m_{ij} = \sum_{n=0}^{n=N_{ij}} A_{ijn} \, (j\omega)^n \, , \tag{6.33}$$

where the coefficients A_{ijn} are real numbers and are obtained by best fitting to the experimental results. This type of parametric synthesis could express experimental data of a much more complex nature than those that can be studied by the scheme of equation (6.30). In particular, equation (6.33) can represent a change of sign in $X(\delta)$.

Applications of Parametric Synthesis

The parametric synthesis proposed in (6.29) and (6.33) are directly applied to the calculation of the stability of the equilibrium state of the hydraulic system. In effect, in order to introduce the complex variable $\tilde{s}$, $j\omega$ must be substituted by $\tilde{s}$ in equation (6.29) or in equation (6.33) in accordance with the parametric synthesis scheme to be used. As an example of this, the most general scheme described by equation (6.33) will be applied to the stability analysis of the case examined in section 6.1.3. It is supposed that the dynamic behaviour of the pump coincides with that found experimentally by Kawata, 1985 [6.15]. As the result, the functions $X(\delta)$ and $Y(\delta)$ are best fitted by the least square method to the following polynomials:

$$X = a_0 - a_2\delta^2 + a_4\delta^4 \, , \qquad Y = a_1\delta \, , \tag{6.34}$$

with the value of the coefficients $a_0 = 0.639$, $a_1 = 20.195$, $a_2 = 0.187 \times 10^3$, $a_4 = 6.974 \times 10^3$. The experimental points and the fitted curves are shown in Figure 6.12.

If $m_{21}(j\omega)$ is calculated, the following result is obtained:

$$m_{21} = -473.59 - 57.15(j\omega) - 2.023(j\omega)^2 - 1.102\times10^{-3}(j\omega)^4 \quad (\mathrm{s/m^2}) \, . \tag{6.35}$$

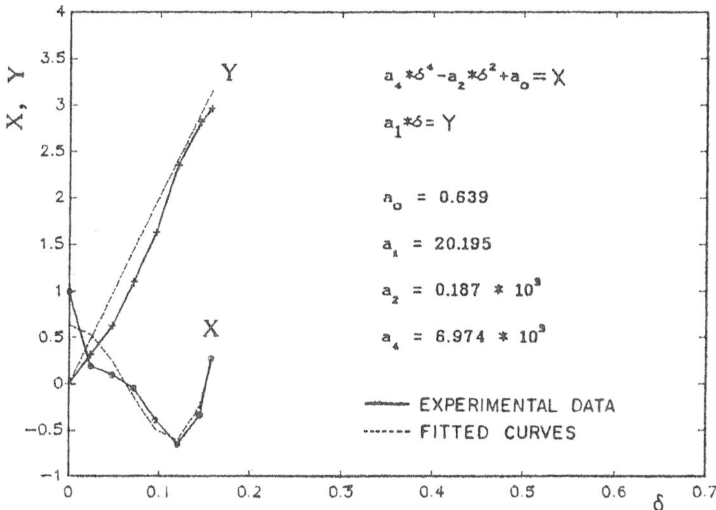

Figure 6.12 Experimental data and fitted curves (Kawata, 1985 [6.15])

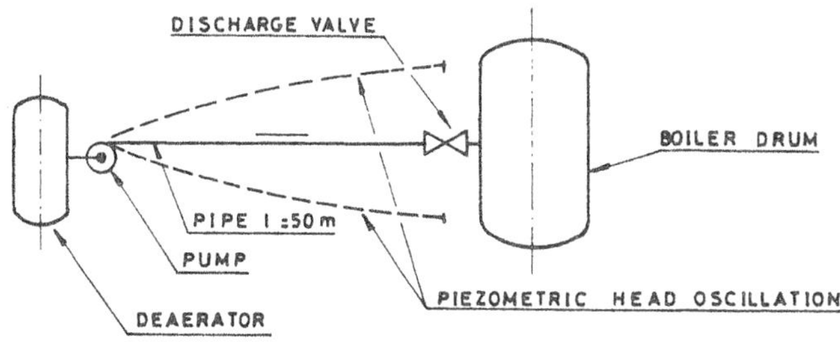

Figure 6.13 Piezometric head oscillation scheme

In the pump which was studied experimentally by Kawata, $m_{11} = 1$. In order to numerically solve the specific system dealt with in section 6.1.3, the following numerical values are assumed $c = 1000$ m/s, $\ell = 50$ m, $Z_c = 1.29 \times 10^4$ s/m^2, $Z = -5.25 \times 10^5$ s/m^2, $n = 41.67$ Hz.

By substituting these values into equation (6.18) and equalizing c_{21} to zero and rearranging the equation, the following result for $\tilde{s}$ is obtained:

$$\tilde{s} = 5 \log_e \left(\frac{\frac{m_{21}}{1.29} \times 10^{-4} + 1}{\frac{m_{21}}{1.29} \times 10^{-4} - 1} \times 0.952 \right) , \qquad (6.36)$$

where m_{21} is the polynomial of $\tilde{s}$ obtained by substituting $\tilde{s}$ for $j\omega$ in equation (6.35).

In order to solve equation (6.36), a scientific subroutine package was used. The result for $\sigma > 0$ is:

$$\sigma = 0.1750 \quad \text{rad/s} , \qquad \omega = 28.8663 \quad \text{rad/s} .$$

This means that the hydraulic system is unstable and departs from the equilibrium state oscillating with a frequency $f = 4.59$ Hz. This frequency almost corresponds to the stationary wave whose wave length is four times the length of the pipe ($\ell = 50$ m). This wave has a frequency of 5 Hz. A diagram of the oscillatory component of the piezometric head for $f = 4.59$ Hz, is shown in Figure 6.13. In the real system, when the departure from the equilibrium state is produced, the amplitudes in the piezometric head and discharge oscillations increase according to $e^{\sigma t}$.

This phenomenon will result in either the breakdown of the hydraulic system or the increase of the dissipation due to the growth of the amplitudes in the oscillations of the discharge. In the latter case, it results in a phenomenon of self-oscillation with a frequency that is close to the one found in the calculation of the stability. Kawata showed the presence of self-oscillation phenomenon by means of an installation similar to the one illustrated in Figure 6.13. The δ corresponding to the frequency $f = 4.59$ Hz is found to be $\delta = 0.1102$. In Figure 6.12 it is observed that for this value of δ the corresponding value of X is negative. As $m_{21}(0)$ is negative, it can be concluded that the corresponding value of $\Re(m_{21})$ is positive.

Theoretical Model of Dynamic Behaviour

The diversity of the dynamic behaviours that are shown in the measurements of m_{21}, illustrated in Figure 6.7, proves the necessity for a theoretical model that allows the interpretation of experimental data, the organization of measurements and the prediction of the dynamic behaviour of the turbomachines with respect to its fundamental characteristics. Ohashi, 1968 [6.21], and Fanelli, 1972 [6.8], have tried to formulate a theoretical model taking into account the fundamental hydrodynamic phenomena which are produced in a vane cascade under oscillatory flow.

Ohashi developed an elaborate fundamental work directed at the calculation of the dynamic behaviour of a staggered flat plate cascade subjected to a fluid whose far upstream velocity either contains an oscillatory sinusoidal component in the direction of the cascade or an oscillatory sinusoidal and progressive component in the space. Ohashi applies the outlined theory to the calculation of the dynamic behaviour of a single-stage axial pump and a centrifugal pump. For the latter case he carries out measurements. It must be pointed out that these are the first measurements of the dynamic behaviour to be performed in a hydraulic machine.

The results obtained are of great interest as they show a very good agreement between the theoretical prediction and the performed measurements. In order to carry out the theoretical analysis in a particular case, a simplified model is applied to the dynamic behaviour of the pump, without including the conduit effect of the pump. Ohashi remarks that this effect is indifferent to the pumping action itself and merely resulted from the conduit effect of the pump casing. In consequence, the author analyzes the behaviour of the following term:

$$x = \frac{m_{21}(\omega_R) - m_{21}(\omega_R = \infty)}{m_{21}(\omega_R = 0)}, \tag{6.37}$$

where $\omega_R = \cos\lambda_R \cdot f/N_R\phi n$ = the reduced frequency and λ_R = stagger angle, ϕ = flow coefficient (discharge divided by the product of the outer diameter, exit width and the peripheral velocity of pump impeller), N_R = rotor blade number. $\omega_R = \infty$ means $n = 0$ (no pumping action). Figure 6.14 shows the amplitude and phase of x that depends on ω_R and the result of the theoretical model appears superposed. The amplitude and phase of x was measured for different flow coefficients ϕ and rotational speeds n. $\phi = \phi_D$ means the ϕ value for the pump design point and $\phi_{D/2}$ means $\phi = \phi_D/2$. The amplitude of x is given in decibels and the phase in degrees.

An interesting result which arises from the experimental data illustrated in Figure 6.14, is that for $\omega_R > 0.1$, the approximation of $m_{21}(\omega_R)$ by $m_{21}(\omega_R = 0)$ (impedance resulting from the static characteristic of the machine) is not possible.

Without any knowledge of the work undertaken by Ohashi, Fanelli developed a theoretical model applying unsteady aerodynamics for an isolated airfoil introduced by Von Karman and Sears. These results are applied to a two-dimensional thin airfoil in non-uniform motion. In order to apply these results to airfoil cascades, Fanelli uses the correction coefficient proposed by Weinig.

In this way the incidence angle variation can be linked with the lift variation on the blades. A runner with radial flow is studied in the paper

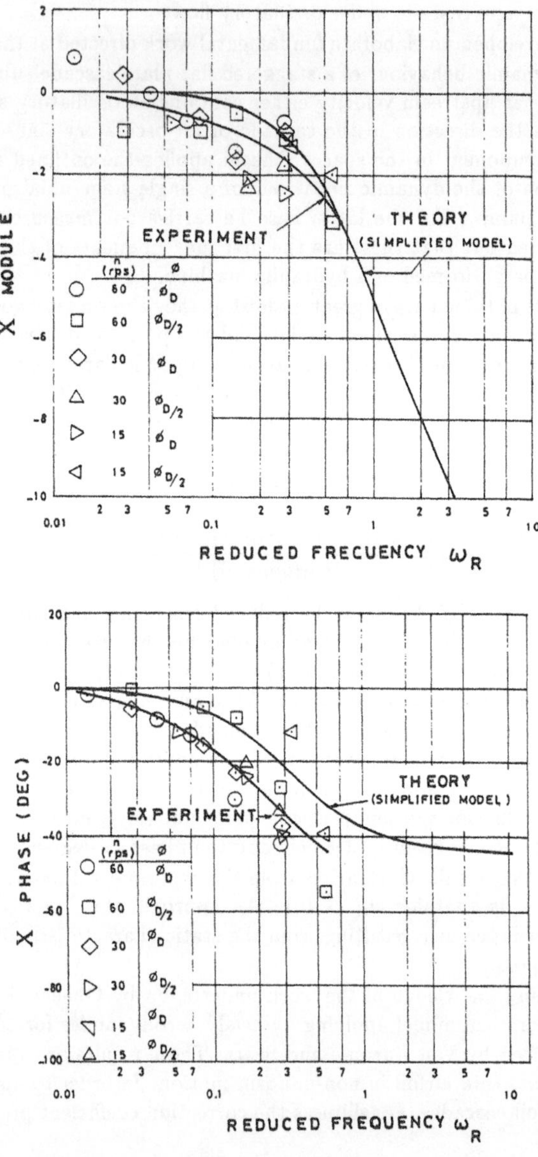

Figure 6.14 Theoretical and experimental results of Ohashi (1968 [6.21])

but the theoretical analysis could also be applied to axial flow machines. In his original paper Fanelli does not give any experimental data that allow the comparison between the theoretical model and the dynamic behaviour of a real machine. This comparison, shown in Figure 6.15, is presented in a later paper (Fanelli, 1972 [6.9]). It is seen that the results herein are also promising.

It is probable that the complexity of both models has been the determining factor in refraining other authors from employing them in the cases they have studied.

6.1.6 Hysteresis Phenomena

The phenomena, previously dealt with, allow an analysis of a linear type to be performed. This analysis permits also a linear formulation of the problem of the stability of the equilibrium states of the hydraulic installations where turbomachines operate. However, there are phenomena that escape from the outlined linear treatment. These are phenomena that occur in the pump-turbines when they operate as turbines. These phenomena have been described by Yamabe, 1971 [6.28]. In Figure 6.16, $Q_1 = Q/\sqrt{H}$ versus $n_1 = n/\sqrt{H}$ relationship for different Guide Vane Opening (GVO) and Thoma cavitation factor $\sigma = (H_a - H_v - H_s)/H = (\text{NPSH})/H$ is shown, where Q = measured discharge (m^3/s), n = measured speed (rpm), H = measured net head (m), H_a = barometric pressure head (m), H_v = vapour pressure head at the temperature of water (m) and H_s = static suction head (m) for a particular machine operating as a turbine. The hysteresis phenomenon during turbine operation is described as follows.

Let the operation point be the point **a** where $GVO = 110$ %, $\sigma = 2.39$, $n_1 = 320$ (rpm/$\sqrt{\text{m}}$). When n_1 is increased gradually, keeping $\sigma = 2.39$, the operation point suddenly shifts to the point **c** at the point **b**. Then when n_1 is increased or decreased, it moves along the curve **d**∼**e** and operation changes abruptly to the point **f** on the curve **a**∼**b** at the point **e**. Here Q_1 and output (characteristic curve was omitted) always show similar change. The operation points just before the sudden change of operational condition have little relationship with experimental head, while it is regulated by the cavitation factor σ.

In his paper the flow conditions inside the runner are described when the operation point shifts between the different operation curves. The work however does not include an analysis of the causes of the hysteresis phenomenon nor does it formulate ways of correcting it.

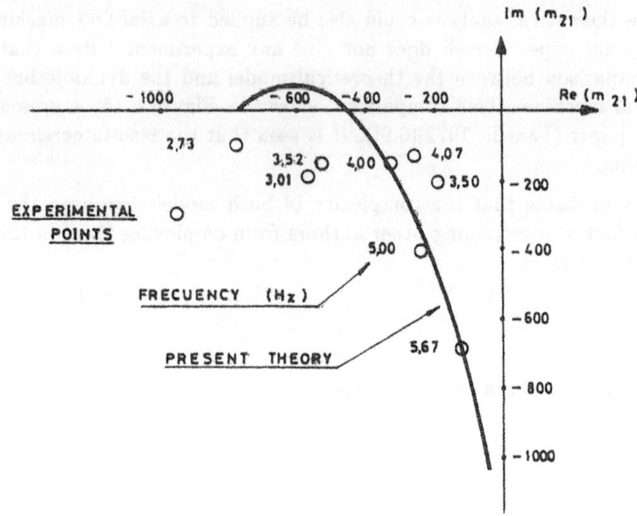

Figure 6.15 Theoretical and experimental results of Fanelli (1972 [6.9])

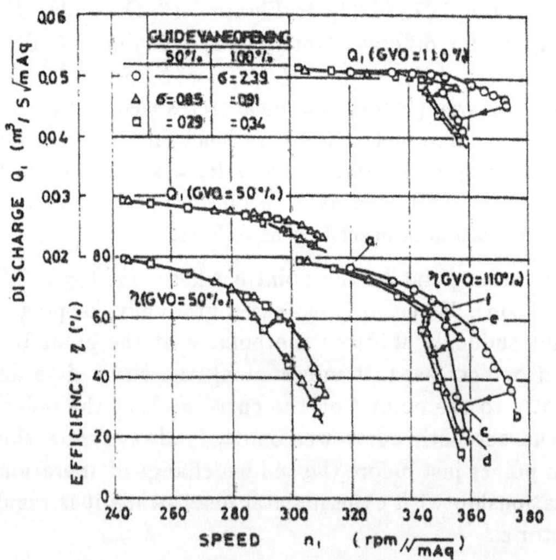

Figure 6.16 Hysteresis characteristics of a model pump-turbine (Yamabe, 1971 [6.28])

6.2 Power Swing

6.2.1 Qualitative Description of the Phenomena

Hydroelectric power units equipped with Francis turbines occasionally experience severe oscillations of hydraulic, mechanical and electrical quantities when operating at off-design load. These phenomena can manifest themselves both at partial load and at overload, albeit with different characteristics. At partial load the draft tube is the seat of a complex flow endowed with a more or less pronounced rotational component.

One, or sometimes two, localized vortex filaments, roughly helicoidal in shape as shown in Figure 6.17 (a), can often be evidenced on model test rigs, especially when the pressure is lowered until the vortex core is cavitating as the photograph of Figure 6.17 (b) shows. The velocity field at the runner outlet is far from being mainly axial and uniform as it is the case at design load for well-conceived machines, and it possesses strong circumferential components. The axial velocity distribution in this section is very much skewed with zones of reverse flow. The main effective flow is rejected in an annular section outlying the helicoidal vortex. Radial and circumferential components can also be detected in the straight part of the draft tube, which is the seat of a purely axial flow at design load.

The pressure field has likewise a complex structure. Pressure sensors positioned under the runner in the straight part of the draft tube as illustrated in Figure 6.18 (a) will record a rotating pressure oscillation, which is precessing in the direction of runner rotation. The angular velocity ω_T of this precession is variable with the degree of partial load but is roughly of the order of magnitude of $0.25 \sim 0.3\ \omega_0$, where $\omega_0 = 2\pi n$ is angular velocity of runner rotation (n is rotational frequency). The sensors arranged in different positions around the circumference of the section will display the same response with phase shifts corresponding to their angular position.

Under particular conditions with cavitated vortex core and resonance of draft tube mass oscillations for example, the mean pressure p_s of a cross-section, i.e. the mean value of wall pressures, $p_1, ..p_4$, experiences also an oscillation with frequency $\omega_T/2\pi$. This synchronous pressure oscillation can be observed globally in the penstock, the machine itself and the draft tube, whilst the rotating pressure field is restricted locally within a cross section.

Figure 6.18 (b) is used to explain two different pressure oscillations, that is, rotating and synchronous ones. The vector $\overline{OO'}$ is rotating around O with angular velocity ω_T and the real part (projection to x-axis) of this vector represents the change of mean pressure p_s, i.e. synchronous oscillation. The rotating fluctuation is represented by the vector $\overline{O'\ p_1}$, for example, which rotates around O′ also with angular velocity ω_T. The real part of vector $\overline{Op_1}$ gives rise to the combined rotational and synchronous pressure change p_1 at

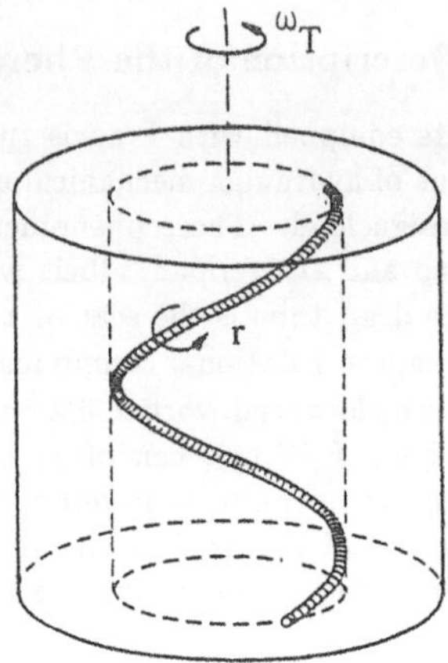

(a) Schematic reference sketch

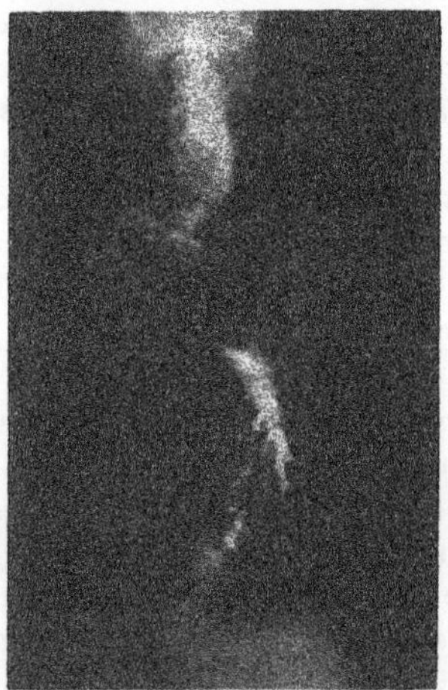

(b) Photograph of actual cavitating vortex rope of a model runner

Figure 6.17 Vortex rope in the straight part of the draft tube (Courtesy of Hydroart)

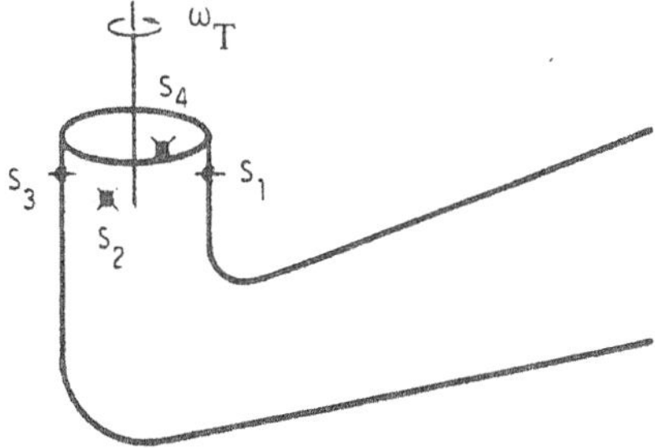

(a) Location of pressure sensors, S_1,..S_4, arranged at 90 deg. angle difference

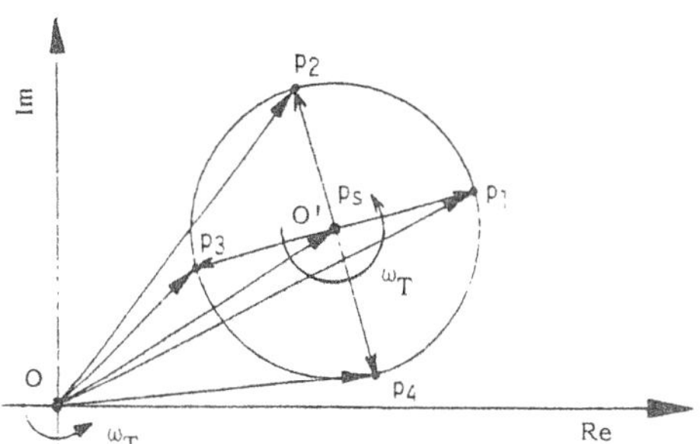

(b) Superposition of rotating and synchronous pressure oscillation

Figure 6.18 Rotating pressure field

sensor location S_1.

The synchronous oscillations are generally associated with the discharge oscillations in the draft tube (*draft tube surge*) and sometimes also in the penstock and through the machine. The hydraulic synchronous oscillations described above can generate the following oscillations in electro-mechanical quantities:

(a) radial and axial thrust on the runner and shaft bearings, Zanetti, 1982 [6.37],

(b) angular velocity of the unit (ω_0),

(c) torque,

(d) electric power,

(e) regulator action, hence opening of guide vanes of the distributor.

Radial thrust oscillations can affect guide bearings and labyrinths; axial thrust oscillations can affect support bearings. Pressure oscillations can attain values that are a considerable fraction of the working head, thus affecting the mechanical safety of the installation; moreover, even if low-valued, the number of cycles can be very large, thus creating a potential for fatigue damage.

Power oscillations (*power surge*) can prejudice correct operation of the unit at partial load, so that the phenomenon has to be taken care of quite often. Remedial measures can be taken with a high probability of beneficial effects, if reckoned on the basis of existing experience. The interventions frequently adopted as the remedy consist of:

(a) injection of air immediately under the runner (usually through the hub of the wheel); however, the discharge of compressed air (or of air being naturally sucked in by the negative pressure, if such is the case) must be carefully controlled, because in a certain range – difficult to assess a priori – there is the risk of amplification of the phenomena (see further section 6.2.3),

(b) installation of radial baffles at the walls or in the centre of the draft tube,

(c) installation of cylindrical ducts or obstacles in the draft tube, etc.

At overload the rotational structure of the flow is less clearly defined,[1] but a cavity of roughly cylindrical shape is formed at the centre of the draft tube (immediately below the runner) and pulsates at a certain frequency in synchronism with pressure oscillations, power swings, etc. It is to be noted that these circulatory patterns of motion and the attending pulsatory phenomena are produced not only at off-load working points in nominally steady operations but also during transients, that is, under time-varying rotational speed.

---

[1]The rotation is in this case in the opposite sense with respect to rotational speed of the runner (retrograde precession).

This entails that the frequency of the pulsatory phenomena is under these conditions altered and time-varying as well. This should be kept in mind in evaluating the possibilities of amplification by the response of different parts of the installation, in particular of the draft tube, albeit amplification under steady- state operation is undoubtedly as a rule more worrying. If manoeuvres are quite frequent as in pumped storage plants, the amplification under transient conditions could eventually lead to harmful effects.

6.2.2 Qualitative Interpretation of the Phenomena

A qualitative interpretation of the above mentioned phenomena is not difficult, at least for partial load. It is known that near the design point the rotational component of flow coming out of the runner should be minimal. This operational condition is called *zero swirl* and the corresponding flowrate is denoted by Q_0. As the discharge is reduced by acting on the opening of distributor wicket gate at constant head and rotational speed, the velocity triangles at outlet change shapes. In particular the outlet velocity triangle will necessarily show a definite rotational component of the absolute velocity as shown in Figure 6.19.

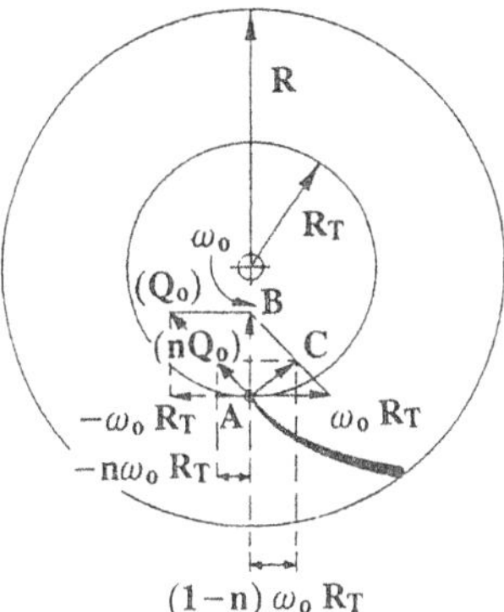

Figure 6.19 Ideal velocity triangles at runner outlet ($\overline{AB}$: absolute outlet velocity at zero-swirl condition $Q = Q_0$, $\overline{AC}$: absolute outlet velocity at $Q = nQ_0$ $(n < 1)$ with circumferential component $(1 - n)\omega_0 R_T$)

By the law of conservation of moment of momentum, the flow in the draft tube will also have a rotational component, which will increase as discharge ratio $n = Q/Q_0$ is reduced. The coexistence of an axial flow and a circumferentially rotating flow in a hypothetical straight and cylindrical draft tube is compatible with the existence of a precessing helical vortex filament within the cylindrical boundary, at least over a certain range of discharge (see Fanelli, 1989 [6.36]).

Near the axis of the vortex filament a field of centrifugal body forces gives rise to a negative pressure gradient as the axis is approached from the more distant regions of the flow field. Thus, when the average pressure in the section is low, the absolute pressure in the vortex core can reach the cavitation threshold p_v, and a cavity filled with a mixture of water vapour and gas (air coming out of solution) will appear all around the helical axis of the vortex filament.

The volume and pressure of this gaseous cavity will be dependent on the degree of partial opening, which influences the intensity of the centrifugal pressure gradient, and on Thoma number[2] σ, which influences the average pressure in the section, as well as on water quality (temperature, dissolved gas content, cavitation nuclei content,..), which influences the cavitation threshold p_v.

Now in a straight cylindrical draft tube this situation could essentially reproduce itself in every successive section, with the only modification of a rigid rotation and a variation in cross-section of the cavity due to the hydrostatic pressure gradient. The only effect that should necessarily accompany this situation would be the rotating pressure field, since there are no synchronous pressure oscillations in a straight draft tube. It is known, however, that the situation can alter drastically when, as is usually the case, the draft tube incorporates an elbow.

Several mechanisms of interaction of the peculiar type of flow described above with the elbow can be identified:

- The presence of a roughly helical vortex filament also in the bend of the draft tube induces pressure fields that, owing to the bend curvature, differ from the pressure fields in the straight part, the difference varying periodically with the rotation of the vortex filament.

- The vortex filament at the exit of the bend interacts with the well-known rotational components of the secondary flow in the final, diverging part (diffuser) of the draft tube. Interaction is markedly different according to the position of the vortex filament, because the rotational sense of the two interacting motions is concurrent in one-half of the diffuser section and is at variance in the other half.

  The interaction is therefore oscillating in nature, with frequency $\omega_T/2\pi$ corresponding to the angular velocity of the vortex filament. Apart from the local interaction, it is probable that the global performance of the diffuser in terms of pressure recovery efficiency is also affected in a periodic way.

---

[2]Thoma's cavitation number defined by (NPSH)/H.

- The pressure fields at inlet and outlet of the bend will be different, again owing to the curvature of the bend, except in the ideal case of toroidal bend of constant circular cross-section containing an even number of half-pitches of a regular helix. Also this difference is oscillating in nature with the same frequency $\omega_T/2\pi$.

All these effects can be seen as the engine or excitation mechanism of the hydraulic oscillations. If it should happen that the frequency of mass oscillation in the draft tube, taking into due account the compliance of the gas pocket contained in the cavitated vortex core, coincides with the frequency $\omega_T/2\pi$ of the excitation, then the response of the system could be appreciably amplified. If, moreover, the dynamic response of the hydraulic-mechanical-electrical system at frequency $\omega_T/2\pi$ is such as to entail appreciable oscillations of torque T, then power swings are to be expected.

In the most extreme case very severe power swings up to 10- 50% of average power can be experienced. Interaction with the regulator can lead to appreciable oscillations of the wicket gate opening. Pressure oscillations in the penstock can eventually be amplified, especially if one of the penstock eigenfrequencies should happen to lie near $\omega_T/2\pi$ or a multiple thereof, because the excitation sometimes contains harmonics of this fundamental frequency owing to the different non- linearities present in the system.[3]

The influence of the electric system is well demonstrated by Rheingans, 1940 [6.39]. In case the generator is connected to a system of infinite capacity, the natural frequency $\omega_n/2\pi$ of the generator rotor oscillations is given by;

$$\omega_n = \sqrt{T/J}\,. \tag{6.38}$$

where T being the restoring torque per radian of displacement from the steady-state angular position of the rotor and J the polar moment of inertia of the rotating masses.

If ω_n is close to ω_T, the hydraulic torque oscillations at frequency $\omega_T/2\pi$ will act as an external excitation on a resonant electro-mechanical system formed by the rotating masses and by the *electrical spring* of rotating torque T, and consequently a large amplification will occur leading to power swings. The installation will not be suitable for operation at the particular reduced load at which resonance sets in. Case history recordings of pressure and power swings for an actual installation and for a reduced-scale model are presented in Figure 6.20 (a) and (b).

6.2.3 Quantitative Relationship

The quantitative relationship between the different variables involved are not yet well established. Some empirical correlations are currently accepted

[3] As confirmed by ad-hoc model tests by Magri, 1987 [6.38]

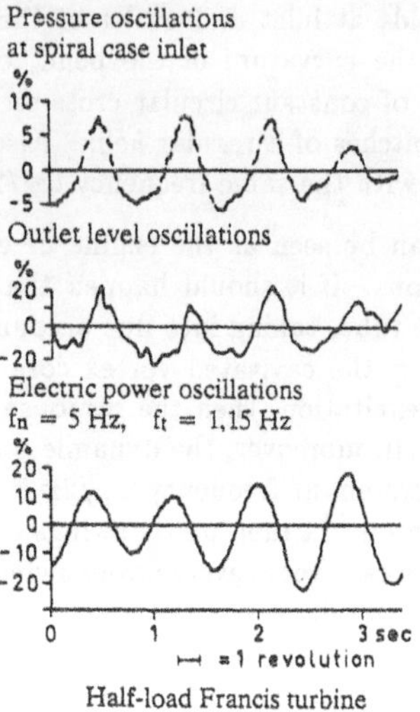

(a) Actual installation

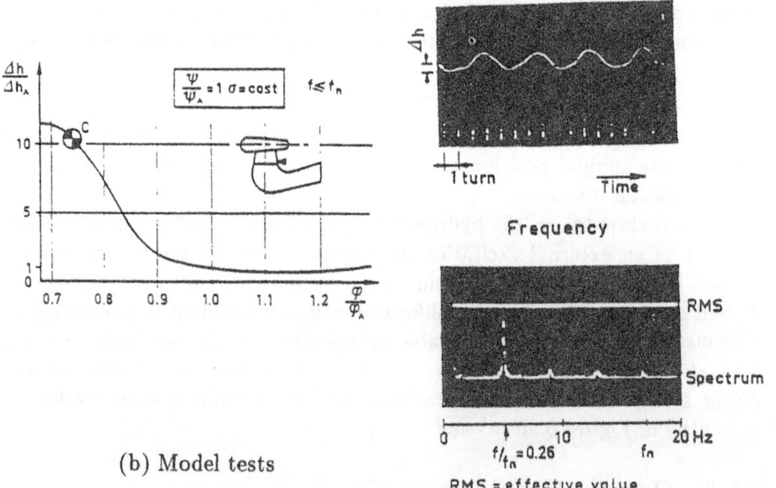

(b) Model tests

Figure 6.20 Record traces of pressure and power swings (courtesy of ENEL)

but more reliable and complete laws will have to wait, in all probability, for a comprehensive theory of the phenomena (see Dörfler, 1982 [6.32] and Fanelli, 1989 [6.36]). Figure 6.21 shows in non- dimensional form the zones of operation of Francis turbines more likely to give rise to the phenomena in question.

One of the first correlations to have been established is the variation of the rotational precession velocity ω_T in relation to the angular velocity of the runner ω_0 as a function of partial discharge rate n. Figure 6.22 shows three of the more often used relationships.

Other correlations have been identified between the σ value and the intensity of the phenomena. There seems to exist a critical value of σ which corresponds to a maximum amplitude of the oscillations at a given discharge. Figure 6.23 shows the general trend of these results. Of course, testing of a machine under a widely variable range of σ is usually possible only in a model test rig, so that the question arises as to how to transpose the results of model tests to prototype for these oscillatory phenomena.

Followings are semi-quantitative relationships which concern the sensitivity of the amplitude of oscillations to σ number on one hand, and to air injection on the other hand:

- If the overall pressure level of a machine is lowered, which is operating at partial load but cavitation-free, the region surrounding the helicoidal axis of the vortex rope is progressively occupied by a cavity, of a roughly circular cross-section, filled with water vapour and air.

  The threshold of cavitation inception seems to correspond to a pressure inside the cavity that is somewhat higher than vapour pressure, so that one may suppose that internal pressure corresponds to the sum of vapour pressure and the partial pressure of air more or less in equilibrium with dissolved air in water. If the value of σ is progressively lowered beyond the value σ_K corresponding to inception of rope cavitation, the amplitude of pressure oscillations grows and reaches a maximum, corresponding to a resonance value σ_R. Simultaneously, the relative importance of the synchronous component of oscillations grows with respect to the rotating component.

  At $\sigma = \sigma_R$ the pressure oscillations are almost completely synchronous and one observes also torque and power oscillation accompanied sometimes by pressure oscillations on the high-pressure side of the runner. The value of σ_R is a function of the machine specific speed. For σ values still lower than σ_R, the amplitude of synchronous oscillations begins to decrease again, see Figure 6.23.

- If air injection is effected under the runner usually through a duct in the hub axis, this produces an effect which mimics a decrease in

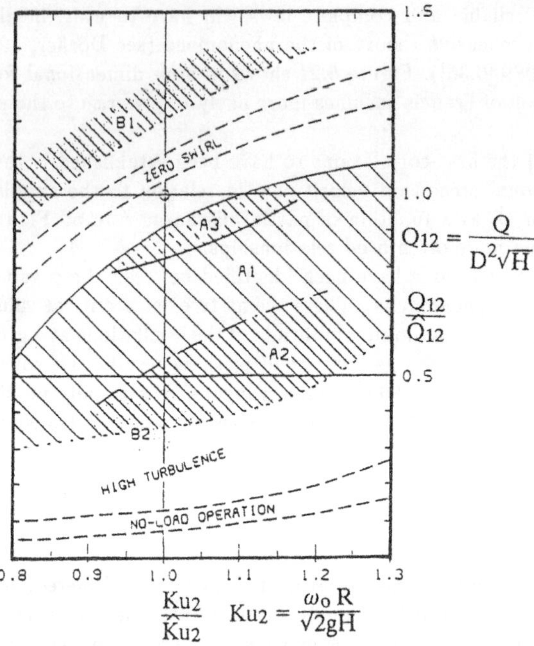

Figure 6.21 Operational zones of Francis turbines (A_1, A_2: rotating helical vortex rope, upper band of A_2: double helix, A_3: unstable helical vortex rope, B_1: central pulsating rope) (courtesy of Escher-Wyss)

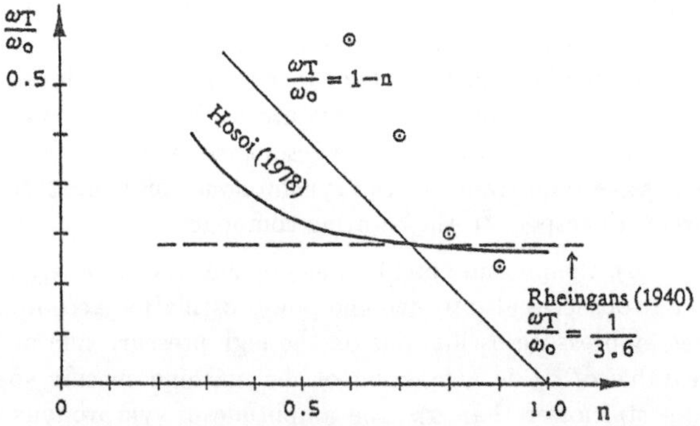

Figure 6.22 Variation of apparent precessional angular velocity of vortex rope ω_T against $n = Q/Q_0$ (Experimental points ⊙ and three correlation curves commonly used are shown.)

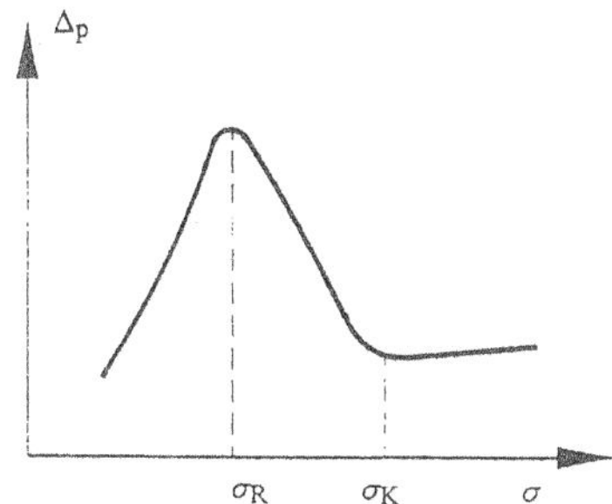

Figure 6.23 Amplitude of synchronous pressure pulsations against σ

pressure. At $\sigma > \sigma_R$ a moderate air injection results in the increase of synchronous oscillations and more substantial quantities of air are needed to reduce them. On the contrary at $\sigma < \sigma_R$ even the injection of a moderate quantity of air leads to a reduction in the amplitude of synchronous oscillations.

In conclusion, both very low values of σ and moderate to substantial discharge of air (up to 2% or more of the full-load water discharge) lead to almost complete elimination of the synchronous oscillations. To these conditions there corresponds a large cavity which usually no longer has a clearly defined helicoidal shape.

Another difficulty which confronts manufacturers, designers and plant operators is that the transposition of test results obtained on reduced-scale models to full-scale prototypes under swinging conditions is fraught with uncertainties. Whereas the ratio of the frequency of the phenomena to the rotational frequency is tied essentially to the degree of partial opening and can be scaled-up with good reliability, the amplitude of the phenomena, its relationship to σ variations and/or to the discharge of injected air, etc. cannot usually be forecast with any degree of certainty from model tests, see Magri, 1987 [6.38].

The reasons for this can be summarized as follows:
(i) The amplification of the response to the excitation engine is strongly dependent on the dynamic properties of the whole hydraulic installation, which are not usually reproduced in the test rig of the machine model. (ii) The similitude of σ values cannot usually be realized, and likewise criteria of similitude for injected air are not clearly defined. (iii) The action of the electrical system, rotating masses and the speed regulator cannot usually

be reproduced at the correct scales (regulator action is totally absent in the model tests).

So, a rational way out of these difficulties could be found eventually only in the development of an acceptable, comprehensive mathematical model of the whole phenomenon (excitation, response, dynamic reaction of the different components of the system). If this model were available and validated, then it could be used to simulate the whole test rig and to isolate the excitation from the model test results by identification procedures. This excitation could be scaled- up according to similitude laws and the response amplification then could be calculated for the full-scale installation taking into account the actual dynamic characteristics of the real components.

Unfortunately, mathematical models of the swinging behaviour at partial load or at overload are still in a very primitive stage of development. One can make a distinction between global models, which do not attempt to go into the detailed dynamics of the flow field, and more ambitious ones, which aim at a reasonably realistic representation of the flow field in the draft tube. For a reasonable mathematical model, a consistent set of equations should be derived, connecting the flow field to pressure variations (and cavity volume in the case of cavitated vortex rope), to the excitation mechanism, to response of the single components, to compatibility of boundary conditions at the junction of consecutive components, to reaction of mechanical and electrical system, etc. Part of the effort has been undertaken by Fanelli, 1989 [6.36], but the final goal is still a long way off.

6.2.4 Conclusions and Further Developments

Summing up the present situation, one can state the following:

(a) Power swing is a harmful phenomenon, which can prevent flexible operation of Francis turbines and/or produce, in the more slight cases, accumulation of damage by fatigue in mechanical pieces through pressure oscillations, radial and axial thrust variations on the bearings and hydroelastic interactions. It is a subject of active research.

(b) Usually it is possible to intervene on an installation presenting power swing with remedial measures, typically injection of air through the runner hub. However, such measures are costly. Not only the cost of the compressor and the attending energy consumption is incurred, quite often it is necessary to stop the installation in order to carry out the alterations needed, so incurring loss of production.[4] Moreover, an appreciable decline of efficiency is often incurred, resulting again in a continuous loss of production whenever the installation is operated in the range where power swing tends to set in.

[4] All the more so, if the alterations involve the draft tube such as in the case of radial baffles installation.

(c) The phenomenon is difficult to predict at least quantitatively at the stage of installation design, and model test results suffer uncertainties when scaled-up to full-size prototype, due to incomplete similitude and different response of the hydraulic-mechanical-electrical system attached to the model and prototype machines.

(d) Statistical data on installations presenting the power swing phenomenon are hard to collect. Some manufacturers have collected partial data, referring mostly to machines of their own production, Dörfler, 1982 [6.32]. Under these conditions, transfer of knowledge and experience is difficult.

In conclusion, there is still room and need for further investigations, both at the theoretical and the experimental level:

(i) On the theoretical side; more detailed three-dimensional (3-D) and realistic representation of the flow field can be sought. The final aim of this effort is to establish physically sound relationships among global quantities (discharge, synchronous pressure, cavity volume oscillations in relation to excitation and response characteristics of the whole hydro-mechanic-electrical system), to be used as an investigation tool which allows to extract the intensity of the excitation mechanism from measurements on reduced-scale models, and to scale-up this intensity to full-scale installation and subsequently to compute the response of the actual hydro-mechanical-electrical system. This goal is deemed to be attainable in the long run.

Complete theoretical derivation, including excitation intensity, radial and axial thrust oscillations, etc., would necessitate 3-D simulation of the flow field in the draft tube and its reaction on the flow field inside the runner, including also boundary layer effects and vorticity decay due to viscosity. This more comprehensive goal is decidedly not within reach of the present state of the art and will probably remain out of reach for some time in the near future. However, the fast pace of progress in complex 3-D hydrodynamic simulation by supercomputers leaves open a possibility of complete simulation in a more long- range perspective.

(ii) On the experimental side; more complete and detailed investigation of the 3-D instantaneous flow field on reduced-scale models, both in the draft tube and in the runner, could help better understanding of the hydrodynamics of the phenomena and is probably within the pale of present possibilities, using for example 3-D laser-doppler anemometry. A considerable progress in test rig conception could consist in providing the external circuit with some degrees of flexibility, for example, by varying the hydraulic impedances of the high-pressure and low-pressure circuits adjoining the model of scroll case-distributor-machine draft tube.

More accurate measurements of torque, axial and radial thrust oscillations are also necessary. The possibility of simulating the regulator action and the electrical system influence on the model (e.g. by means of an on-line, real-time computer interaction) would also constitute an important progress. New shapes of draft tube should be investigated, hoping to find less sensitive configurations. A synergistic use of numerical and physical modeling is recommended in any case, as a powerful means of extracting the maximum information from both tools.

For designers and purchasers of installations, it is recommended that the possibility of air injection through the runner hub, and the facilities for air compressor installation, be foreseen from the beginning, so that, should the necessity arise, no alteration or lengthy stops are required to correct the situation. Also, pressure taps with attending signal cables should be liberally installed in the full-scale unit, especially in areas such as the draft tube, which present difficulties of access. In this way a thorough knowledge/documentation of the phenomena, should they arise, can be obtained with a minimum of additional effort.

Guarantees on efficiency at part load or at overload should take into account the possibility that injection of air could become necessary, and thus a slight reduction in expected efficiency values in these ranges of load could have to be accepted. What to do in this event (penalties or other financial terms) should be clearly defined in specifications and final tendering.

Finally, it is to be kept in mind that the phenomenon presents the maximum amplification for a certain σ value, which is low but in the range of normally accepted values[5]. Under this respect, the adoption of σ installation values somewhat higher than current practice (i.e. higher submergence of the runner) could be highly beneficial, and the higher costs incurred for civil works (excavations) would probably be more than justified by the absence of troubles.

Finally, a systematic collection of measurement data on actual installations is to be recommended to owners and manufacturers alike, possibly in a standardized form aimed at establishing an International Data Base on these phenomena.

6.3 Pressure Surge Due To Vortex Core In Draft Tube

6.3.1 Cavitating Vortex Core

Violent pressure fluctuation called *pressure surge* frequently occurs in the draft tube of a Francis turbine at part load operation. The amount of swirl

[5] In fact, machines with high submergence (e.g. reversible pump- turbines working in the generating mode) are less prone to troubles of this kind.

left in the discharge flow from a runner increases as the guide vane opening is increased or decreased from the design condition. Thus, it is understandable that the strength of swirl at the draft tube inlet is related to the occurrence of pressure surge (Falvey, 1971 [6.40]).

Photographs in Figure 6.24 show various vortex cores observed in the swirl flow inside the elbow draft tube of a model Francis turbine where runner rotation is in the clockwise direction relative to the downstream. The operating points are represented by two parameters: one is the Thoma number σ and the other is the discharge ratio Q/Q_0, where Q_0 denotes *zero-swirl* discharge. Figure 6.24 (a) is the case at overload, and (b), (c) and (d) correspond to part load cases. It is observed that the shape of the cavitating core, sometimes called *vortex rope*, varies with the discharge ratio. The wall pressure measurement showed that large pressure fluctuation occurred at conditions shown in Figure 6.24 (d). This pressure surge at part load is the subject of the present section.

6.3.2 Swirl Flow Model and Similarity Law

One Parameter Model

Figure 6.25 shows time averaged velocity distributions of swirl flow at part load measured at the inlet cone of a draft tube. The top graph corresponds to the axial component of velocity v_a and the bottom to the peripheral component v_u. Both of them are non- dimensionalized by the area-mean axial velocity derived from the discharge Q. The experimental data represented by solid lines indicate that the v_u distribution looks like a Rankine vortex and the v_a distribution corresponds to one having a wake-like velocity defect at the centre of uniform flow.

These features in velocity distributions are recognized in the discharge flow from a runner at various part load operating conditions. Thus, the discharge flow is approximated by the following simplest model that the irrotational swirling main flow surrounds the central dead water region with radius R_d (see Nishi et al., 1982 [6.33]).

$$v_a = \begin{cases} 0 & (0 \leq R < R_d) \\ \bar{v}_a = Q/[A(1-r_d^2)] & (R_d \leq R \leq R_w) \end{cases} \tag{6.39}$$

$$v_u = \begin{cases} 0 & (0 \leq R < R_d) \\ v_{uw} R_w / R & (R_d \leq R \leq R_w) \end{cases} \tag{6.40}$$

where R : radial distance
R_d : radius of dead water region
R_w : radius of draft tube
r_d : dimensionless dead water radius $= R_d/R_w$

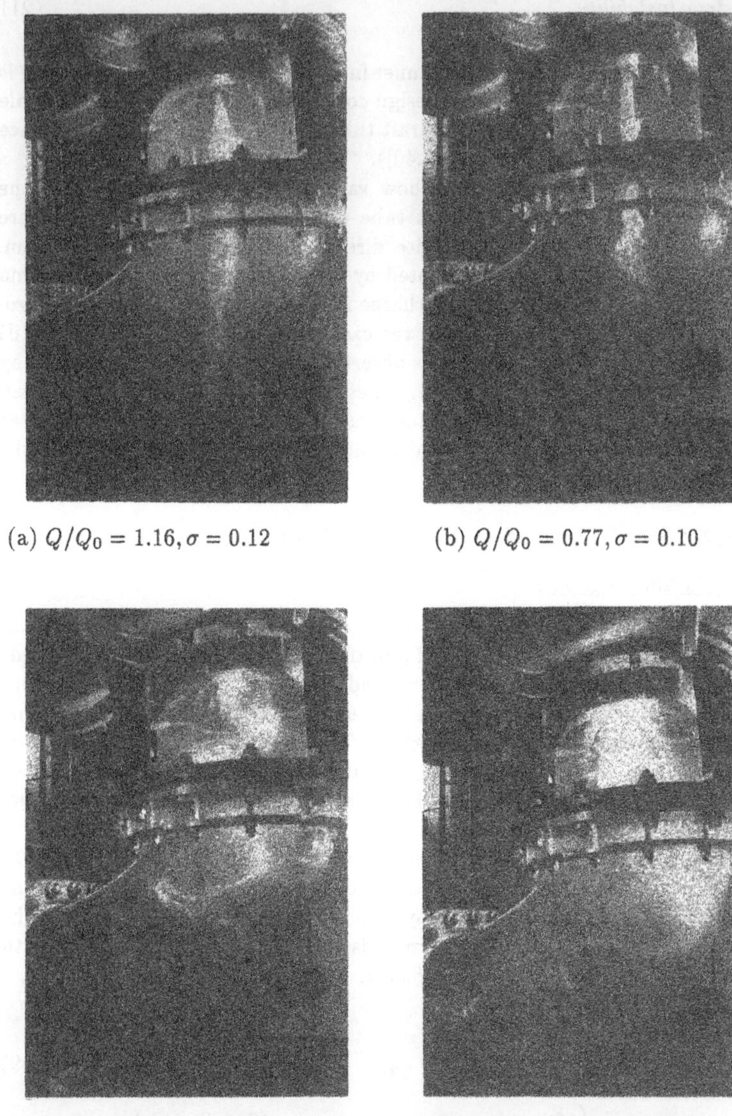

(a) $Q/Q_0 = 1.16, \sigma = 0.12$

(b) $Q/Q_0 = 0.77, \sigma = 0.10$

(c) $Q/Q_0 = 0.65, \sigma = 0.10$

(d) $Q/Q_0 = 0.46, \sigma = 0.10$

Figure 6.24 Cavitating vortex core in the draft tube of model Francis turbine (Courtesy of Fuji Electric Co., Ltd.)

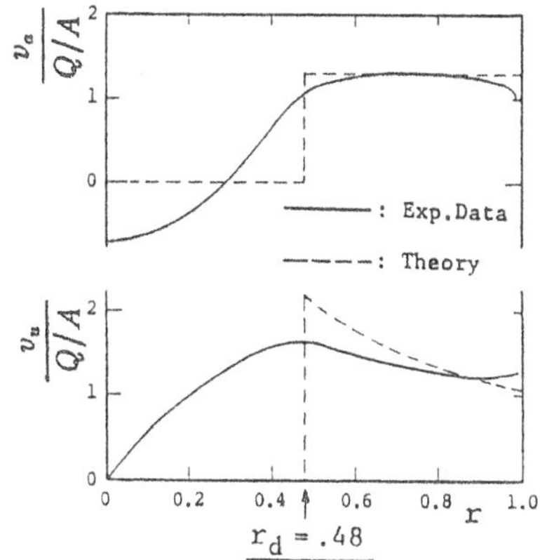

Figure 6.25 Velocity distribution of swirl flow at the draft tube inlet of a model Francis turbine

A : cross-sectional area $= \pi R_w{}^2$

v_{uw} : swirl velocity at draft tube wall determined by free-vortex formula

Concerning the parameter to evaluate the strength of swirl at the draft tube inlet, the following equation is adopted to define the swirl rate:

$$m = \int_0^{R_w} v_a v_u R^2 \, dR \; / \; R_w \int_0^{R_w} v_a{}^2 R \, dR \tag{6.41}$$

Substituting equations (6.39) and (6.40) into equation (6.41) and integrating, yields

$$m = v_{uw} \; / \; \bar{v}_a \; . \tag{6.42}$$

If the principle of minimum kinematic energy proposed by Shogenji and Shimoyama, 1933 [6.41], is applicable to swirl flow, the following relationship is derived between m and r_d;

$$m = \sqrt{2r_d{}^2(1 - r_d{}^2)/(1 - r_d{}^2 + 2r_d{}^2 \; \log_e r_d)} \; . \tag{6.43}$$

Comparison of equation (6.43) with experimental data is shown in Figure 6.26 (Nishi et al., 1982 [6.33]). The dash-dot line shows the following curve fit relationship;

$$m = r_d + 3\, r_d{}^3 \; . \tag{6.44}$$

While the correlation may be fair, it is acceptable to assume that the velocity profile of swirl flow is represented by one parameter family, swirl

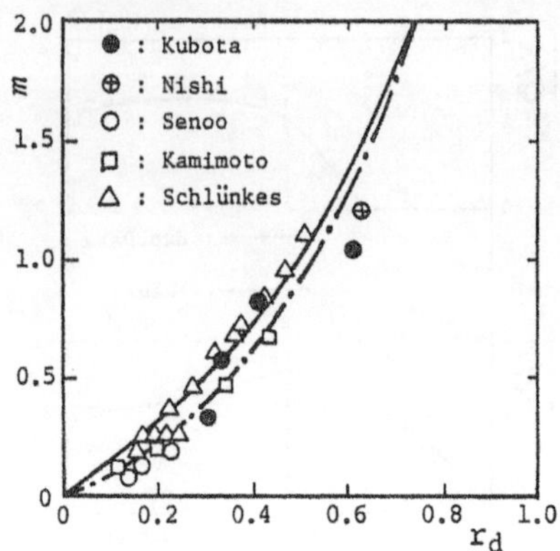

Figure 6.26 Relationship between the radius of dead water region and swirl rate (The solid line indicates equation (6.43) and the dash-dot line equation (6.44).)

rate m. Dotted lines in Figure 6.25 show the velocity model computed from the experimental value of swirl rate.

Similarity Law

From dimensional analysis the following relations and parameters are obtained, which affect the amplitude and frequency of pressure fluctuation at part load, providing that the geometry of a draft tube is prescribed:

$$\Delta\psi = \text{func}(m,\ K,\ R_e,\ F_r)\,, \tag{6.45}$$

$$S_t = \text{func}(m,\ K,\ R_e,\ F_r)\,, \tag{6.46}$$

where $\Delta\psi$: dimensionless amplitude;

$$= 2g\Delta H/v_i^2 \tag{6.47}$$

ΔH : peak to peak value of pressure head fluctuation
v_i : mean velocity at draft tube inlet $= Q/A_i$
K : cavitation parameter;

$$= 2g(H_a - H_s - H_v)/v_i^2 = 2g(\text{NPSH})/v_i^2 \tag{6.48}$$

H_a : absolute pressure head acting on the tail water
H_s : static draft tube head
H_v : vapor pressure head
R_e : Reynolds number;

$$= 2R_{wi}v_i/\nu \tag{6.49}$$

F_r : Froude number;

$$= v_i/\sqrt{2gR_{wi}} \tag{6.50}$$

S_t : Strouhal number or dimensionless frequency;

$$= 2fR_{wi}/v_i \tag{6.51}$$

f : fundamental frequency (Hz).

If the experiment is conducted at a Reynolds number larger than 10^5, its influence on the swirl flow becomes negligible (Falvey et al., 1970 [6.42]). The effect of Froude number arises when the cavitating core is observed in the draft tube at low pressure level. As it is difficult to fit the similitude of Froude number for model tests, attempts have been made to include the effect in the static draft tube head or the reference elevation from the tail water surface which determines the cavitation parameter (Ulith et al., 1974 [6.43]).

Consequently, equations (6.45) and (6.46) are respectively approximated by

$$\Delta\psi = \text{func}(m,\ K) \quad \text{and} \quad S_t = \text{func}(m,\ K)\,. \tag{6.52}$$

Thus, we are able to attain the identical pressure fluctuation for the model test, if both parameters m and K for the model and the prototype are equal to one another (Nishi et al., 1980 [6.44]). It also indicates that we are able to study the pressure surge using the test apparatus without a turbine runner (cf. Nishi et al., 1980 [6.44], Guarga et al., 1986 [6.34]). It has been verified that similar velocity distributions represented by a single parameter, swirl rate m, are observed in swirl flows which are generated by a set of stationary guide vanes.

For the sake of convenience, the above equations are replaced by the following for the model test of a Francis turbine (Kubota et al., 1982 [6.45]).

$$\Delta H/H = \text{func}(Q/Q_0,\ \sigma) \tag{6.53}$$

$$f/n = \text{func}(Q/Q_0,\ \sigma) \tag{6.54}$$

where H is the effective head and n rotational speed of the runner (Hz). It should be noticed that σ=const does not correspond to K = const.

According to Hosoi's work, 1978 [6.31], it is necessary to consider the effect of the specific speed n_s of the Francis runner on the dimensionless frequency f/n for low n_s machines. It suggests that one parameter model is not always sufficient to represent swirling flows at the draft tube inlet. If the velocity model shown in the preceding section is applicable to the flows, the other parameter is the nondimensional radius of dead water region r_d. In such cases both $\Delta\psi$ and S_t are regarded as a function of three independent variables, m, r_d and K. But all the consideration that follows are restricted to the case where the velocity model with the single parameter is applicable.

6.3.3 Flow Regimes in a Draft Tube

Behaviour of Vortex Core and Flow Regimes

Each vortex core shown in Figure 6.24 (c) and (d) has the spiral shape similar to a corkscrew and rotates around the axis of the draft tube. This spiral core is a rolled up vortex sheet which originates between the central dead water region with flow reversal and the swirling main flow as shown in Figure 6.27 (Nishi et al., 1982 [6.33]). The core trails in the opposite direction to the swirling main flow (see Dörfler, 1980 [6.46]) and the pitch of the spiral increases with the strength of swirl.

Swirl flows in a draft tube are classified into four major flow regimes, which are characterized by size and location of the dead water region. Figure 6.28 shows a flow regime chart for an elbow draft tube having inlet cone angle of 9 degrees and exit-to-inlet area ratio of 4.1 (Nishi et al., 1982 [6.33]). The ordinate m_{th} denotes the theoretical value of swirl rate computed by the quasi three-dimensional inviscid flow analysis, cf. Senoo et al., 1971, 1972 [6.47, 48].

As shown in Figure 6.28, each flow regime primarily depends on the strength of swirl. Brief descriptions of these regimes are given below:

(a) Flow regime I: A vortex core is observed as a straight rope, since a stagnation point does not occur at the centre axis due to weak swirl.
(b) Flow regime II: Various shapes of vortex core appear irregularly due to the transitory nature of the dead water region (Fisher et al., 1980 [6.49]).
(c) Flow regime III: A single spiral vortex core similar to the corkscrew appears stably where the dead water region almost always exists near the inlet, and the core rotates continuously in the draft tube. Figure 6.29 shows sketches of the spiral vortex core in two typical cases during a pressure fluctuation; Figure 6.29 (a) corresponds to the case where the core is situated near the small-radius wall at the bend inlet, while figure b to the opposite case where the core is located near the large-radius wall. The tail of the vortex core together with the main flow follows one side wall. It indicates that the flow may separate near the foot inlet and the opposite wall is covered by a large separated region or the hatching zone in the illustrations.
(d) Flow regime IV: Two vortex cores usually appear due to strong swirl, since the dead water region extends to the upstream of a draft tube.

The same general feature is to be expected in different draft tube geometry, but the values of swirl rate to divide these regimes are affected by the

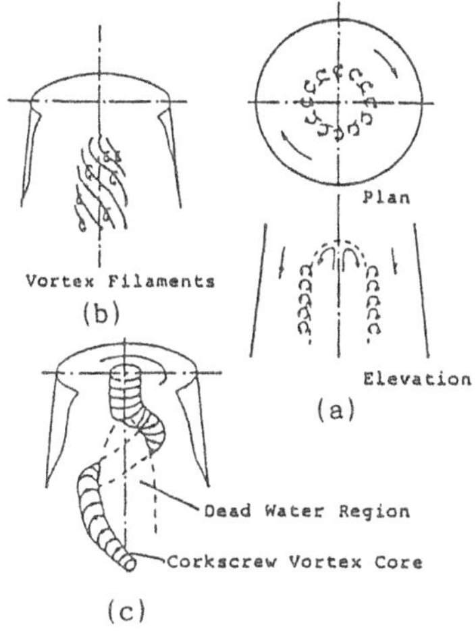

Figure 6.27 Origination of a spiral vortex core

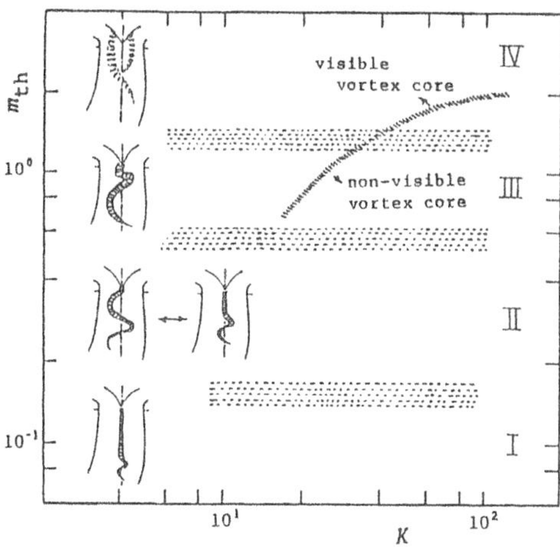

Figure 6.28 Flow regime chart

geometry (Nishi et al., 1986 [6.50]).

Pressure Fluctuations in Flow Regime III

Since the flow regime III corresponds to the half- load surge, the cavitation characteristics of pressure fluctuation in this regime are shown in Figure 6.30. $\Delta\psi_{rms}$ denotes the rms value and K_i is the cavitation parameter which is defined at the draft tube inlet. Pressure fluctuations were measured at two taps L1 and L2 shown in Figure 6.29. The dimensionless frequency S_t is almost constant for a high cavitation parameter and tends to decrease gradually with K_i due to the effect of inhomogeneous flow.

Concerning the amplitude parameter $\Delta\psi_{rms}$, there exists the critical value of cavitation parameter, where the amplitude increases considerably. The pressure wave at $K_i = 8.5$ is shown in Figure 6.31 (b). Pressures oscillate almost synchronously, and their amplitudes are different at the measuring points. Considering the quasi-synchronous nature, this pressure surge is called QS type. The result in Figure 6.31 (a) is an example of QR type pressure fluctuation. This type is observed in the range of high cavitation parameter, where $\Delta\psi_{rms}$ and S_t are regarded constant. In the case of QR type, the waveform varies point by point and time lag between the pressure troughs of L1 and L2 data roughly corresponds to the circumferential difference of the two measuring points.

6.3.4 Pressure Surge Physics

Frequency

Assuming that the flow in a draft tube is homogeneous, a simple theory is derived to predict the fundamental frequency of pressure surge which agrees with that of vortex core rotation. If the rotational speed of the spiral vortex core v_v is regarded as one half the velocity difference between the peripheral components of swirling main flow and axi-symmetric dead water region, v_v is expressed as

$$v_v = \frac{k\, v_{uw}}{2\, r_d}\,, \tag{6.55}$$

where the correction coefficient k is introduced to correlate the discrepancy between the peripheral component of actual velocity and the analytical model at radial position $R = R_d$. The frequency is given by

$$f = \frac{v_v}{2\pi R_d}\,. \tag{6.56}$$

Consequently, the following equation which indicates the dependence of the dimensionless frequency on swirl rate m, is derived:

$$S_t = \left[\frac{k\, m}{2\pi r_d^2(1 - r_d^2)}\right]_i \tag{6.57}$$

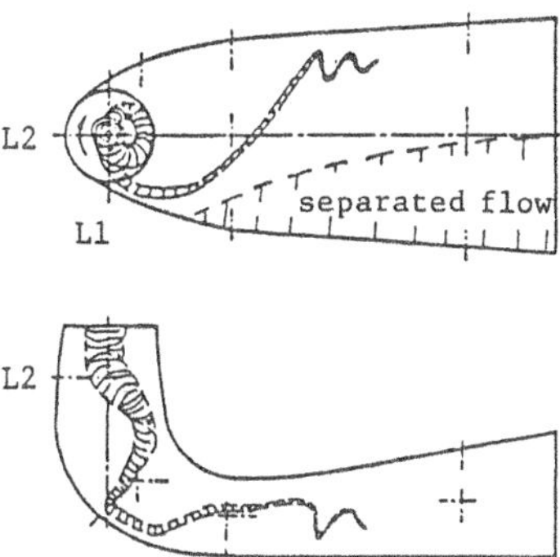

(a) A thick vortex core at the minimum of inlet pressure

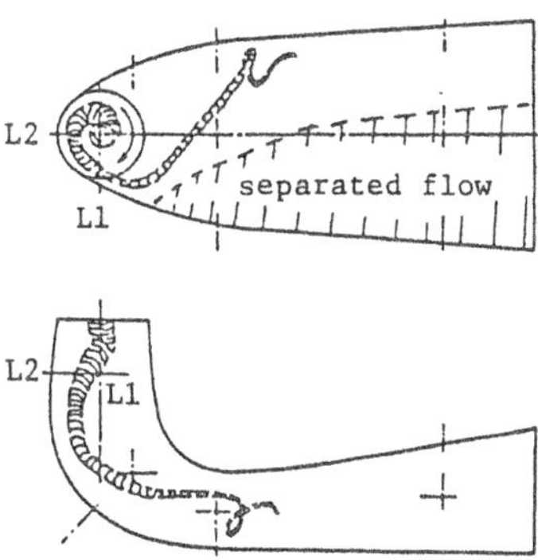

(b) A thin vortex core at the maximum of inlet pressure

Figure 6.29 Sketches of a spiral vortex core in a draft tube during pressure surge

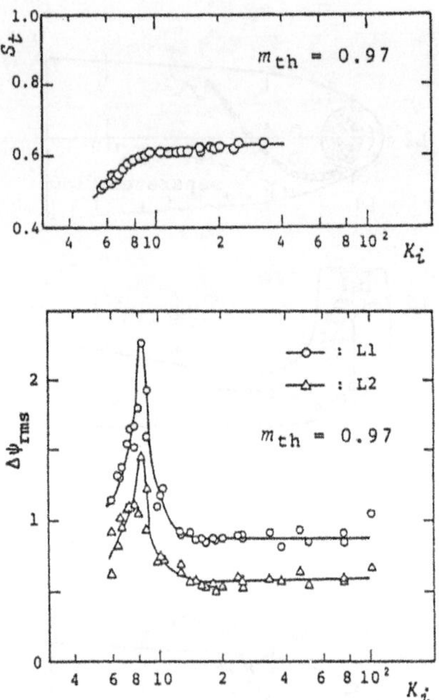

Figure 6.30 Cavitation characteristics of wall pressure fluctuations

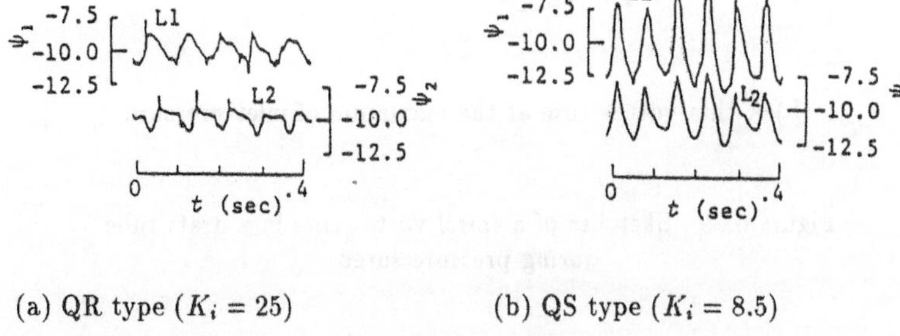

(a) QR type ($K_i = 25$) (b) QS type ($K_i = 8.5$)

Figure 6.31 Waveforms of pressure fluctuations

Various marks in Figure 6.32 show the relationship between S_t and m determined using experimental data shown in literature (Nishi et al., 1982 [6.33]). This data is reasonably represented by the solid line, which is estimated using equation (6.57) with $k = 3/4$ and equation (6.44).

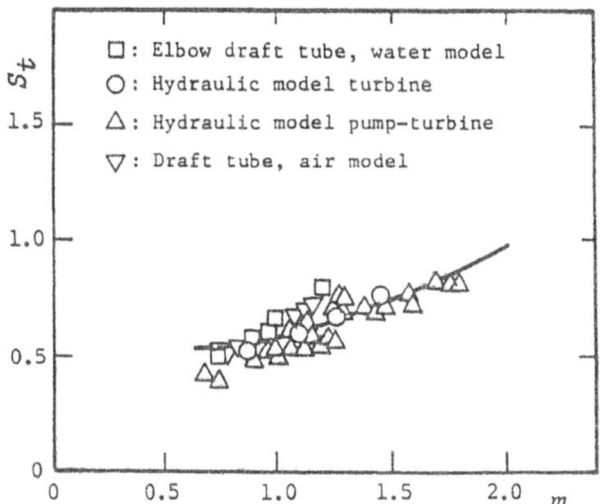

Figure 6.32 Relationship between swirl rate m and frequency parameter S_t

Pressure Model and Application

It is understandable that the flow field is distorted considerably by the vortex core. The contour illustrated in Figure 6.33 is regarded as a snapshot of pressure distribution measured in the cross-section of the conical diffuser, Nishi et al., 1988 [6.51]. Thus, this pressure pattern rotates around the axis of the draft tube together with the spiral vortex core, and the pressure detected at the wall varies periodically. This has a convective nature as shown in Figure 6.34, and is called rotating pressure fluctuation (RO type).

It is recognized in Figure 6.29 that the size of the separated region in the foot portion varies with the position of the core. Thus, rotation of the core in an elbow draft tube induces the periodical variation of pressure recovery. Pressures detected in the inlet section of the draft tube fluctuate in unison, even if flow entering the draft tube does not have an oscillatory nature. They become the lowest at maximum pressure recovery and the highest at minimum recovery due to the diffuser effect. This corresponds to the synchronous fluctuation (SY type) shown in Figure 6.34.

Consequently, the periodic component $\tilde{\psi}$ of the dimensionless pressure is represented by the sum of the rotating component $\tilde{\psi}_{ro}$ and the synchronous

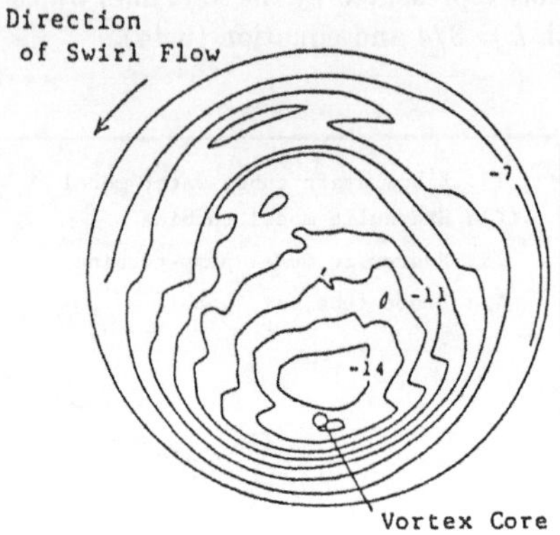

Figure 6.33 A snapshot of pressure distribution (m_{th} = 1.03, K_i = 22. Numbers shown in the contour correspond to the dimensionless piezometric pressure based on the dynamic pressure calculated from mass-averaged velocity.)

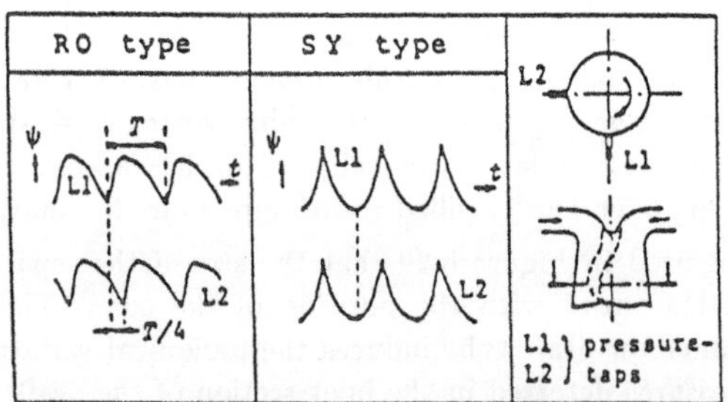

Figure 6.34 Classification of pressure fluctuation

component $\tilde{\psi}_{sy}$, both of which have the same period of time T.

$$\tilde{\psi} = \tilde{\psi}_{sy} + \tilde{\psi}_{ro} \tag{6.58}$$

If two pressure taps L1 and L2 are installed in the section of the inlet cone making an angle of 90 degrees as shown in Figure 6.34, both components $\tilde{\psi}_{ro}$ and $\tilde{\psi}_{sy}$ can be determined using the following relationship with experimental data $\tilde{\psi}_1$ and $\tilde{\psi}_2$;

$$\tilde{\psi}_{ro}(t) - \tilde{\psi}_{ro}(t - \Delta t) = \tilde{\psi}_1(t) - \tilde{\psi}_2(t) \tag{6.59}$$

$$\tilde{\psi}_{sy}(t) - \tilde{\psi}_{sy}(t + \Delta t) = \tilde{\psi}_1(t) - \tilde{\psi}_2(t + \Delta t) \tag{6.60}$$

where subscripts 1 and 2 denote the measuring point L1 and L2 respectively. $\Delta t = T/4$ in this case, since 90 degrees corresponds to a quarter of the period T. It is noted that the Fourier series up to the third harmonic is usually sufficient to approximate the periodic pressure, see Casacci et al., 1982 [6.52].

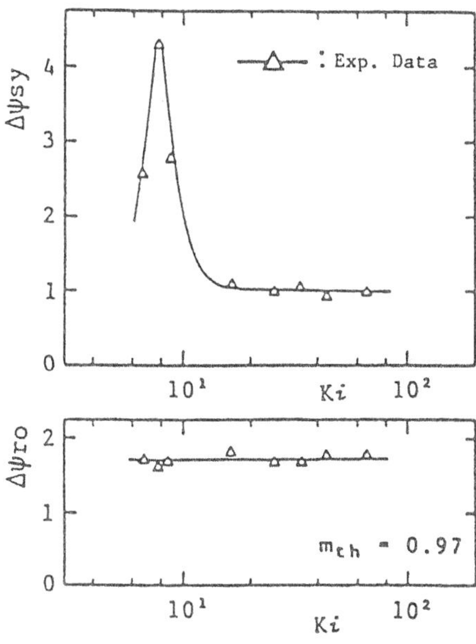

Figure 6.35 Cavitation characteristics of amplitude parameters $\Delta\psi_{sy}$ and $\Delta\psi_{ro}$

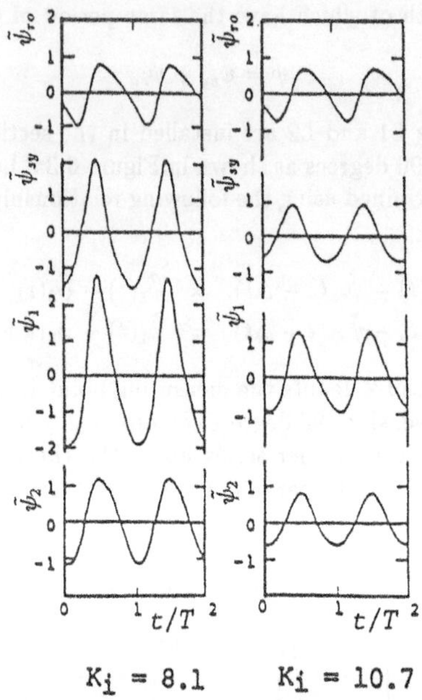

Figure 6.36 Waveforms of pressure fluctuations $\tilde{\psi}_{sy}$ and $\tilde{\psi}_{ro}$ together with $\tilde{\psi}_1$ and $\tilde{\psi}_2$ ($m_{th} = 0.97$)

Figure 6.35 shows the cavitation characteristics for the amplitude parameters $\Delta\psi_{sy}$ and $\Delta\psi_{ro}$. These results were obtained using the above method with measured data $\tilde{\psi}_1$ and $\tilde{\psi}_2$ corresponding to those in Figure 6.30. The rotating component $\Delta\psi_{ro}$ is regarded as constant in the range of the measurement, since the cavitation parameter has little effect on the distorted pressure field. This kind of data presentation is desirable in order to evaluate the pressure fluctuation because the differences among pressure fluctuations detected at various measuring points may be removed. Typical waveforms of both components $\tilde{\psi}_{sy}$ and $\tilde{\psi}_{ro}$ together with measured data $\tilde{\psi}_1$ and $\tilde{\psi}_2$ are illustrated in Figure 6.36, where $t = 0$ corresponds to the minimum value of $\tilde{\psi}_1$.

Resonance

In Figure 6.35 considerable increase in $\Delta\psi_{sy}$ is observed around $K_i = 8$ where a QS type pressure surge occurred. Such large amplitudes cannot be attributed to the diffuser effect. From the visual observation of the behaviour of spiral vortex core at these conditions, it was found that the cavitating core varied its size periodically, i.e., the thick core was observed at the minimum of inlet pressure (see Figure 6.29 (a)) and the thin core was at the maximum of the pressure (see Figure 6.29 (b)).

This indicates that oscillatory flow occurred in the draft tube. If the natural frequency of an elbow draft tube system is close to the frequency of the pressure-recovery fluctuation at a certain value of cavitation parameter, the resonance may occur and the oscillation of flow appears in the draft tube. The flow oscillation does not occur when the natural frequency of the system varies with the gas volume in the cavitating core by changing the pressure level and deviates from the rotating frequency to some extent. Consequently, the large synchronous component in the corresponding case is attributable to the resonance entrained by the fluctuation of pressure recovery.

In conclusion, the following remarks are applicable in order to reduce violent $\tilde{\psi}_{sy}$ or the QS type pressure surge:

(a) Determine the draft tube shape so as to secure sufficient pressure rise in the inlet conical diffuser portion.

(b) Choose the area ratio of foot portion so as to reduce the zone of separated flow.

It is noted that various remedies to alleviate the pressure surge are summarized in the literature (Grein, 1980 [6.53]).

6.4 Cavitation Induced Oscillation And Two-phase Flow Instability

Severe oscillations of pressure and flow rate are sometimes observed in a pump operating under cavitating condition or two-phase flow condition, even when the operating point is set in the stable range of a steady-state head-capacity curve. These undesirable phenomena should be avoided since they deteriorate the reliability of pumping systems. Typical examples of such instabilities are described briefly in the following sections.

6.4.1 Cavitation Induced Oscillation

Observation of Instabilities in a Cavitating Inducer

Figure 6.37 shows a typical inducer with three helical blades. In the case of a turbopump of liquid-propellant rocket engines, such an inducer is installed ahead of a main-impeller to improve the suction performance considerably. An example of the performance for two inducers, dimensions of which are listed in Table 6.2, is shown in Figure 6.38 obtained by Kamijo et al., 1977 [6.54]. Three curves of head coefficient ψ versus cavitation number K are plotted for constant values of flow coefficient ϕ.[6] for definition. As shown in the figure, two kinds of instabilities were observed, rotating cavitation and low cycle oscillation. The latter occurred in the negative slope region of ψ with respect to K.

In order to deepen our understanding on the former instability, variation of cavity length on the blades with time and waveforms of pressure fluctuations detected at both upstream and downstream of the inducer are reproduced in Figures 6.39 (a) and (b) respectively. Measurements of cavity length in Figure 6.39 (a) were made using high speed motion films at two radial positions, the blade tip and the middle of the blade height. According to those pictures, cavitating zones at the tip and middle of the blade rotated together around the inlet centre and their rotational frequency was estimated to be 147 Hz, i.e., 1.16 times the inducer rotation. However, wall pressure at the inlet fluctuated with the frequency of 60 Hz as shown in Figure 6.39 (b). This value agreed fairly well with the frequency of cavity volume variation at the inducer inlet, provided that the volume was represented by the total length of those cavities in Figure 6.39 (a).

Table 6.2 Dimensions of test inducers

Tip diameter (mm)	127.0	Blade thickness at tip (mm)	2.5
Inlet hub diameter (mm)	37.0	Blade thickness at hub (mm)	3.5
Outlet hub diameter (mm)	65.0	Inlet blade angle at tip (deg.)	10.0
Radial tip clearance (mm)	0.5	Outlet blade angle at tip (deg.)	12.0
Number of blades	3	Hub taper angle (deg.)	21.0
Solidity at tip	2.5	Cant angle (deg.) Inducer A	0
		Inducer B	10.5

When the instability occurred at higher cavitation number than that shown in Figure 6.39, the wall pressure fluctuated with the rotating frequency of the cavitation pattern. While it was seen that the cavitation number affected the relationship between the rotating frequency of the cavitation and the frequency of the wall pressure fluctuation at the inducer inlet, the cause of the pressure fluctuation was not made clear.

Concerning the low cycle oscillation, Figure 6.40 shows the feature of this instability. The cavities on the blades vary their length in unison as shown in Figure 6.40 (a). The corresponding frequency was about 10 Hz and it agreed well with the frequency of pressure fluctuations measured at

[6] $K = 2\Delta p/\rho w_i^2$, $\phi = v_{ai}/u_t$, $\psi = g\Delta H/u_t^2$
where v_{ai} = mean axial velocity at inducer inlet, ΔH = inducer static head, Δp = difference between inducer inlet static pressure and vapor pressure, u_t = inducer tip speed at inlet and w_i = relative velocity at inducer inlet.

Figure 6.37 An inducer with three helical blades (Courtesy of National Aerospace Laboratory)

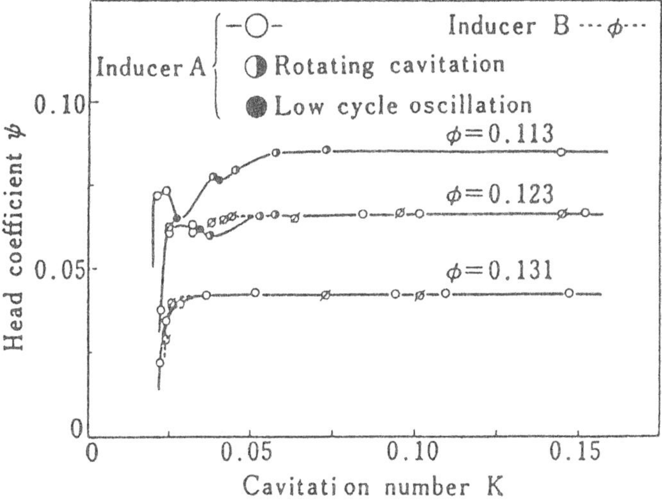

Figure 6.38 Suction performance of inducers

both inlet and exit of inducer. It is suspected from the waveforms shown in Figure 6.40 (b) that the oscillatory flow occurred in both pipes upstream and downstream of the inducer. Thus, the low cycle oscillation is regarded as a cavitation-induced system oscillation.

Case of a Centrifugal Pump

The suction performance of a centrifugal pump is improvable to some extent, if the number of blades is decreased to reduce the blockage effect due to cavities. However, the instability may occur in such a pump, when it operates at partial capacities under cavitation conditions. For instance, oscillograms in Figure 6.41 show two kinds of instabilities observed in a cavitating centrifugal pump (Yamamoto, 1980 [6.55]). ψ_1 and ψ_2 represent time histories of dimensionless pressures measured at suction and delivery pipes, and ϕ_1 and ϕ_2 are corresponding flow coefficients. Figure 6.41 (a) corresponds to the rotating cavitation. In this case, the frequency of suction pressure fluctuation is estimated to be 62.5 Hz and the rotational speed of the impeller corresponds to 50 Hz. The high speed motion films taken at this condition revealed that the cavitation pattern at the inlet had rotated at a speed of 1.25 times the impeller velocity. Thus, it is recognized that the suction pressure fluctuated with the frequency of the rotating cavitation.

In the case of Figure 6.41 (b), which corresponds to the condition of lower cavitation number, it is observed that both pressure and flow in the suction pipe fluctuated violently in unison. The high speed motion films revealed that the size of the cavitating zone in the reversed flow in the suction nozzle varied with respect to the oscillation of pressure and interaction occurred between the reverse flow and the incoming flow at the impeller inlet. The frequency in the latter case depends on the suction pipe length, as shown in Figure 6.42, while the length has no effect on the frequency of rotating cavitation.

Consideration of Rotating Cavitation

If the hydrodynamic radial force is produced by the distortion of the flow pattern due to the rotating cavitation, cavitating turbopumps may suffer undesirable shaft deflections and bearing loads (Rosenmann, 1965 [6.56]).

Figure 6.43 shows the sketches of cavitation zone on the inducer blades at the rotating cavitation drawn by Kamijo et al., 1977 [6.54], considering the results of Figure 6.39 (a). In (a), a large size cavity appears on suction side of No.1 blade and tip cavities are observed on pressure sides of the other two blades. Such positive and negative angles of incidence at this condition indicate that there is a severe non-uniformity in the inlet flow with preswirl. Sketches in (b) and (c) illustrate that the suction side cavity propagates toward the rotational direction relative to the blade row, resulting in an

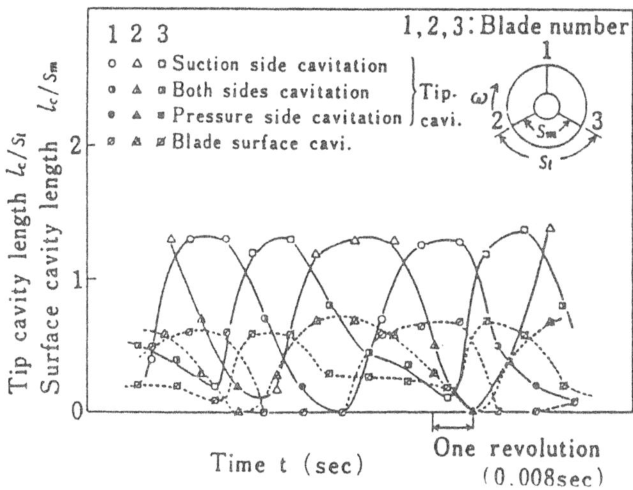

(a) Variation of cavity length with time
(s_t, s_m : blade spacing at tip and at middle of blade height)

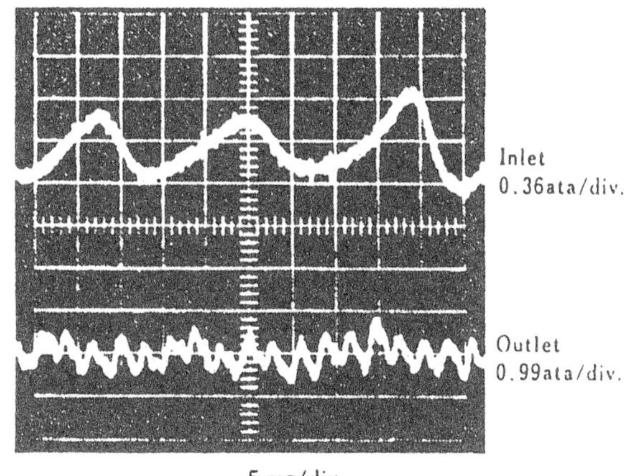

(b) Waveforms of pressure fluctuations
($n = 7,500$ rpm, $K = 0.045$, $\phi = 0.094$)

Figure 6.39 Instability due to rotating cavitation

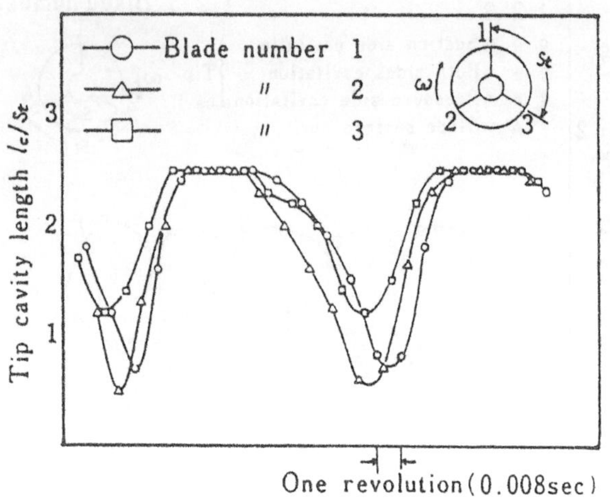

(a) Variation of cavity length with time

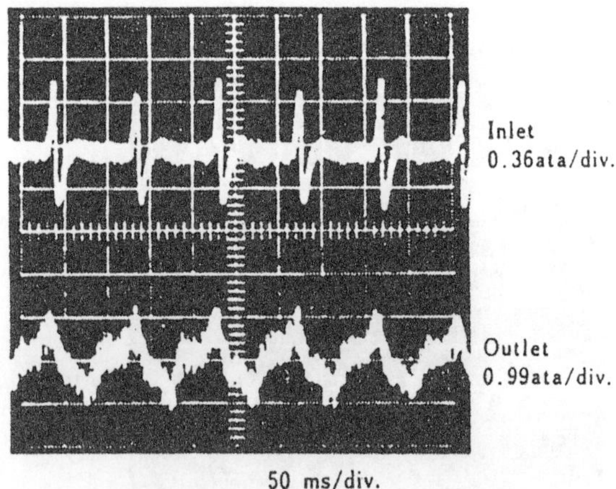

(b) Waveforms of pressure fluctuations
($n = 7,500$ rpm, $K = 0.026$, $\phi = 0.104$)

Figure 6.40 Instability due to low cycle oscillation

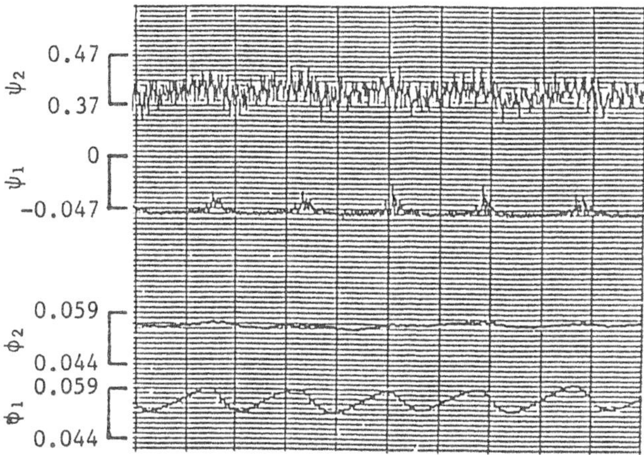

(a) In the case of rotating cavitation ($K = 0.21$)

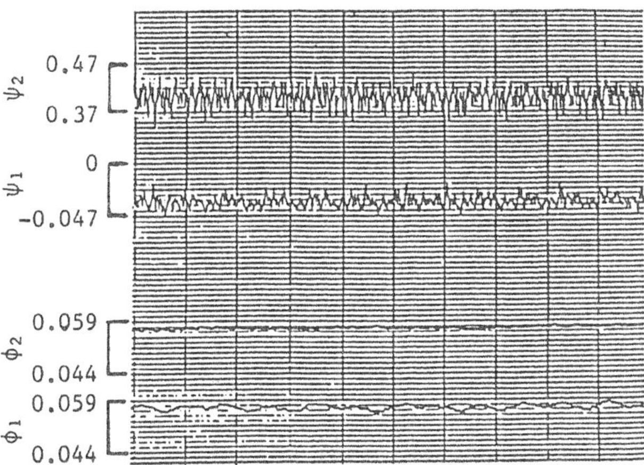

(b) In the case of low cycle oscillation ($K = 0.10$)

Figure 6.41 Oscillograms for a centrifugal pump

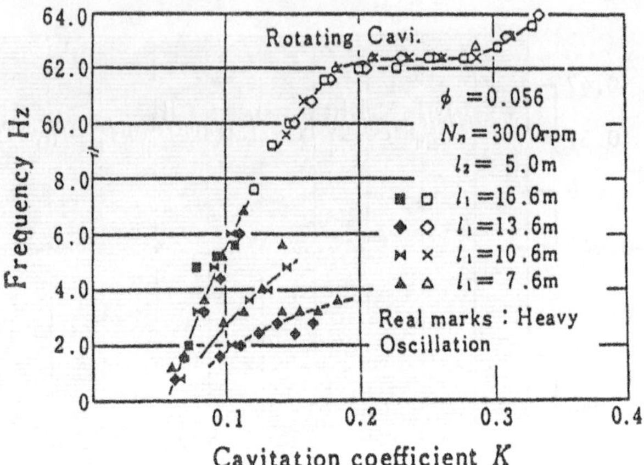

Figure 6.42 Effect of suction pipe length on the frequency

absolute propagation faster than the blade speed.

The mechanism of this phenomenon is not clear at this stage. In usual rotating stall at partial discharge, the stall zone on suction surface of a blade propagates in the opposite direction to the blade rotation, thus resulting in slower propagation than the blade speed. The propagation of cavitating zone is, however, rather similar to the rotating stall caused by pressure surface flow separation reported by Madhaven et al., 1985 [6.57]. A qualitative interpretation of the rotating cavitation could be given as follows: when once a large negative incidence causes stall on the pressure surface of No.3 blade in (a), a flow blockage occurs in the passage between No.2 and No.3 blades, and the incoming flow is diverted to No.1 and No.2. Due to an increased angle of incidence on No.1, cavitation is observed on the suction surface. Since the negative angle of incidence on No.2 becomes larger, the blade now tends to stall, and the onset of the stall will suppress the suction side cavitation on No.1. Consequently, stall on the pressure surface will propagate forward together with the suction-side cavitation as shown in (b) and (c).

Description of System Oscillation

As described in the previous section, the instability which induces a severe fluctuation in mass flow through a cavitating turbopump is known as the low cycle system oscillation. In order to avoid this cavitation surging and to secure its smooth operation at cavitating conditions, it is necessary to

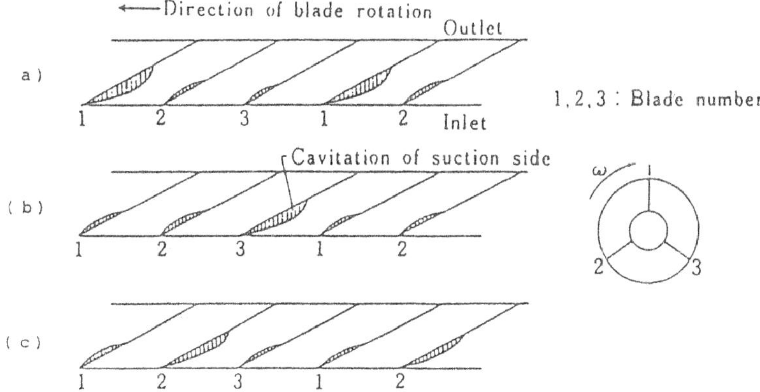

Figure 6.43 Sketches of cavitation zone on inducer blades at the rotating cavitation

examine the stability of the pumping system in advance. For this purpose, using a simple example shown in Figure 6.44, a method to analyze the system stability is presented in this section.

If it is assumed that cavities are only observed within a pump and one-dimensional quasi-static analysis is applicable to each element, the system in Figure 6.44 is represented by the transmission matrices in Figure 6.45 which describe the relationship between small perturbations in heads and flow rates at inlet and outlet of those elements. These perturbations are represented by $\tilde{H}_i$ and $\tilde{Q}_i$ $(i = 0 \sim 3)$ respectively, where i denotes the station number given in the figure. The dynamic behaviour of a cavitating turbopump operating at constant rotational speed is expressed by the following transmission matrix (see Watanabe et al., 1978 [6.58]);

$$\left\{ \begin{array}{c} \tilde{H}_2 \\ \tilde{Q}_2 \end{array} \right\} = \left[\begin{array}{cc} T_{11} & T_{12} \\ T_{21} & T_{22} \end{array} \right] \left\{ \begin{array}{c} \tilde{H}_1 \\ \tilde{Q}_1 \end{array} \right\}$$

$$= \left[\begin{array}{cc} 1 + \tilde{\mu}_h - \tilde{C}_c s(\tilde{R}_h - \tilde{L}_h s) & (\tilde{R}_h - \tilde{L}_h s)(1 - \tilde{M}_b s) \\ -\tilde{C}_c s & 1 - \tilde{M}_b s \end{array} \right] \left\{ \begin{array}{c} \tilde{H}_1 \\ \tilde{Q}_1 \end{array} \right\} \quad (6.61)$$

where pump inertance:

$$\tilde{L}_h = \int_{pump} (1/g\,A)\,dL \quad (6.62)$$

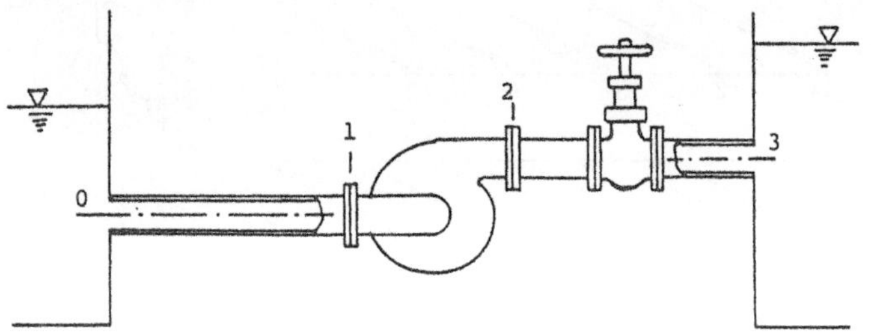

Figure 6.44 A pumping system for analysis
(0: upstream tank, 1: pump inlet, 2: pump outlet, 3: downstream tank)

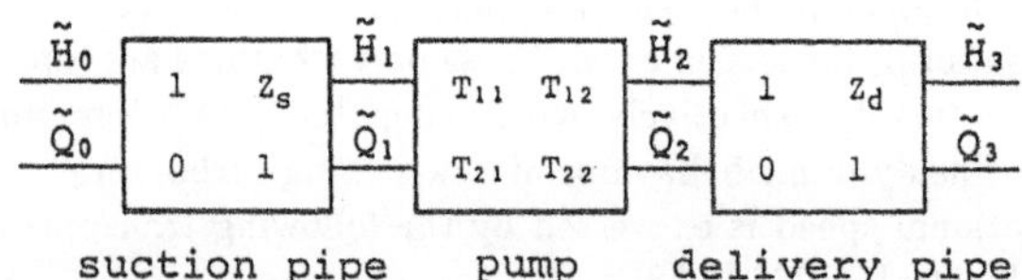

Figure 6.45 Transmission matrix representation

slope of $H-Q$ curve:

$$\tilde{R}_h = \partial H/\partial Q_2 \tag{6.63}$$

slope of cavitation performance curve:

$$\tilde{\mu}_h = \partial H/\partial H_1 \tag{6.64}$$

cavitation compliance:

$$\tilde{C}_c = -\partial V_c/\partial H_1 \tag{6.65}$$

mass flow gain factor:

$$\tilde{M}_b = -\partial V_c/\partial Q_1 \tag{6.66}$$

and
s : Laplace operator
A : cross-sectional area
g : acceleration due to gravity
H : pump head rise
H_1, H_2 : total heads at pump inlet and outlet
L : length
Q_1, Q_2 : flow rates at pump inlet and outlet
V_c : volume occupied by cavitation.

It is noted that parameters $\tilde{\mu}_h, \tilde{C}_c$ and $\tilde{M}_b$ are those directly linked with the presence of cavitation. The hydraulic impedance, inertance and resistance of water in each short pipeline can be written as

$$\text{pipe impedance:} \quad Z_i = -(\tilde{R}_i + \tilde{L}_i s)\ , \tag{6.67}$$

$$\text{pipe inertance:} \quad \tilde{L}_i = L_i/gA_i\ , \tag{6.68}$$

$$\text{pipe resistance:} \quad \tilde{R}_i = 2H_{ri}/Q_i\ , \tag{6.69}$$

where H_{ri} is head loss in the respective pipe.

The subscripts $i=s$ and $i=d$ denote the suction pipeline and the delivery pipeline respectively. Consequently, the matrix for the overall system is given by

$$\begin{Bmatrix} \tilde{H}_3 \\ \tilde{Q}_3 \end{Bmatrix} = \begin{bmatrix} T_{11}+Z_dT_{21} & Z_s(T_{11}+Z_dT_{21})+T_{12}+Z_dT_{22} \\ T_{21} & Z_sT_{21}+T_{22} \end{bmatrix} \begin{Bmatrix} \tilde{H}_0 \\ \tilde{Q}_0 \end{Bmatrix} . \tag{6.70}$$

Since both tanks are assumed to be very large, $\tilde{H}_0 = \tilde{H}_3 = 0$. Thus, we obtain the following characteristic equation;

$$Z_s(T_{11}+Z_dT_{21})+T_{12}+Z_dT_{22} = 0$$

or

$$F_0s^3 + F_1s^2 + F_2s + F_3 = 0\ , \tag{6.71}$$

where

$$\left.\begin{aligned}F_0 &= \tilde{C}_c\tilde{L}_s(\tilde{L}_h + \tilde{L}_d)\\ F_1 &= \tilde{C}_c[\tilde{R}_s(\tilde{L}_h + \tilde{L}_d) - \tilde{L}_s(\tilde{R}_h - \tilde{R}_d)] - \tilde{M}_b(\tilde{L}_h + \tilde{L}_d)\\ F_2 &= \tilde{L}_s(1 + \tilde{\mu}_h) + (\tilde{L}_h + \tilde{L}_d) - (\tilde{C}_c\tilde{R}_s - \tilde{M}_b)(\tilde{R}_h - \tilde{R}_d)\\ F_3 &= \tilde{R}_s(1 + \tilde{\mu}_h) - \tilde{R}_h + \tilde{R}_d\end{aligned}\right\} . \tag{6.72}$$

The Routh-Hurwitz criterion for stability condition is written as

$$F_1,\ F_2,\ F_3 \ \text{and}\ (F_1F_2 - F_0F_3) > 0\,, \tag{6.73}$$

where F_0 is a positive number. If those quantities of equation (6.72) at operating points of interest satisfy the condition (6.73), the pumping system is regarded as stable.

From the above relationship the qualitative effect of each parameter on the system stability can be observed: to increase the resistance of the pipeline contributes to the stabilization of the system. A steeper negative slope of head-capacity curve is desirable to stabilize the system. However, $\tilde{\mu}_h$ or the slope of cavitation performance curve should be positive. This remark corresponds to the result in Figure 6.38, which reveals that the low cycle oscillation occurred in the region where the cavitation performance curve had the negative sloped part. Equations (6.72) and (6.73) show that the mass flow gain factor $\tilde{M}_b$ affects the system stability most. If $\tilde{M}_b$ becomes large, the system will become unstable. Concerning the cavitation compliance $\tilde{C}_c$, it is the primary parameter to affect the resonance frequency which is derived from the imaginary part of the roots of equation (6.71), though an increase of the compliance will contribute slightly to stabilization of the system.

To determine $\tilde{C}_c$ and $\tilde{M}_b$ in the above method, it is necessary to predict the total volume of cavitation bubbles inside a pump as a function of the operating point. The blade cavitation and the additional cavitation due to backflow and tip vortex are its main sources, if the volume of dissolved or entrained gas in the cavitation region is negligible.

The analyses based on quasi-steady free-streamline cascade theory have been utilized as the first approximation to evaluate the fully developed and attached blade cavities, see Ghahremani et al., 1972 [6.59] and Brennen et al., 1976 [6.60]. It is noted that these theoretical results are lower than the experiments indicate. For the practical use, the following curve-fit equation (see [6.59]) is sometimes adopted to represent cavitation compliance data in the case of rocket pumps;

$$(\tilde{C}_c/A_1) \times 10^5 = -22.4 + 286(S/S^*) - 274(S/S^*)^2 + 92.6(S/S^*)^3 \tag{6.74}$$

where A_1 is pump inlet cross-sectional area and $S = n\sqrt{Q}/(\mathrm{NPSH})^{3/4}$ the suction specific speed. Subscript * denotes the condition at cavitation breakdown point.

Photographs of Figure 6.46 show the cavitation patterns at various operating conditions of an inducer. It is seen that large amounts of cavities extend upstream from the suction side of the blades at violent system oscillations. Thus, consideration of backflow cavitation is essential to estimate the cavitation compliance and the mass flow gain factor. Watanabe and Kawata studied a mathematical model of backflow cavitation, and they analyzed the stability of a cavitating inducer system using the method described above. Their comparison of calculated results with experiment is shown in Figure 6.47, where the abscissa is the ratio of flow coefficient ϕ to the tangent of blade tip angle β_t, and the ordinate is the net positive suction head (NPSH) for the oscillation inception.

Further Consideration

More precise information of the dynamic performance of a cavitating pump is required to achieve more quantitative prediction of system stability. Since the transmission matrix in equation (6.61) is useful to describe the dynamic performance, attempts have been made to determine the elements as function of frequency and mean operating condition theoretically (Brennen, 1978 [6.61], Nishiyama et al. 1984[6.62]) and experimentally (Ng et al., 1978 [6.19], Yamamoto, 1986 [6.63]). The dynamic characteristic of a cavitating rocket pump has been measured using cryogenic fluid by Shimura et al., 1985 [6.64]. The results are used to examine the possibility of the POGO instability, which is a hazardous longitudinal structural vibration of a liquid propelled rocket caused by dynamic interaction of structure with the propulsion system.

Few contributions are found concerning the flow patterns inside a turbopump at cavitating condition where the instability occurs, due to the difficulties in measuring such unsteady two-phase flow accurately. One good example of velocity distributions at the inlet of an inducer suffering low cycle oscillation is shown in Figure 6.48, where the pressure-probe measurements were accomplished using the phase-locked ensemble averaging technique. In the figure, dimensionless axial velocity ϕ_1 and dimensionless peripheral velocity λ_1 have been plotted against dimensionless radial distance r for various values of dimensionless time t. It is seen that the strong backflow zone is limited to the tip region at all times during one oscillation cycle, while alternate occurrence of the normal flow and the backflow was observed at the impeller inlet during the low cycle oscillation, see Kitamura et al., 1978 [6.65]. Further studies are really needed to clarify the unsteady behaviour of such flows.

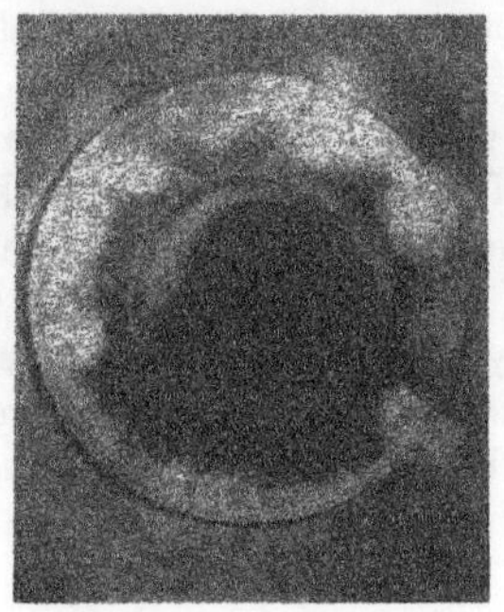

(a) $g\,(\mathrm{NPSH})/{u_t}^2 = 0.0791$

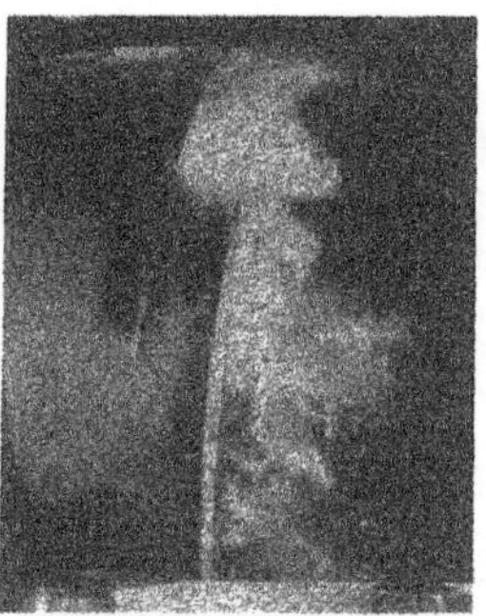

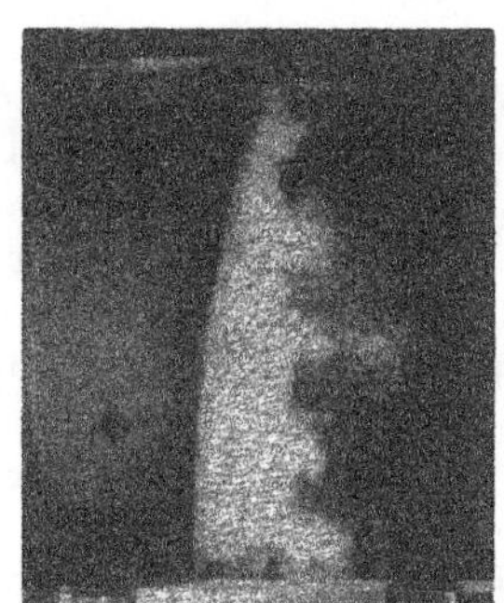

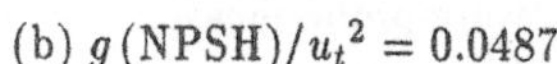

(b) $g\,(\mathrm{NPSH})/{u_t}^2 = 0.0487$

(c) $g\,(\mathrm{NPSH})/{u_t}^2 = 0.0278$

Figure 6.46 Cavities observed in a helical inducer at $Q/D^2u_t = 0.0425$ (the direction of flow from right to left) (Courtesy of Mitsubishi Heavy Industries, Ltd.)

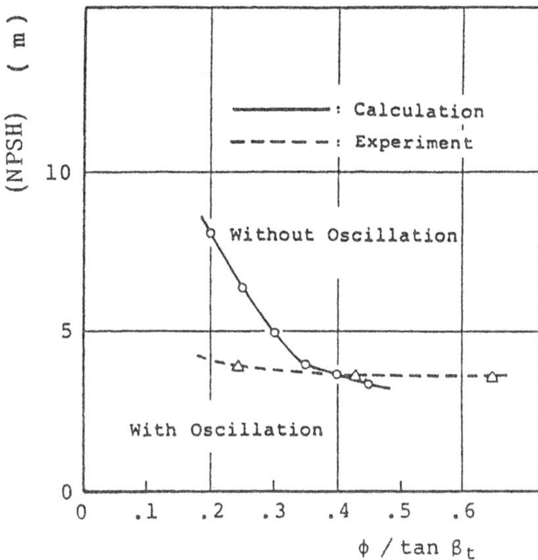

Figure 6.47 Comparison of prediction with experiment (ϕ= flow coefficient, β_t=blade angle at the tip, NPSH=net positive suction head)

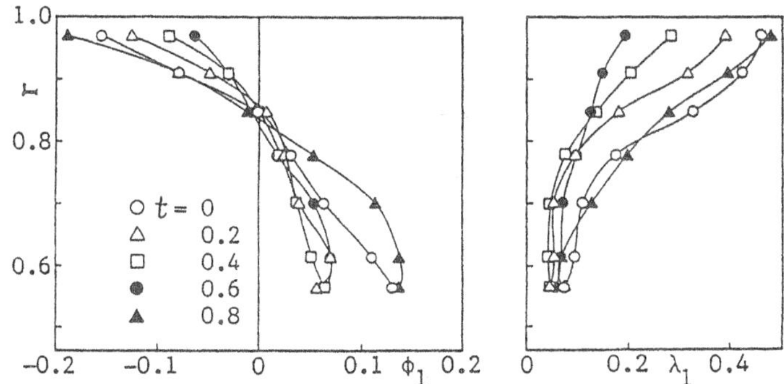

Figure 6.48 Velocity distributions at the inducer inlet (r: dimensionless radial distance, $r = 1$: the radius, t: dimensionless time, $t = 1$: the period, ϕ_1: dimensionless axial velocity at inlet $=v_{a1}/u_t$, λ_1: dimensionless peripheral velocity at inlet $= v_{u1}/u_t$) (Courtesy of Kyushu University)

6.4.2 Two-Phase Flow Instability

Two-Phase Flow Pump Performance

Figure 6.49 shows two-phase flow pump performance data on the model of a PWR primary coolant pump, obtained by changing void fraction at the pump inlet (Rothe et al., 1978 [6.66]). In the figure, the head coefficient ψ is plotted against the flow coefficient ϕ for various values of inlet void fraction α (approximately represented by gas volumetric flow divided by total volumetric flow at pump inlet). The rapid degradation in head coefficient is observed at low flow coefficients with increasing void fraction and the performance curves for inlet void fractions above 0.06 have a positive slope, although the pump has a stable negative-slope characteristic for a single-phase flow. Since this positively sloped characteristic is associated with surge (see Greitzer, 1981 [6.12]), serious flow oscillations may occur in a two-phase flow pumping system. It is noted that the marked change in the pump characteristic is qualitatively explainable if the drag of gas bubbles due to the slip velocity between gas and liquid in the pump is considered (Patel et al., 1978 [6.67]).

Air Injection Instability

A large amplitude oscillation may be observed in a pumping system when air in the atmosphere is naturally sucked into the pipe upstream of the pump through cracks or holes in the wall caused by some accident. The instability of this type was studied by Kaneko and Ohashi, 1982 [6.68], using a scale model of a BWR recirculation pump. The model pump test loop is illustrated in Figure 6.50, where an air supply device is mounted on the suction pipe to generate bubbly air-water two-phase flow.

Figure 6.51 shows time histories of dimensionless quantities $\beta_1, \beta_s, \phi_\ell$ and ψ_ℓ during oscillation. β_1 and β_s denote void fractions at the air supply point and at the pump suction inlet. The definitions of flow coefficient ϕ_ℓ and head coefficient ψ_ℓ are;

$$\phi_\ell = Q_\ell/\pi D_2 B_2 u_2\ , \quad \psi_\ell = 2gH_\ell/u_2^2\ ,$$

where u_2 is the tip speed of the impeller with tip diameter D_2 and exit width B_2. Q_ℓ is the liquid flow rate and H_ℓ is the pump head rise. Figure 6.51 demonstrates the following flow physics.

When large air voids reach the impeller inlet, the pump head is drastically reduced. It forces the flow to decelerate and then to increase the pressure level at the pump suction due to reduced head loss, resulting in no air supply to the suction pipe. After the flow in the pipe becomes single-phase, the pump gradually recovers its head rise and flow rate, and the suction pressure starts to decrease again. This gives rise to an increased air supply into the

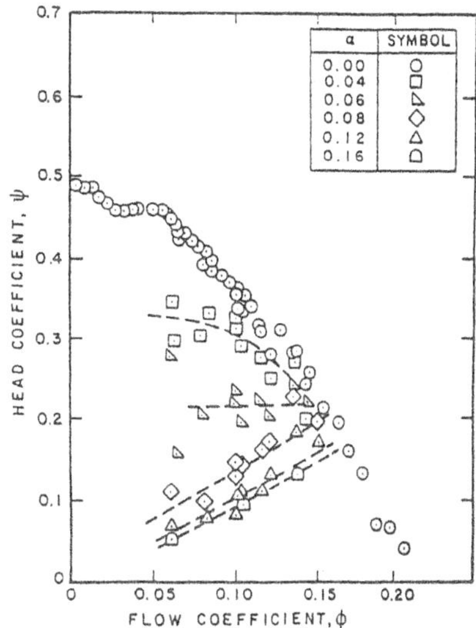

Figure 6.49 Pump performance for air-water two-phase condition

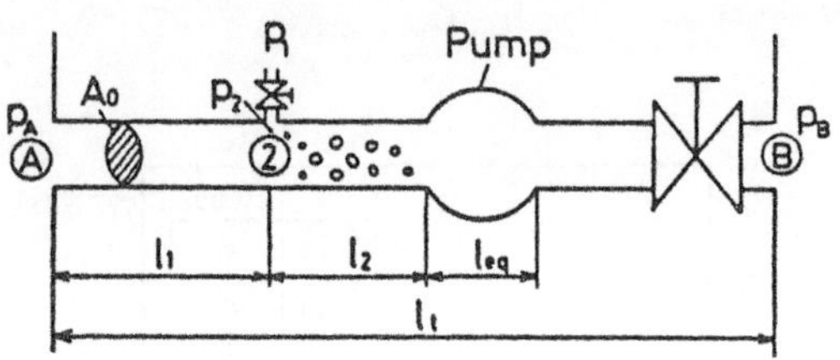

Figure 6.50 Schematic of the model pumping system

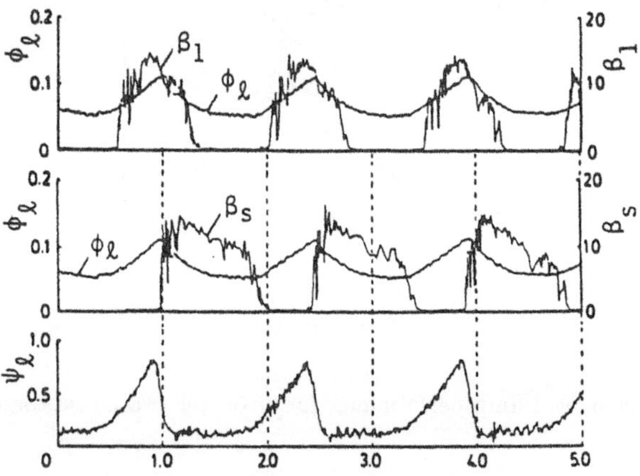

Figure 6.51 Time histories of $\beta_1, \beta_s, \phi_\ell$ and ψ_ℓ

suction pipe. Thus, the cause of this self-excited oscillation is attributed to the delay time required for the transport of air bubbles from the supply point to the impeller inlet. Experiments also reveal that the period of oscillation is nearly equal to twice the delay time. Further, it is clarified that a simple linear analysis is useful to predict the stability limit of the system shown in Figure 6.50.

References

6.1 **Anderson, D.A. et al.(1971). Response of a radial-bladed centrifugal pump to sinusoidal disturbances for non cavitating flow, NASA-TN-6556.**

6.2 Arshenevskii, N.N. et al. (1981). Stable operation of stations with axial-flow pumps, Gidrotekhnicheskoe stroitel'stvo, No. 11, pp. 25-27.

6.3 Black, H.F., et al. (1975). 'Stability of oscillations in boiler feed pump pipe-line systems', Proc. of the Conf. on Vibrations and Noise in Pump, Fan and Compressor Installation, Inst. of Mech. Eng., pp. 99-106.

6.4 Borel, L. (1960). Stabilité de réglage des installations hydroelectriques, Dunod, Paris, pp. 1-237.

6.5 Brown, F.T. et al. (1969). Small-amplitude frequency behaviour of fluid lines with turbulent flow, Trans. ASME, December, pp. 578-593.

6.6 Carolus, T. (1985). 'Experimental investigation of the transient behaviour of centrifugal fans', 8th Conf. on Fluid Flow Machinery, Budapest, September, pp. 1-8.

6.7 Chaudry, H. (1970). 'Resonance in pressurized piping systems', *ASCE Journal of the Hydraulics Division*, HY9, pp. 1819-1837.

6.8 Fanelli, M. (1972). 'Further considerations on the dynamic behaviour of hydraulic turbo-machinery', *Water Power*, June, pp. 208-222.

6.9 Fanelli, M. (1972). 'Current studies on instationary behaviour of hydraulic machinery'. IAHR Symposium, Rome, September.

6.10 Fanelli, M. et al. (1980). Réponse d'une turbomachine hydraulique a des fluctuations des paramètres dynamiques du circuit, La Houille Blanche, Na. 1/2, pp. 115-121.

6.11 Fanelli, M. et al. (1982). Determinazione sperimentale della impedenza idraulica di una pompa in regime pulsante, ENEL-DSR-CRIS, pp. 1-27.

6.12 Greitzer, E.M. (1981). 'The stability of pumping systems', *ASME Journal of Fluids Engineering*, Vol. 103, June, pp. 193-242.

6.13 Jacob T. et al. (1988). 'Matrices de transfert d'une pump-turbine', *XX Journées de L'Hydraulique de la Soc. Hyd. Francaise*, Lausanne, September, pp. 1-8.

6.14 Katto, Y. (1960). 'Some fundamental natures of resonant surge', *Bulletin of JSME*, Vol. 3, No. 12, pp. 484-495.

6.15 Kawata, Y. et al. (1985). 'The dynamic behaviour of centrifugal pump and its effect on the system instability', IAHR, Work Group on the Behaviour of Hydraulic Machinery Under Steady Oscillatory Conditions, 2nd Meeting, Mexico.

6.16 Kawata, Y. et al. (1987). 'Experimental research on the measurements of the dynamic behaviour of multistage centrifugal pump', IAHR, Work Group on the Behaviour of Hydraulic Machinery Under Steady Oscillatory Conditions 3rd Meeting, Lille.

6.17 Knapp., R. (1937). Complete characteristics of centrifugal pumps and their use in the prediction of transient behaviour, Trans. ASME, November, pp. 683-689.

6.18 Ng, S.L. (1976). 'Dynamic response of cavitating turbomachines', Ph. D. thesis, California Institute of Technology, Pasadena, pp. 1-190.

6.19 Ng, S.L. et al. (1978). 'Experiments on the dynamic behaviour of cavitating pumps', Trans. ASME, *J. Fluids Eng.*, Vol. 100, pp. 166-176.

6.20 Ogata, K. (1970). *Modern control engineering*, Prentice-Hall Inc.

6.21 Ohashi, H. (1968). 'Analytical and experimental study on dynamic characteristics on turbopumps', NASA-TN-D-4298.

6.22 Pfleiderer, C. (1955). *Die Kreiselpumpen für Flüssigkeiten und Gase*, 4 ed., Springer-Verlag.

6.23 Sedille, M. (1967). *Turbo-machines hydrauliques et tecniques*, Masson et Cie Editeurs.

6.24 Simpson, H.C. et al. (1974). 'Pulsations in power station feed pump systems', *Conference Publication 4*, Inst. of Mech. Eng., pp. 32-40.

6.25 Stepanoff, A.J. (1957). *Centrifugal and axial flow pumps*, 2nd ed., John Wiley and Sons, New York.

6.26 Stirnemann, A. et al. (1985). 'Experimental determination of the dynamic transfer matrix for a pump', ASME, Winter Annual Meeting, Miami Beach, November, pp. 1-8.

6.27 Stirnemann, A. (1987). 'Using the transfer matrix notation to describe the stability of hydraulic systems', IAHR, Work Group on the Behaviour of Hydraulic Machinery Under Steady Oscillatory Conditions, 3rd Meeting, Lille, pp. 4/1-4/11.

6.28 Yamabe, M. (1971). 'Hysteresis characteristics of Francis pump-turbines when operated as turbine', Trans. ASME, March, pp. 80-84.

6.29 Levy, H. and Forsdyke, A.G. (1928). 'The steady motion and stability of a helical vortex', Proc. of the Royal Society of London, Ser. A, Vol CXX, October, pp. 689.

6.30 Lamb, H. (1974). *Hydrodynamics*, Cambridge University, pp. 203-204.

6.31 Hosoi, Y. (1978). 'Contributions to model tests of draft tube surges of Francis turbines', IAHR Symposium, Fort Collins, Vol. 1. pp 141-150.

6.32 Dörfler, P. (1982). 'Pulsations in the draft tube as excited by the vortex rope in systems including Francis turbines working at part load' (in German), Thesis, Technical University of Vienna, pp. 3-183.

6.33 Nishi, M., Matsunaga, S., Kubota, T. and Senoo, Y. (1982). 'Flow regimes in an elbow-type draft tube', IAHR Symposium, Amsterdam, Vol.II, paper 38.

6.34 Guarga, R., Torres, J.J., Solorio, A. and Rodal, E. (1986). 'Comparative study of La Angostura draft tube vortex oscillatory behaviour', 13rd IAHR Symposium, Montreal, Vol. I, paper 6.

6.35 Fritsch, A. and Maria, D. (1987). 'Dynamic behaviour of a Francis turbine at partial load', 3rd Round Table of the IAHR WG on Behaviour of Hydraulic Machinery under Steady Oscillatory Conditions, September, Lille, pp. 10.1-10.14.

6.36 Fanelli, M. (1989). 'The vortex rope in the draft tube of Francis turbines operating at partial load: a proposal for a mathematical model', *Journal of Hydraulic Research*, Vol. 17, No. 6, pp. 769-807.

6.37 Zanetti, V. (1982). 'La pousse radiale dans les machines hydrauliques: experiences de laboratoire', *La Houille Blanche*, No. 2/3, pp. 237-241. and Henry, P., Zanetti, V. and Wegner, M. (1982). 'Charactristiques des pousses axiales', Ibidem, pp. 227-234.

6.38 Margi, L. (1987). *Problematiche delle turbomacchina idrauliche per produzione di energia elettrica*, Pitagora Editrice Bologna, pp. 177-211.

6.39 Rheingans, W.J. (1940). 'Power swings in hydroelectric power plants', Trans. ASME, April, pp. 174.

6.40 Falvey, H.T. (1971). Draft tube surges: A review of present knowledge and annotated bibliography, Report No. REC-ERC-71- 42, U.S.B.O.R., Denver.

6.41 Shogenji, K. and Shimoyama, Y. (1933). 'On the flow of water through the draft tube of a water turbine', *Journal of the Faculty of Engineering*, Kyushu Imperial University, pp. 145-183.

6.42 Falvey, H.T. and Cassidy, J.J. (1970). 'Frequency and amplitude of pressure surges generated by swirling flow', Trans. IAHR Symposium, Stockholm, Part 1, E-1.

6.43 Ulith, P., Jaeger, E.U. and Strscheletzky, M. (1974). 'Contribution to clarifying the inception of nonstationary flow phenomena in the draft tube of high specific speed Francis turbines operating at part load', IAHR Symposium, Vienna, III 4.

6.44 Nishi, M., Kubota, T., Matsunaga, S. and Senoo, Y. (1980). 'Study on swirl flow and surge in an elbow type draft tube', Proc. IAHR 10th Symposium, Tokyo, Vol. 1, pp. 557-568.

6.45 Kubota, T. and Yamada, S. (1982). 'Effect of cone angle at draft tube inlet on hydraulic characteristic of Francis turbine', Proc. IAHR 11th Symposium, Amsterdam, Vol. 2, Paper 53.

6.46 Dörfler, P. (1980). 'Mathematical model of the pulsations in Francis turbines caused by the vortex core at part load', *Escher Wyss News*, 1/2, pp. 101-106.

6.47 Senoo, Y. and Nakase, Y. (1971). 'A blade theory of an impeller with an arbitrary surface of revolution', *ASME Journal of Engineering for Power*, Vol. 97, No. 4, pp. 454-460.

6.48 Senoo, Y. and Nakase, Y. (1972). 'An analysis of flow through a mixed flow impeller', *ASME Journal of Engineering for Power*, Vol. 94, No. 1, pp. 43-50.

6.49 Fisher Jr., R.K., Palde, U. and Ulith, P. (1980). 'Comparison of draft tube surging of homologous scale models and prototype Francis turbines', Proc. IAHR 10th Symposium, Tokyo, Vol. 1, pp. 541-556.

6.50 Nishi, M., Matsunaga, S., Kubota, T. and Senoo, Y. (1986). 'Effect of draft tube shape on the characteristics of pressure surge and swirl flow', Proc. IAHR 13th Symposium, Montreal, Vol. 1, Paper 7.

6.51 Nishi, M., Matsunaga, S., Okamoto, M., Uno, M. and Nishitani, K. (1988). 'Measurement of three-dimensional periodic flow in a conical draft tube at surging condition, Flows in Non- Rotating Turbomachinery Components', *ASME FED*, Vol. 69, p. 81.

6.52 Casacci, S., Wegner, H., Henry, P. and Graeser, J. (1982). 'Examen experimental de la stabilite des turbines Francis sur modele et sur prototype a charge partielle', Proc. IAHR 11th Symposium, Amsterdam, Vol. 2, Paper 41.

6.53 Grein, H. (1980). 'Vibration phenomena in Francis turbines: their causes and prevention', Proc. IAHR 10th Symposium, Tokyo, Vol. 1, pp. 527-539.

6.54 Kamijo, K., Shimura, T. and Watanabe, M. (1977). An experimental investigation of cavitating inducer instability, ASME Publication, 77-WA/FE-14.

6.55 Yamamoto, K. (1980). 'An experimental study on instability in a cavitating centrifugal pump with a volute suction nozzle', Proc. IAHR 10th Symposium, Tokyo, Vol. 1, pp. 303-312.

6.56 Rosenmann, W. (1965). 'Experimental investigations of hydrodynamically induced shaft force with a three bladed inducer', ASME Symposium on Cavitation in Fluid Machinery, pp. 172-195.

6.57 Madhaven, S. and Wright, T. (1985). 'Rotating stall caused by pressure surface flow separation on centrifugal fan blades', *ASME Journal of Engineering for Gas Turbine and Power*, Vol. 107, pp. 775-781.

6.58 Watanabe, T. and Kawata, Y. (1978). 'Research on the oscillation in cavitating inducer', Proc. Joint Symposium on Design and Operation of Fluid Machinery, Fort Collins, Vol. 2, pp. 265-277.

6.59 Ghahremani, F.G. and Rubin, S. (1972). Empirical evaluation of pump inlet compliance - Final report, NASA CR 123963.

6.60 Brennen, C. and Acosta, A.J. (1976). 'The dynamic transfer function for a cavitating inducer', *ASME Journal of Fluids Engineering*, Vol. 98, No. 2, pp. 182-191.

6.61 Brennen, C. (1978). 'Bubbly flow model for the dynamic characteristics of cavitating pumps', *Journal of Fluid Mechanics*, Vol. 89, No. 2, pp. 223-240.

6.62 Nishiyama, T. and Nishiyama, H. (1984). 'Dynamic responses of partially cavitating hydrofoil cascade to axial gust in bubbly water', *ASME Journal of Fluids Engineering*, Vol. 106, No. 3, pp. 312-318.

6.63 Yamamoto, K. (1986). 'Experimental study on the dynamic behavior of cavitating centrifugal pump', Proc. IAHR 13th Symposium, Montreal, Vol. 1, Paper 19.

6.64 Shimura, T. and Kamijo, K. (1985). 'Dynamic response of the LE-5 rocket engine liquid oxygen pump', *Journal of Spacecraft and Rockets*, Vol. 22, No. 2, p. 195.

6.65 Kitamura, N. and Kubota, N. (1978). 'Cavitation performance of tandem impeller at partial capacities', Proc. Joint Symposium on Design and Operation of Fluid Machinery, Fort Collins, Vol. 2, pp. 621-630.

6.66 Rothe, P.H., Runstadler, P.W., Jr. and Dolan, F.X. (1978). 'Pump surge due to two-phase flow', Proc. Symposium on Polyphase Flow in Turbomachinery, ASME, pp. 121-137.

6.67 Patel, B.R. and Runstadler Jr., P.W. (1978). 'Investigations into the two-phase flow behavior of centrifugal pumps', Proc. Symposium on Polyphase Flow in Turbomachinery, ASME, pp. 79-100.

6.68 Kaneko, M. and Ohashi, H. (1982). 'Self- excited oscillation of a centrifugal pump system under air/water two-phase flow condition', Proc. IAHR 11th Symposium, Amsterdam, Vol. 2, Paper 36.

Chapter 7

Noise of Hydraulic Machinery

E. Egusquiza

7.1 Fundamentals of Noise

7.1.1 Introduction

Noise, the *unwanted sound*, is an unavoidable by- product of machines. It is the sensation caused in the human ear by pressure oscillation of surrounding air produced by vibration of a structure or by fluctuation in a fluid. In all processes there are always some unsteadinesses that are transmitted to the surfaces of the machines from which they radiate sound.

New trends in the construction of hydraulic machinery are towards units with smaller dimensions and weight, but with a larger power concentration. A better knowledge of the flow through the machine, has resulted in an increase in its speed. New materials and more accurate machine-tools have allowed the reduction of size and weight of pumps and turbines. Unfortunately, the increase of the hydrodynamic load in the blades may enhance the unsteady flow and, consequently, vibration and noise.

The sequential process of noise is generation, transmission and reception. In a hydraulic installation, noise is generated especially in the hydraulic machine, as shown schematically in Figure 7.1. Fluid-borne noise generated inside the machine is transmitted to the piping system and transferred to the casing where it produces mechanical vibrations (structure-borne noise) which radiate air-borne noise.

Generation of fluid-borne noise can be caused by many types of hydrodynamic sources. The noise produced is generally a mixture of periodic noise and random noise. Periodic noise is produced by the rotor movement and random noise by turbulence and cavitation. Mechanical vibrations due to misalignment, imbalance, bearings, etc. are also produced.

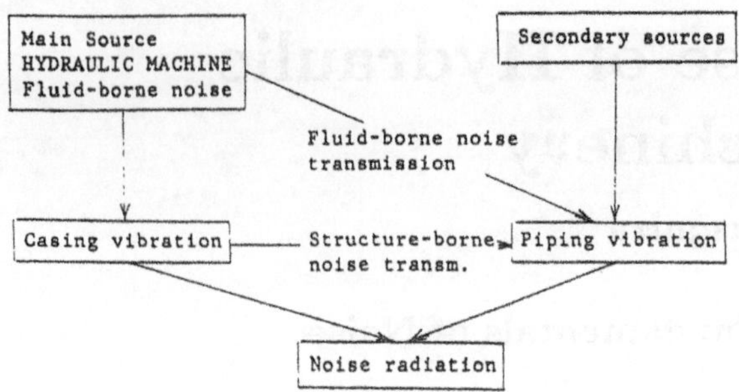

Figure 7.1 The noise chain in hydraulic systems

The generated noise is propagated in all possible directions through a wave motion pattern. When the amplitude of the waves is small, the problems are dealt with as acoustic phenomena. In that case, several simplifications in the fundamental equations can be made. To express the intensity of noise, special parameters that consider the subjective response of the human ear are used. These are evaluated on logarithmic scales related to a reference value of pressure and power.

The problems in acoustics can be approached mainly in three basic ways; wave acoustics, ray acoustics and energy acoustics. Wave acoustics treats the wave propagation in homogeneous media with simple boundaries. The three-dimensional approach is the general one used for complicated sources. For simple problems or in cases where propagation is mainly in one direction, a one-dimensional approach is enough. When there is a reverberant enclosure (diffuse field), wave acoustics are no longer useful and statistical methods (energy acoustics) are necessary. These are used for noise in rooms of any geometry. On the other hand, the propagation of sound over long distances suffers the consequences of the inhomogeneity of the medium and has to be approached in a different way (ray acoustics).

During the propagation, when a wave strikes a boundary, a part of the wave is transmitted and a part reflected, suffering some dissipation. The sound energy dissipation when striking on a surface (absorption) is an important part in the noise control methods.

7.1.2 Fluid-Borne Noise Generation

Noise is produced when a fluid flows over a body or inside a conduit. In both cases, some fluid dynamic processes may result in unsteady pressure fields. Fluctuating pressures are a source of noise by itself or because they excite structural vibrations which radiate noise. Intermittencies of large-scale, mainly due to vortex-like structures more or less periodic, and fine-scale fluctuations of random characteristics are important sources of fluid-borne noise.

There are various basic flow mechanisms leading to fluid excitation. Examples are; wake type flows, shear layer instabilities and cavitation.

Wake Type Flows

When a fluid passes over a body, a boundary layer and a wake are developed. For bluff bodies, the boundary layer separates, creating a zone with vortex and turbulence generation. This zone has an unsteady nature, leading to a fluctuation of pressure on the body. In general, all vortex- dominated flows have an unsteady nature. Travelling eddies produce unsteady pressures and later are destroyed by viscous stresses with a high amount of turbulence production.

Wake flows, especially behind bluff bodies, produce vortex shedding, generally with non-symmetric characteristics, leads to fluctuations of drag and lift forces. The typical case is the vortex shedding in cylindrical structures (Von Karman vortex street) which causes the singing of wires in a cross wind. These vortices may be considered impulsively generated by the periodic deceleration and acceleration of the fluid which accompanies the alternating circulation around the cylinder.

The Strouhal number S_t is usually used for scaling vortex shedding from bluff bodies

$$S_t = fD/v \tag{7.1}$$

where f is the vortex shedding frequency, D a characteristic dimension of the body and v the upstream fluid velocity. The frequency f, for a determined geometrical shape, depends on several parameters, mainly on the Reynolds number ($R_e = vD/\nu$, where ν is the kinematic viscosity of the fluid). Diagrams of Strouhal number versus Reynolds number, have been obtained from experimental research (see Blevins, 1977 [7.1], for example).

However, the most important wake-type flows involve wakes that are considerably more complicated than the wake of two-dimensional bodies. Three-dimensional effects due to finite aspect ratio, effects of confining walls, submergence and cavitation can modify considerably the two-dimensional results. Vortex shedding can also occur behind trailing edges of streamlined bodies like blades. Strong vibrations have been found in stay vanes and blades of turbines and propellers when this excitation matches one of the natural frequencies of the body (i.e. propeller singing).

Longitudinal vorticity may introduce fluctuations of pressure. Examples of longitudinal vorticity are the tip vortices in turbomachinery blades and horse-shoe vortices. In general, swirling flows may be a source of pressure fluctuations, especially if vortex breakdown occurs. In swirling flows of large angular momentum, the streamwise vortices develop into a spiral form which finally bursts into intense turbulence. In draft tubes of hydraulic turbines, this process is generally accompanied by cavitation and leads to pressure pulsations.

Shear layer roll-up is another mechanism of vortex formation. Shear layers are inherently unstable and generally involve a transfer of energy from the mean flow to flow disturbances. These disturbances are nearly periodic for a certain range of Reynolds numbers becoming random for higher values.

Other types of separated flows can produce severe intermittent fluctuations and are known as bi-stable flows. Characteristics of these flows are that there is a change between two stable states with a movement of the mean separating streamline. Examples are the transitory stall in diffusers and force fluctuations in protruding walls (Naudascher et al., 1974 [7.2]).

Boundary Layer

In the turbulent boundary layer there are always some fluctuations of pressure produced by semi-periodic eddy formation as shown by Corcos, 1964 [7.3]. Low frequency components associated with large eddies are convected by the mean flow at a speed of $0.6 \sim 0.85$ times the flow velocity v. Eddy structures of smaller size are generated in the laminar buffer zone and are convected downstream with a smaller velocity. The vortices are converted into fine-scale turbulence which generates noise of high frequency. The noise generated has a broad band spectrum and produces wall pressure fluctuations. The wall pressure spectrum peaks at a Strouhal number of about 0.2 ($S_t = f\delta_m/v$), where f is the frequency and δ_m the momentum thickness of the boundary layer.

Cavitation

Cavitation occurs when the liquid pressure at some point drops to around its vapour limit. Then, vapour bubbles tend to form around any free-gas nuclei. When they are carried downstream to regions of higher pressure, they collapse in a very short time producing extremely high pressures and temperatures. The implosions produce a characteristic noise.

There are several regimes of cavitation that produce different pressure fluctuations. In the incipient stage, bubbles are very small and the fluid-borne noise is of high frequency. For well developed cavitation, in addition to the high frequency pressure fluctuations, there are fluctuations of lower frequency of the developed cavity. This may lead to force fluctuations in the boundaries close to the cavitation region, especially when the cavitation pocket is filled with a compressible liquid-vapour mixture.

Fluid Resonance

All perturbations can be amplified considerably either by fluid resonance or by structure resonance. In cavities, for instance, Heller and Bliss, 1975 [7.4], showed that the shear layer instability may be amplified by the resonance of the fluid inside the cavity. Problems of resonance in hydraulic power plants and methods of solution have been reported several times, for example Jaeger, 1977 [7.5], and Fanelli, 1975 [7.6].

Fluid oscillators that can lead to the amplification of an existent perturbation may be of resonator type (i.e. Helmholtz) or of standing wave type. Standing waves can be of one-dimensional character (long pipes), or two- and three-dimensional like standing waves in basins or boxes. In both cases the frequencies of resonance are the important parameters to know.

In a circuit, the wave modes that can appear depend on boundary conditions, wave propagation speed and dimensions. For simple cases, the frequency of resonance is given by

$$f_n = c/\lambda \tag{7.2}$$

where c is the propagation speed (see section 7.2.2) and λ is the wavelength. For a pipe of length L' (including end correction) it holds the relation that $\lambda = 2L'/n \quad n = 1, 2, \ldots n$ for close-close or open-open ends and $\lambda = 4L'/(2n-1)$ for close-open ends.

7.1.3 Structural Response

The unsteady pressure field produced by any of the mechanisms described above, generates fluid-borne noise and produces a force fluctuation on the structure. Once the force fluctuation is known, it is necessary to know the mechanical admittance of the structure in order to obtain the structural response. This function can be evaluated theoretically by several methods (finite elements, etc.) or experimentally by modal analysis. A characteristic mechanical admittance function is represented in Figure 7.2. At low frequencies stiffness dominates and excitation and response are nearly in phase. At the resonance condition (f_n) the response reaches the maximum amplitude, which depends on system damping. At higher frequencies the inertia dominates.

It has to be noted that the mechanical admittance is greatly affected in hydraulics by the added mass, added damping and added stiffness of the fluid surrounding the structure. These effects depend on flow conditions, vibration characteristics of the structure and fluid properties. So, the response of the structure in water flow may be considerably different from the one obtained in still air.

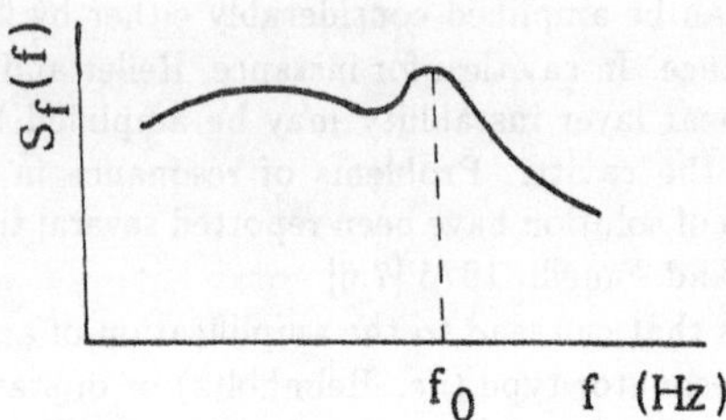

(a) excitation power spectrum $\cdots S_f(f)$

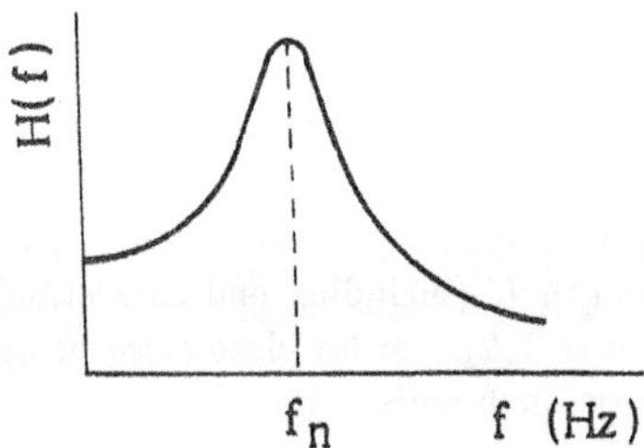

(b) mechanical admittance function $\cdots H(f)$

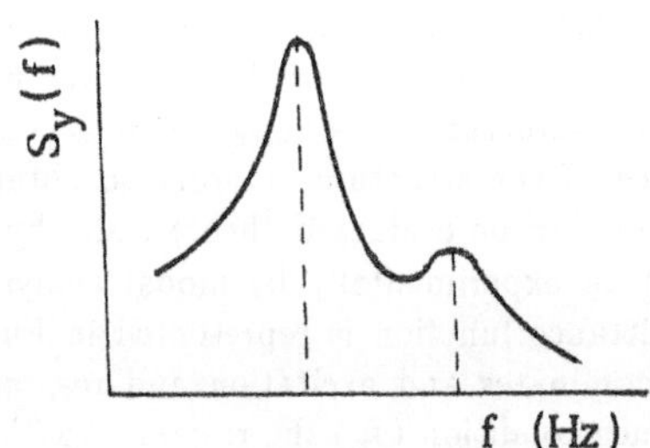

(c) structure-borne noise $\cdots S_y(f)$

Figure 7.2 Transference of fluid-borne noise to structure-borne noise

It is convenient for the study of the structural response, to classify flow induced vibrations into three categories (cf. Naudascher, 1982 [7.7]) depending on the characteristics of the excitation.

Extraneous Excitation

In this case vibration is produced by an exciting force which is independent of the vibrating structure and it is not affected by its motion (forced vibrations). The loading is more or less random and therefore, it is best described by a power spectrum. The power spectrum of a fluctuating variable (velocity, force, etc.) reveals the energy content of that variable in the frequency domain. If the force is strictly periodic, a discrete spectrum with merely a spike at a frequency would have been obtained. Knowing the load spectrum and the mechanical admittance, the response of the structure can be calculated (see Figure 7.2). Turbulence buffeting is a common case of forced vibration due to extraneous excitation.

Excessive response of the structure which occurs when its resonance frequency is within the range of the excitation spectrum, can be alleviated by several methods. Changes in the structural stiffness can move the resonance peak to avoid excessive responses at a determined frequency. An increase in damping is more effective, regarding mitigation of structural vibrations. The current trend towards more flexible and lightly damped structures results in a decrease of the damping ratio and a decrease in natural frequency. This makes the fluid structures more prone to flow induced vibrations.

Movement-Induced Excitation

The exciting force depends on body motion, that is, fluid forces depend on the structural displacement and the flow periodicity ceases if the structure is brought to rest. The effects of added mass, damping and rigidity can play an important role in this case. Negative added damping and negative added stiffness normally induce this kind of vibration. These are self-excited vibrations which may lead to catastrophic failures in a short time. Examples are airfoil flutter and fluid-elastic instability in tube bundles.

Flutter is a well known phenomenon in blades, when the excitation frequency matches one of the resonant frequencies of the structure leading to failure. Galloping is a problem found in some bluff bodies that can move (Dowell, 1989 [7.8]).

Instability-Induced Excitations

In this case, the flow excitation is related to the vibrating body. It is not induced by the body motion, although it can be greatly influenced by it. If the body is brought to rest, the flow periodicity does not disappear. Vortex-shedding excitations are in this group.

If the excitation frequency is separated from the resonance frequency of the structure, a case of forced vibration occurs. If there is some coincidence between them, then the structural vibration affects the fluid-excitation enhancing it. The lock-in phenomenon of cylinders is a typical result of coincidence (McCroskey, 1977 [7.9]). The movement of the body enhances the vortex strength and the vortex-shedding coherence along it.

7.2 Propagation and Radiation

7.2.1 Introduction

Any perturbation produced in a continuous medium is propagated in all possible directions. The propagation of the perturbations occurs at a high speed and the process is accompanied by dissipation due to viscous stresses. During the propagation there is a change in density, temperature and particle velocity of the fluid.

The speed of propagation depends basically on the medium's characteristics, disturbance size and even frequency in certain cases. The wave form changes especially for high disturbances due to the variation in the acoustic velocity and the wave front may become a shock wave. The speed of propagation should not be confused with the particle velocity, that is the velocity induced on the fluid particles during the passing of the wave.

The waves reflect when they strike on a solid boundary or when there is a change in the state or characteristics of the medium. The reflection can change the nature of the waves, for example at an open end of a duct.

It is possible to classify the waves in several ways. When their amplitude is important and the non-linear effects cannot be disregarded, these waves are called finite amplitude waves. When the amplitudes are small, non-linear effects can be neglected which implies an easier theoretical treatment.

Waves travel in all directions, so they are three- dimensional. However, in certain cases when they are restricted, they travel only in one direction (inside a pipe). In that case we have a one-dimensional wave and there is a great simplification of the problem. If only one wave form is travelling, we have a simple wave. With more waves travelling in different directions we have a compound system. Another possible classification is with regard to how they propagate. In that case, the waves can be classified as progressive or standing waves. If the wave is repeated continually, we have a periodic sound.

7.2.2 Acoustic Waves

Acoustic Wave Equation

The basic equations of fluid mechanics can be used to derive the linear acoustic wave equation (cf. Reynolds, 1981 [7.10]). It can be assumed that all points of the wave propagate with the same velocity and the wave profiles remain unchanged during the propagation. For a one-dimensional case:

$$\frac{d^2p'}{dt^2} - c^2\frac{d^2p'}{dx^2} = 0\,, \tag{7.3}$$

where $p'(x,t) = p(x,t) - p_0$ (p' is the fluctuating part of the pressure and p_0 the time mean and c the wave speed). Alternatively, the wave equation could have been constructed in terms of the associated fluid particle velocity v'.

Wave Speed

It can be easily demonstrated that for a plane wave of acoustic amplitude, the propagation speed c is given by

$$c^2 = \frac{dp}{d\rho} = \frac{B}{\rho}\,, \tag{7.4}$$

where B is the bulk modulus of the medium. As the medium becomes more incompressible c approaches infinity. c is greater in liquids than in gases due to the differences in compressibility of both media. For air under normal atmospheric conditions, $B = 1.42 \times 10^5$ Pa, for water $B = 2.13 \times 10^9$ Pa. To calculate the value of c for a liquid inside a conduit, the elasticity of the pipe has to be taken into account. Several expressions have been developed, see Jaeger, 1977 [7.5].

Acoustics Fields

When acoustic pressure p' and particle velocity v' are in phase, the region is called *far field*. When they are out of phase it is called *near field* and the sound pressure is influenced by the radiation characteristics of the sound source. Its extension depends on frequency of sound, radiation characteristics and dimensions of the sound source.

For a spherical source the near field extends to a distance of around two wavelengths. For an irregular and directional source it extends from 2 to 5 times the largest dimension of it. In the far field sound pressure is inversely proportional to the distance R away from the source. Doubling the distance implies halving the sound pressure. This is called the *inverse square law* (see 7.2.3 for sound pressure level L_p).

$$L_{p1} = L_{p2} + 20\ \log_{10}(R_2/R_1) \tag{7.5}$$

Another possible classification is between free and reverberant field. In a free field there are only direct radiated sound waves moving away from the sound source. There are no reflected sound waves. This condition is approximately materialized in large open spaces or in anechoic chambers, where the wall absorbs most of the incident waves. The reverberant field occurs in closed spaces where the incident waves are almost completely reflected by the walls, as in a reverberant chamber.

Waves of Finite Amplitude

In a finite amplitude wave, particle velocities and pressure changes may be considerably large and acoustic equations cease to apply. Each point of the wave profile travels at a speed that has been influenced by the preceding portion of the wave. For a one-dimensional flow of a gas, it can be demonstrated that the propagation and particle velocity are given by:

$$c = c_0 \left[6 \left(p_0/p \right)^{1/7} - 5 \right] \tag{7.6}$$

$$v' = 5\, c_0 \left[\left(p/p_0 \right)^{1/7} - 1 \right] \tag{7.7}$$

If p is larger than p_0, then there is a compression wave, c is larger than c_0 and v' is positive. If p is smaller than p_0, there is an expansion wave, c is smaller than c_0 and v' is negative.

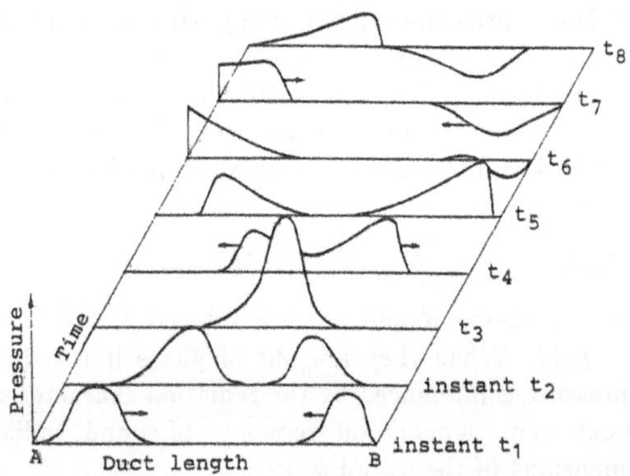

Figure 7.3 Process of propagation of two waves of finite amplitude in a duct (**A** is a closed end and **B** an open end. The steepening and reflection of the wave is clearly seen.)

Because of the different velocities of propagation in any wave, the propagation is accompanied by a change in wave form as shown in Figure 7.3. The steepening of the profile occurs on the leading face of a compression pulse and in the rear face of a expansion pulse. This process leads to the formation of a shock wave in a compression pulse. When a shock wave develops, the system can no longer be considered isentropic, and the effects of heat conduction and viscous dissipation become of dominant importance. The former equations are not valid for the description of the process.

7.2.3 Fundamental Parameters of Acoustics

The human ear can detect a broad band range (in amplitude and frequency) of pressure fluctuations. The smallest pressure fluctuation is of about 2×10^{-5} Pa and pressure amplitudes of more than 100 Pa cause pain and hearing damage. On the other hand, the audible frequency range for an average person with no hearing loss, is from around 20 Hz to 20 kHz. Sound with frequencies lower and higher of these limits are called infrasounds and ultrasounds.

The human ear has a non-linear response to pressure fluctuations and interprets a 10 times higher pressure fluctuation as if the noise were doubled. Moreover, between 2×10^{-5} and 2×10^{2} Pa there is a great difference and, therefore, a logarithmic scale is used for measuring the sound. Subjectiveness is another characteristic of the human ear, which is less sensitive to low frequency noise than to high frequency noise. To account for this sensitivity, weighting scales which are contours of equal loudness (compared with the one at the frequency of 1 kHz) are utilized, see Figure 7.4. "A" weighting corresponds most closely to the response of the human ear.

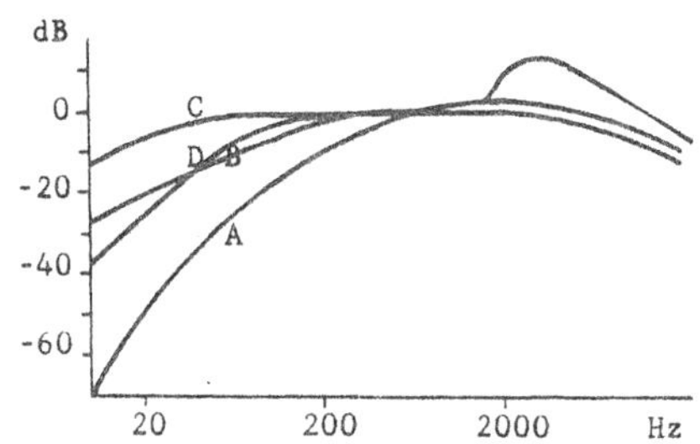

Figure 7.4 Weighting scales A, B, C and D ("A" weighting is normally used.)

Sound Pressure Level

The most common parameter used is the Sound Pressure Level (*SPL* or L_p)

defined by

$$L_p = 20\log_{10}(p'_{rms}/p_{ref}), \tag{7.8}$$

where p_{ref} is a reference pressure. In air, the reference value of 2×10^{-5} Pa is normally used, because it corresponds approximately to the threshold of human hearing at a frequency of 1 kHz. In water the value of p_{ref} is generally of 0.1 Pa. The measurement of L_p can be carried out with a microphone.

Intensity Level

Acoustic intensity I is defined as the time- average rate of sound energy on time interval T through a unit area;

$$I = \frac{1}{T}\int_0^T p'v'\,dt\,. \tag{7.9}$$

The intensity level is defined by

$$L_i = 10\log_{10}(I/I_{ref})\,. \tag{7.10}$$

where $I_{ref} = 10^{-12}$ W/m^2.

For free progressive plane and spherical waves, the following simple relations hold:

$$\begin{aligned} I &= \frac{p'^2}{\rho c}\,, \\ L_i &= L_p + 10\log_{10}\frac{p'^2_{ref}}{\rho c W_{ref}} = L_p - 0.2 \approx L_p \end{aligned}$$

Sound Power Level

In many cases, sound pressure or intensity may not give a true indication of the noise characteristics of the source. Then, the acoustic sound power W' which is independent of both distance and position around a noise source is obtained by integrating the intensity I over a control surface A enclosing the source:

$$W' = \int_A I\,dA \tag{7.11}$$

The Sound Power Level is defined by

$$L_w = 10\log_{10}(W'/W_{ref})\,, \tag{7.12}$$

where $W_{ref} = 10^{-12}$ W. The measurement of acoustic power can be carried out in several ways. One possibility is from L_p measurements using a fictive surface surrounding the source. The control surface is divided into several area segments, and for each area L_p is measured. Several standards (cf.

[7.11-13]) instruct the procedure. In the case of plural sources closely located, this method presents many difficulties. It can only be reliably applied in test beds in which the external sound sources are attenuated.

Intensimetry methods are based on the simultaneous measurement of the acoustic pressures in two points. This method allows the measurement of the noise radiated by a determined source preventing the effects of nearby sources. The normal component of the intensity vector is determined for all the area segments of the control surface.

Acoustic power level can also be calculated from vibration measurements. Vibration velocity and acoustic power can be related as follows:

$$W' = K_r \, \rho \, A \, c \, v_v^2 \tag{7.13}$$

where K_r is the radiation coefficient, ρ the air density, A the surface area and v_v the mean vibration velocity. The problem is to determine K_r, which depends on the frequency and on the structural dimensions. The following equation is suggested by VDI 3743 (cf. [7.14]) for the estimation of K_r of ducts;

$$K_r(f) = 1/[1 + (c/4Df)]^3 \tag{7.14}$$

where D is the pipe diameter and f the frequency.

7.2.4 Radiation Fundamentals

All sounds sources can be classified in accordance with its noise producing mechanisms. These mechanisms can be related to the various ways in which energy may be converted into acoustic energy and radiated away. These ways are: a fluctuation of mass, a fluctuation of momentum and a fluctuation of momentum flux. The first case can be associated with a *monopole*, the second with a *dipole* and the third with a *quadrupole.*

A monopole radiates sound waves that are only a function of radial distance R from the source. It is a point source of mass like the one in irrotational flow, but with a fluctuating mass and the rate of change of mass flux as its strength. Although they are omni-directional, directional radiation patterns can be produced with arrays of monopoles.

In practical terms sound is not generated at a singular point. The region of mass fluctuations may be simulated with a distribution of monopoles. When the scale of the distribution is much less than the wavelength, differences in the retarded times are small and the behaviour is similar to a single monopole. When the scale is greater, interference produces a sound field which has directional properties. As matter cannot be created in the fluid itself, real monopole type sources require a boundary. Examples of monopole type sources are pulsating bubbles (cavitation) and noise produced at the end of engine exhaust ducts.

Fluctuating forces produced by vibration at a boundary are associated with dipoles. A dipole may be envisaged by considering the existence of two monopoles equal and completely out of phase, separated by a distance smaller than the wavelength. If the wavelength is increased, the dipole will approach to a source-sink pair in an incompressible fluid, resulting in a substantial decrease in acoustic power. The dipole has directional properties, the maximum being in the direction of the axis. The dipole strength is defined by the force vector.

The combination of two equal but completely out of phase dipoles, separated by a distance of less than a wavelength constitutes an acoustic quadrupole. The quadrupole may be described in terms of a fluctuating force pair or stress, the stress defining the quadrupole strength. Quadrupoles do not need boundaries for their appearance. The fluctuating stresses in turbulent flow are the most important quadrupole sources of noise.

There are two distinct physical ways in which acoustic source terms can arise. One is a fluctuation of mass, force and stress in a position fixed in space. The other is an oscillation in position of non-fluctuating quantities. Both time and space changes produce acoustic disturbances. Of these two, oscillation in time is more important especially in water. Motion in space must occur at speeds comparable to the speed of sound to be effective.

Radiation Efficiency

Not all the energy is converted into sound. Actual sources create fluid dynamic as well as acoustic fields. For fluid sources, it has been demonstrated by Ross, 1987 [7.15], that radiation efficiency K_r is related to the Mach number M_a, with a dependence that is a function of the order of the multipole. Increasing the order of the multipole m results in an increase in M_a dependence;

$$K_r \sim kL^{(2m+1)} \sim M_a^n \, , \tag{7.15}$$

where k is the wave number $= \omega/v$ (a spatial frequency), L a source dimension, ω the angular frequency and v the flow speed. For a monopole $m=0$, $n=1$, for a dipole $m=1$, $n=3$ and for a quadrupole $m=2$, $n=5$. K_r differs between air and water. Since many sources in liquids have kL smaller than unity, the lower the multipole order the more efficient it is as an acoustic radiator.

Sound power of a monopole depends on source size, strength and frequency. With a small kL the power radiated is small. On the other hand, the larger the size the lower the frequency of efficient radiation (large loudspeakers for low frequency radiation).

7.3 Noise Generation

7.3.1 Introduction

The excitation mechanisms described in section 7.1.2 can appear during the energy transfer process that takes place when a machine is working. The noise generated can be broadly classified into noise of mechanical origin and noise of hydraulic origin. The chain of physical processes that links the hydraulic excitation to the resultant noise radiation is as follows. Internal and external disturbances result in a pressure distribution which produces unsteady forces on the machine elements (blades, casing, etc.). Unsteady forces acting on mechanical surfaces result in mechanical vibrations, which radiate air-borne noise with a certain efficiency.

To evaluate the sound generated by a machine, the noise generating mechanisms have to be identified and from them a distribution of sound sources representative of the actual ones has to be estimated. Although theoretical models depend on data only known by experimental tests, they help us to understand the role of the representative parameters. With the prediction of fluid-borne noise, the mechanical admittance and the radiation efficiency, an estimation of the air-borne noise generated by the machine can be made. As the process is very complex, empirical equations are normally used.

7.3.2 Fluid-Borne Noise Generation

The main mechanisms of random fluid-borne noise generation are: inflow turbulence and its interaction with the runner; boundary layer and wake turbulence; flow separation and interaction with the trailing edge; cavitation; blade tip interaction with casing boundary layer and secondary flows.

Mechanisms of periodic fluid-borne noise are: interaction of runner blades with periodic inlet disturbances; rotation of steady loads in the blades; rotor-stator interactions due to potential field and blade wakes; vortex-shedding (although not related to rotor frequency).

Another sources of fluid-borne noise in a circuit are valves, bends, and other singularities where vortex-like structures and turbulence can be produced. When machines are operating at off-design conditions, more disturbances are generated, depending on the type of machine and they increase when it moves away from the design point. Phenomena like draft tube surge may appear.

Cavitation Noise

The pulsation of gas bubbles in liquids generates noise with monopole characteristics. During the phenomenon of cavitation, volume changes are produced which radiate sound. Due to the large number of bubbles, the cavitation process covers a broad frequency range.

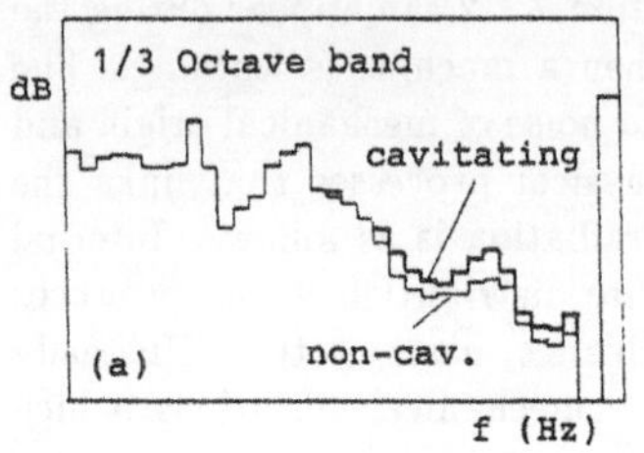

(a) cavitation in the incipient stage

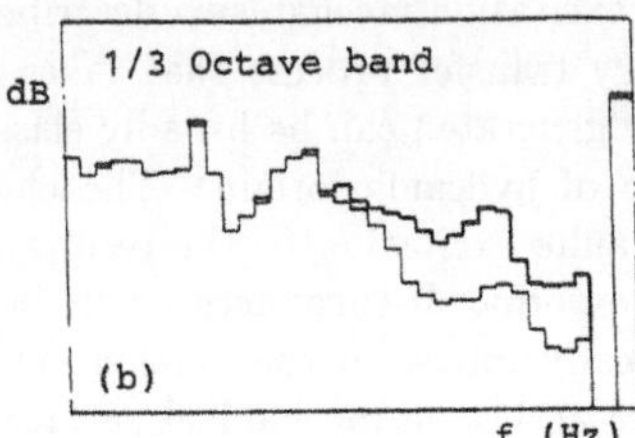

(b) developed cavitation before 3% head drop

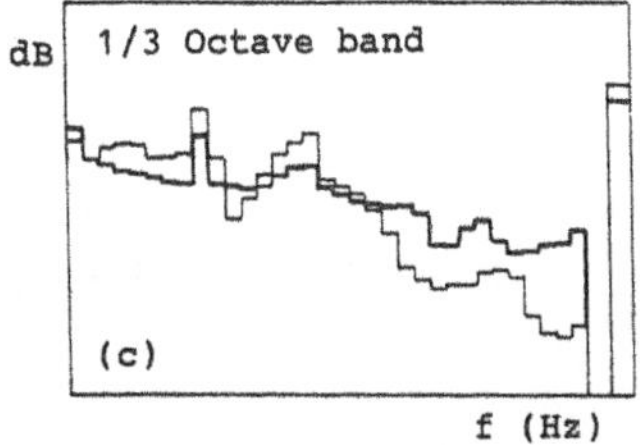

(c) developed cavitation after 3% head drop

Figure 7.5 Comparison of a pump fluid-borne noise with and without cavitation

The amplitude and frequency range of the spectrum varies as a function of the condition of the cavitation. Incipient cavitation in hydraulic machinery has been demonstrated to appear before the 3% head drop. Some results of cavitation noise are indicated in Figure 7.5, where the noise spectra at a pump inlet for several regimes of cavitation are plotted. In the incipient stage, noise has a high frequency content. Decreasing the Net Positive Suction Head (NPSH) decreases the frequency range due to a larger bubble size. Beyond a certain point cavitation noise decreases.

To correlate the sound pressure in pumps, measured at the cavitation inception region, a cavitation inception number I_c was was introduced by McNulty, 1986 [7.16], and defined as the ratio of acoustic power W' (W) to the hydraulic power at the pump inlet $\rho g Q(\text{NPSH})$, where ρQ is the mass flow (kg/s) and (NPSH) the net positive suction head (m). The relationship between I_c and the pump impeller tip speed u_1 (m/s) has been found to be

$$\begin{aligned} I_c &= W'/\rho g Q(\text{NPSH}) = 3.2 \times 10^{-14}\, u_1^{1.8} \,, \\ W' &= p'^2 A/\rho c \,, \end{aligned} \tag{7.16}$$

where p' is the sound pressure at 40 kHz (Pa) measured in the fluid and A is the area of the pipe (m^2). In the same paper a non-dimensional parameter M_c was introduced to judge if a pump is cavitating from the measurement of pressure fluctuation:

$$M_c = p'^2/v(\text{NPSH})u_1^{1.8} = 4.56 \times 10^{-4} \,, \tag{7.17}$$

where v is the velocity of the water (m/s) at the section where the noise is measured. If the M_c calculated is larger than 4.56×10^{-4} there is cavitation. Besides blade cavitation, vortex cavitation can occur in blade tips, in turbine draft tubes and pump inlets at part loads. Vortex cavitation produces noise but not with the intensity of blade cavitation.

Periodic Noise

In the absence of cavitation, noise of dipole origin is predominant in hydraulic machinery. Sound radiated by the rotor is generated by steady rotating forces and unsteady forces. The first are representative of force variations with space, and the second with time. Noise radiated by a rotating static force distribution depends basically on tip Mach number. For low Mach numbers typical of hydraulic machinery, the fluid-borne noise produced by steady rotating forces is negligible compared with that produced by unsteady fluctuations. Application of the dipole equations shows a dependence of the intensity on the sixth power of flow speed.

Blade-inlet vortices interaction is produced during the passage of a blade through a vortex. Sound radiation depends on vortex size. If periodic, the larger the vortex size, the lower the noise frequency.

Noise of discrete frequency is generated by a non-uniform circumferential pressure field, mainly due to some degree of imbalance and asymmetry. Its frequency f_f is the shaft rotation frequency n (Hz),

$$f_f = n \,. \tag{7.18}$$

Rotor-stator interaction generates noise by the effects of the potential flow field and the blade wakes. The spacing between rotor and stator (or rotor and volute tongue) is an important parameter. For close spacing the noise is maximum. The frequency f_b is

$$f_b = z_b\, n \,, \tag{7.19}$$

where z_b is the number of rotating blades. Figure 7.6 shows the periodic noise generated in a pump. The blade passing frequency f_b is clearly seen. Another discrete frequency f_{rs} may appear at

$$f_{rs} = z_b\, z_g\, n \,/\, (\text{highest common factor})\,, \tag{7.20}$$

where z_g is the number of stator blades. Blade passing frequency increases with the speed of rotation.

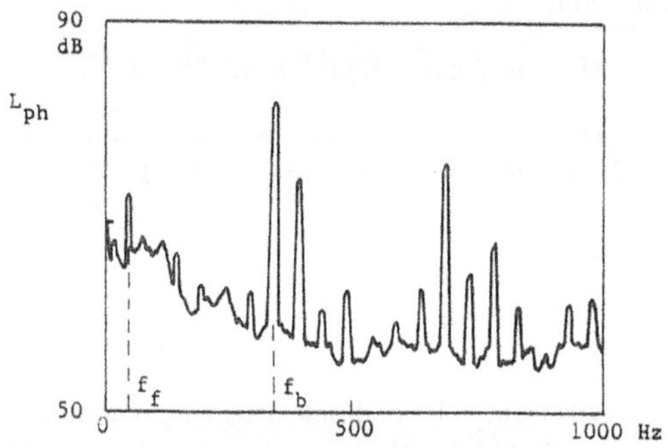

Figure 7.6 Periodic fluid-borne noise measured in a pump (pump with 7 blades at 3000 rpm)

Fluctuating stresses due to turbulence generate fluid-borne noise of quadrupole radiation and are not very important in hydraulics due to the low Mach number of the flows. This mechanism is only important for flows at high velocity as in the jet of aviation engines. Turbulent noise has a wide band spectrum and is generated in boundary layers and wakes. Although direct quadrupole radiation is generally negligible in hydraulics, it can induce wall pressure fluctuations that may radiate noise of dipole nature.

Variation with Flowrate

Noise generated by a hydraulic machine has a minimum around the best efficiency point. Departing from this, especially for low flowrates, the noise increases considerably due to secondary flows and recirculations. Variations in the fluid incidence angle results in a growing of the boundary layer which finally separates. Recirculation induces prewhirl in pumps. In turbines, draft tube surge appears at part loads with a characteristic excitation frequency (see section 6.3).

7.3.3 Prediction of Noise Levels

Hydraulic Noise

Noise generation depends on machine type and its dimensions. Impeller tip speed u_1 and shaft power P seem to be the most important factors in noise generation. Liquid-borne noise in pumps can be estimated by empirical equations in which the coefficients have been found from experimental data of tests on several pumps (cf. McNulty, 1981 [7.17]):

$$W_h' = 2.8 \times 10^{-13}(\rho g Q H) u_1^5 , \tag{7.21}$$

where W_h' is hydraulic sound power (W), ρQ the mass flow (kg/s) and H the pump total head (m). Noise produced by cavitation was not taken into account. Variations of the sound pressure level with frequency and flowrate have been given in graphs in the same paper.

Another expression for estimating the fluid-borne noise produced by the potential field and the wakes of the blades was introduced by Simpson et al., 1967 [7.18], as

$$L_{ph} = K + 20 \log_{10} \frac{3600\, pQ}{N_s\, n_{rpm}\, D_2^2\, b_2} , \qquad (p_{ref} = 2 \times 10^{-5} \text{ Pa}) \tag{7.22}$$

where $K = 150$, $N_s = n_{rpm}(Q)^{1/2}/H^{3/4}$, p (bar), Q (m^3/s), n_{rpm} rotation speed (rpm), D_2 impeller external diameter (m) and b_2 the outlet width (m). Turbulence and separation were not taken into account. These equations are valid for pumps working at the highest efficiency point without cavitation. Important parameters in noise level like cut-water and tip impeller clearance were not considered.

Air-Borne Noise

More information exists for the prediction of air- borne noise from pumps. Empirical expressions give the sound pressure level L_p as a function of the power P (kW), the rotational speed n_{rpm} and the number of stages n_{st}:

$$L_{pa} = A + B \log_{10}(n_{rpm} P / n_{st}). \tag{7.23}$$

The values of the constants A and B may be calculated experimentally for different types of machines or by scaling from model tests in laboratory to prototype. Several values can be found in published literature (cf. [7.19-21]), values A ranging from 31 to 37 and values B from 10 to 11 approximately. Another equation was given by VDI 3743 [7.14] as:

$$L_{pa} = 48 + 12.5 \log_{10} P + 3 \log_{10} n_{rpm} \tag{7.24}$$

For sound power level calculations (L_w with A weighting), similar expressions are used

$$L_{wa} = A + B \log_{10} P \tag{7.25}$$

where A and B are constants that can be found in literature: values of 75.7 and 10 are proposed by Zoog et al. 1983 [7.22], and 71 and 13.5 proposed by VDI 3743 for volute pumps. More recently Zoog et al., 1987 [7.23], carried out the comparison of the values of L_{wa} measured with intensimetry and with classical methods. The results indicated that calculation of L_{wa} from sound pressure measurements in plant gives an overestimation of the actual value and are not valid. However, calculation of L_w from measurements of vibration velocity gives a good approximation except for low frequencies. The values proposed for the constants A and B are 58.8 and 10. Other different values (A =59 (1450 rpm), A =63 (2900 rpm) and B =12) were found by Schmidt and Stoffel, 1985 [7.24], by testing other pumps.

To predict the noise components in octave frequency bands, the following equation is proposed by Erskine, 1977 [7.25]:

$$L_w = 70 + 10 \log_{10} P + K_2 \tag{7.26}$$

where K_2 is a correction factor which yields L_w in octave bands ($K_2 = -20$, −15, −10, −8, −8, −6, −6 for 4k, 2k, 1k, 500, 250, 125 and 63 Hz octave bands). Some manufacturers, assuming that the maximum noise level occurs at the octave band which contains the blade passing frequency f_b, follow the next steps. (a) Calculate f_b and select the octave band f_{obf} which includes f_b (i.e. if f_b=278 then f_{obf}=250). (b) Calculate L_p with an empirical formula in which the coefficients have been determined for each class of machines. (c) Calculate the noise spectrum correcting the L_p calculated with a factor K_{bi} as a function of the ratio f_{obi}/f_{obf} (i.e. $L_{pi} = L_p - K_{bi}$, where f_{obi} is the frequency of the band i). (d) Finally calculate the L_{pa} in each band with the addition of the A weighting scale (i.e. $L_{pia} = L_{pi} - K_{ai}, K_{ai} = +1, +1$, 0, −3, −8, −16 for 4k, 2k, 1k, 500, 250 and 125 Hz octave band).

A similar process assuming that the peak of noise occurred in the octave band containing a *pseudo-vane* frequency f_{sv} (f_{sv} =2 $z_b f_b / z_v$ =integer) is described by Palgrave, 1989 [7.26]. The L_p in that band is determined as a function of the impeller tip speed (average values from 75 dB at 30 m/s to around 88 at 80 m/s). Then a decay rate at each side of the peak noise band is used to calculate the spectrum. Approximate values of $-11, -3, -1, 0, -2, -10, -18$ are given for f_{obi}/f_{obf} ratios of $8, 4, 2, 1, -0.5, -0.25, -0.125$. The considerable scatter of results indicate that much work is still required to predict noise with accuracy.

7.3.4 Noise of Mechanical Origin

All machines have some degree of imbalance, eccentricity and misalignment. During the normal wearing process of the machine, these effects can increase considerably. Imbalance produces a fluctuating force proportional to the square of the angular speed. The characteristic periodic vibration has a stable phase and a frequency that equals the shaft frequency. The amplitude increases as the fourth power of the rotational speed. Eccentricity produces a vibration with the frequency of the shaft rotation and the amplitude is maximum in the direction of eccentricity. Misalignment is another common problem in rotating machinery. It produces high levels of vibration, with a high amplitude at a frequency double that of the shaft rotation, especially in the axial direction. Phase differences between shaft ends are approximately 180 degrees.

Bearings are sources of structure-borne noise. The many types of mechanical defects induce different types of vibrations. In rolling elements bearings, damage in the balls or in the race surfaces, produce high frequency vibrations. Frequencies of interest are;

$$\begin{aligned} f_0 &\simeq nD_i/(D_i + D_o)\,, && \text{(cage)}\,, \\ f_b &\simeq (nD_iD_o)/(D_i + D_o)D_b && \text{(ball damage)}\,, \\ f_{ext} &\simeq nD_i z_{balls}/(D_i + D_o) && \text{(outer race damage)}\,, \\ f_{in} &\simeq nD_o z_{balls}/(D_i + D_o) && \text{(inner race damage)}\,, \end{aligned} \qquad (7.27)$$

where D_o is the outer race diameter, D_i the inner race diameter, D_b the ball diameter and z_{ball} the number of rolling elements.

In journal bearings, vibrations can be due to excessive gaps (frequencies 2 or 3 f_f) and fluid instabilities like the *oil whirling* with frequency:

$$f_{ow} = (0.40 \sim 0.48)\, n \qquad (7.28)$$

Gear boxes are other source of mechanical vibration and noise as shown by Houser, 1988 [7.27]. Excitations at different frequencies are produced by: gear tooth impacts, changes in frictional forces due to gear tooth sliding, air pocketing, lubricant entrainment, mesh stiffness changes and transmission errors. The frequencies of interest are the shaft frequencies and the gear mesh frequency f_{gm} given by:

$$f_{gm} = z_{ti}\, n_i \qquad i = 1,\, 2 \qquad (7.29)$$

where z_{ti} and n_i are number of gear tooth and rotational frequency of mating shafts. Misalignment in gear box bearings produces a vibration of $2n_i$ and two components around the gear mesh frequency $f_{gm} \pm 2n_i$ (modulation). Figure 7.7 shows a spectrum of mechanical noise.

Condition monitoring of hydraulic machinery can be carried out through the analysis of fluid-borne, structure-borne and air-borne noise as demonstrated by Palgrave, 1989 [7.26], and Egusquiza, 1989 [7.28, 29].

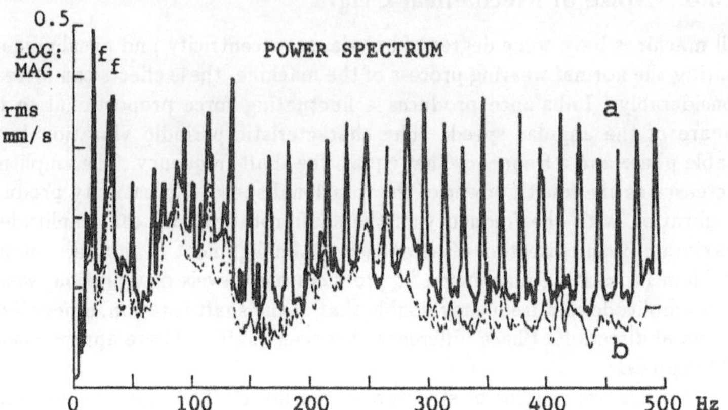

Figure 7.7 Mechanical noise in a bearing of a gear box (solid line **a** for faulty bearing with lubrication problem and broken line **b** for repaired bearing)

7.4 Noise Control

7.4.1 Noise Isolation

The principle of sound isolation is to reduce its transmission by using solid panels. The performance of a panel is given by a sound transmission loss coefficient *TL* defined as (cf. [7.11] and [7.30]):

$$TL = 20\ \log_{10}(p_1'/p_2') \qquad (\text{dB})\ , \tag{7.30}$$

where p_1' and p_2' are the sound pressures before and after the panel. Sound isolation depends on the angle of incidence of the sound waves. For the most common practical cases, random incidence is considered and most experimental data given in literature are for this condition.

Several facts have to be taken into account in reference to sound isolation. First the *TL* is frequency dependent, so the performance of a sound isolating material has to be described at different frequencies. In Figure 7.8, a typical diagram showing the *TL* versus frequency can be seen. It can be observed that low frequencies are very difficult to isolate, while high frequencies are easily isolated.

At low frequencies, the *TL* is controlled by the wall stiffness. Increasing the frequency, there are several resonances because the frequency of the incident sound coincides with natural frequencies of the panel, with a considerable reduction of the isolation. This behaviour is most marked at low frequencies. For the next frequency range, *TL* is controlled by mass. At

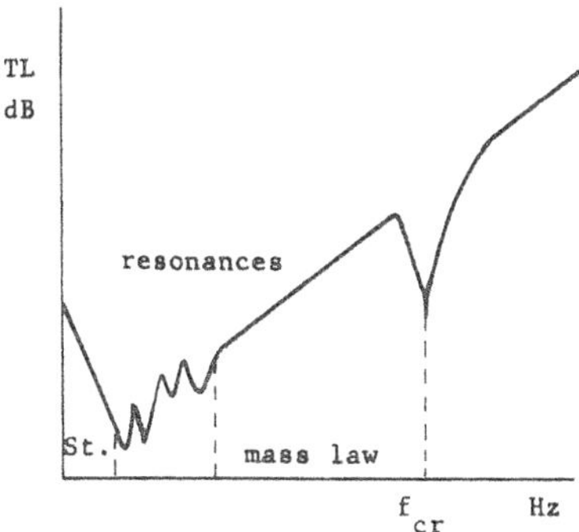

Figure 7.8 Transmission loss of a single panel

a certain frequency the phase of incident sound will tend to coincide with the phase vibrations of the panel. This phenomenon (coincidence effect) reduces sound isolation considerably. The frequency of coincidence f_{cr} for thin panels can be estimated as

$$f_{cr} = c^2/(1.8\, h\, c_m) \tag{7.31}$$

where c_m is the flexural wave speed (m/s) and h the panel thickness (m). $f_{cr} = 12/h$ for metal plates in air and $f_{cr} = 275/h$ for metal plates in water. In the absence of empirical data, for the mass controlled region the TL can be approximated by (mass law)

$$TL = 20 \log_{10}(M f) - K \,, \tag{7.32}$$

where M is the mass of the wall per unit area (kg/m^2), and f the frequency. To have good isolation, the critical frequency of coincidence f_{cr} should be away from the frequency range to isolate. If more isolation is required, more mass is necessary. Care should be taken because while TL is proportional to thickness (for a given surface), f_{cr} is inversely proportional to thickness, so increasing mass makes f_{cr} decrease. If f_{cr} can not be avoided in the operating range, damping has to be increased to diminish the coincidence effect.

7.4.2 Sound in Enclosed Spaces

The behaviour of sound in enclosed spaces is rather complex because of the successive reflections that it follows before it is damped out. In small rectangular enclosures with negligible sound absorption the standing wave patterns and its natural resonant frequencies can be calculated. For large and irregular enclosures statistical methods are normally used. The existence of a near field and a reverberant field, with the acoustic waves impacting the wall at different angles, makes the problem very difficult to be approached by wave acoustic theories.

Sound Absorption

As sound waves travel through the air, some absorption occurs due to viscosity. This absorption is small except for long distances or at high frequencies. In rooms, absorption is provided by the walls, by some internal damping process or by allowing some transmission through them. Sound absorbing materials absorb sound by converting acoustic energy into heat and vibration which have porous characteristics. To measure the absorption characteristics, the sound absorption coefficient c_a, defined as the ratio of the sound energy absorbed to the sound energy incident, is used;

$$c_a = W'_a/W'_i \,. \tag{7.33}$$

When there are several surfaces of different materials the average absorption coefficient is

$$c_{am} = (c_{a1}A_1 + ... + c_{an}A_n)/(A_1 + ... + A_n) \,. \tag{7.34}$$

Reverberation Time

When an acoustic source is turned off in a room, a few seconds pass before the sound completely disappears. This time depends on the acoustic absorption of the walls and other objects in the room. The more absorption the more rapidly the sound disappears. To quantify the absorption characteristics of an enclosed space, the reverberation time T_r is used. It is defined as the time that a sound decays by 60 dB. To relate T_r and c_a the Sabine formula can be used;

$$T_r = 0.16V/A_b \,, \tag{7.35}$$

where A_b (absorption units in m^2)$= c_{a1}A_1 + \ldots + c_{an}A_n$ at a determined frequency band, and V the volume of the room (m^3).

Directionality

The directivity factor Q_d is used to describe the directionality of a noise source. Q_d is defined as the ratio of the sound intensity produced by a source in a certain direction at a given distance to the average sound intensity $\bar{L}_p$ at the same distance. Another parameter used is the directivity index Di defined by

$$D_i = L_p - \bar{L}_p = 10 \log_{10} Q_d \,. \tag{7.36}$$

In a free field, the L_p at a given distance from the source can be related to L_w by

$$L_p = L_w - 20\log_{10} R - 11 + D_i\,, \tag{7.37}$$

where R is the distance from the source (m). Di has different values ranging from 0 when situated at the centre of a room to +9 when situated at a corner junction of three planes.

To calculate the L_p in an enclosed environment, there are contributions from both the direct and reverberant fields as:

$$L_p = L_w + 10\log_{10}\left(\frac{Q_d}{4\pi R^2} + \frac{4}{A_b}\right)\,. \tag{7.38}$$

In the direct field $(4/A_b) < (Q_d/4\pi R^2)$. With this expression, knowing L_w and the reverberation time T_r, the L_p can be calculated.

Enclosures

Enclosures are usually the best way to reduce noise. Their walls have to be adapted as a function of the noise spectrum. Machinery enclosures should be made as stiff as possible for low frequency isolation, as large as possible for the medium frequency range and with high stiffness and damping for frequencies above f_{cr}.

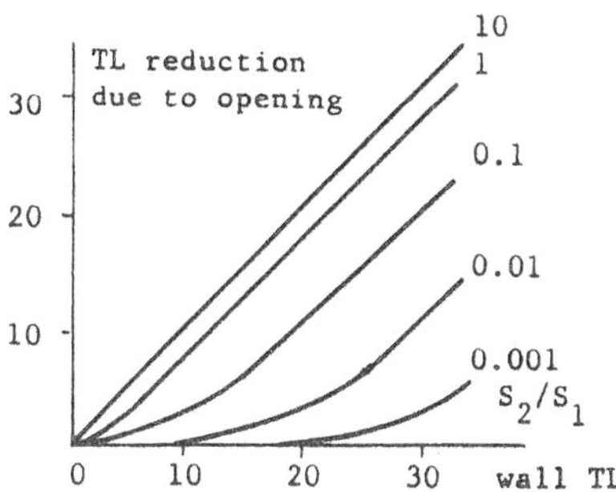

Figure 7.9 Isolation reduction due to openings (S_2 is the opening area and S_1 the wall area)

It is important to coat the internal part of the enclosure with absorbing material to avoid an increase of the noise source intensity by reverberant field effects. The prevention of acoustic leaks is necessary to avoid an important decrease in isolation as is shown in Figure 7.9. An opening of 1% in the enclosure reduces the TL from 25 to 19 dB. From another point of view, radiation properties of the enclosure should be taken into account. Large enclosures are more efficient radiators of low frequency sound.

The difference between noise levels before and after the enclosure installation (insertion loss IL) is

$$IL = TL - 10\log_{10} A + 10\log_{10} A_b \tag{7.39}$$

where A is the total area of the enclosure walls and A_b is the total absorption inside the enclosure. After a certain point, it is better to use double walls than single ones. The attenuation is larger with the same mass for a double wall with a space filled with porous materials.

Barriers

Solid barriers are effective when their dimensions are large compared with the noise wavelength λ. TL can be calculated with the equation

$$TL = 20\log_{10}(K/\tanh K) + 5 \quad (\text{dB}), \tag{7.40}$$

where $K = (2\pi N_f)^{1/2}$ and $N_f = \pm 2/\lambda(A + B - d)$ (see Figure 7.10).

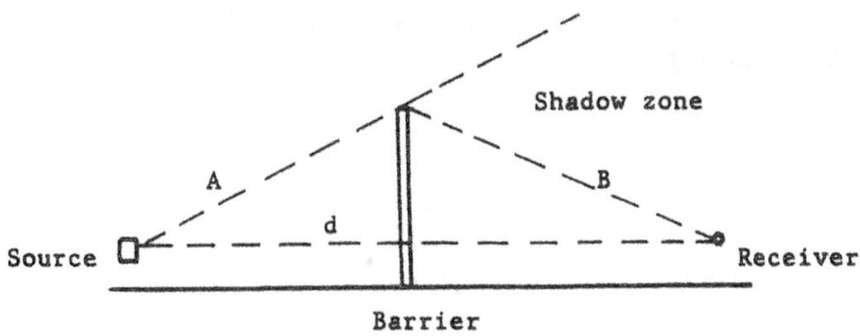

Figure 7.10 Zones of attenuation in a barrier

Vibration Isolation

Vibration isolation is fundamental for noise reduction. Vibrations produced in the source can be propagated through the structure and radiate noise to other points away from the source.

Anti-vibration mountings can be used to avoid vibration transmission from the machine to the foundations. For good isolation, the resonant frequency of the mounting should be much lower than the frequency to be isolated. The resonant frequency f_n is:

$$f_n = \frac{1}{2\pi}\left(\frac{k}{M}\right)^{0.5} = \frac{0.5}{x^{0.5}}, \tag{7.41}$$

(7.41) where k is the spring stiffness, M the mass (kg) and x the static deflection (m).

A wide variety of materials can be used. At 25 Hz and above, cork and rubber pads are recommended. For the 5 ~ 30 Hz frequency range, rubber and elastomers. Springs are used in the low frequency range. Although very effective for low frequencies, they allow the passing of high frequencies and they have a very low damping. To avoid these inconveniences springs are fitted with other materials. Flexible couplings and bellows are appropriate to avoid vibration transmission through pipes.

7.4.3 Silencers

Silencers, when feasible, can reduce the noise amplitude in ducts considerably at selected frequencies. Expansion box elements (a box between two pipes), side resonators and intake silencers are often used for silencing. The frequencies of maximum and minimum attenuation are related to their resonance frequency. Many published works can be found on this topic, for example [7.30] and [7.31]. Absorption silencing is obtained by coating the inner part of a duct with sound- absorbent lining. For low speed flows with wavelengths larger than the diameter D the following equation can be used:

$$TL = 4.2(L/D)c_a^{1.4} , \tag{7.42}$$

where TL is the transmission loss, L the length of the duct (m) and c_a the absorption coefficient.

Active Control

Active control systems are another possibility that are emerging. Arrays of loudspeakers reduce noise by anti-noise production, see Epstein, 1984 [7.32]. Application to turbomachinery reduces not only noise but also the instability itself. Flutter of aircraft wings can be controlled by this system. Accelerometers located on the airfoil control the vibration. When this arises, the system generates a signal so that the loudspeakers can generate a compensating force. Compressor surge can be reduced if a loudspeaker introduces a perturbation into the system that cancels the original surge perturbation, see Archibald, 1975 [7.33].

7.4.4 Control of Noise in Hydraulic Machinery

Noise control has to be studied in the planning stage for maximum effectiveness and minimum cost. Pumping stations and hydro-power plants may have different acoustical characteristics so the approach for noise control may be different in each case. As legislation is very strict and noise levels allowed are decreasing, all possible noise control methods should be considered. Noise regulations indicate the maximum time of exposure during a day. In areas of higher noise levels, where noise is more difficult to avoid, personnel should be exposed for the minimum time possible.

Noise control should follow the chain of noise, that is: generation, transmission and reception. Noise can be controlled at the source, along the path or at the receiver.

Reduction of Fluid-Borne Noise

The present trend in construction is towards units with more power concentration, so reduction of fluid-borne noise generation by design is most difficult. As seen in section 7.3, overall noise levels are a function of rotational speed and power, therefore, quieter units are slower and consequently of larger dimensions, which is not economically justifiable. However, design efforts are being made to minimize noise generation by a better profiling of the water passages and paying attention to items such as blade trailing edge shape.

To reduce the rotor-stator interaction noise, the clearance between blades and stator vanes should be increased but unfortunately this is detrimental to machine efficiency. However it seems that a little increase in clearance signifies a considerable reduction in noise and only a negligible reduction in efficiency. Increasing the blade number will decrease the blade passing frequency amplitude. Replacement of long inlet guide vanes set close to the impeller by many short blades with larger stator-rotor clearance, has produced a reduction of 10 dB in outlet noise, see McNulty, 1981 [7.17]. However, an increase in the blade number will produce noise in a frequency where the ear is more sensitive. The blade trailing edge is important in minimizing the effect of vortex-shedding. Different shapes have been tested by Dupont et al., 1987 [7.34] and can be used to reduce the excitation on the blade, so that failures can be avoided.

Operating conditions and installation greatly affect noise generation. Operating at highest efficiency point is one way to maintain noise at a minimum level. If the machine has to operate at off-design conditions, and cavitation surge appears, control methods (air injection, fins, etc.) have to be used to reduce the excitation.

Ingested turbulence can considerably affect fluid-borne noise generation. Inlet flow should be free of vortices and flow asymmetries. Cavitation noise will depend on machine design and installation. The elimination of all cavitation would be impractical due to the high values of NPSH required.

Structure and fluid resonances should be avoided by design. Forced vibrations produced by the machine at shaft and blade frequencies (and harmonics) should not coincide with any fluid or structural, natural frequency. Apart from the normal excitation frequencies, if the machine has to operate at off-design point, this signifies that new excitations at new frequencies will appear and have to be taken into account. An example is the possibility of penstock resonance by draft tube surging.

Since fluid-borne noise can propagate though the duct system where noise can be radiated with a better efficiency than in the machine itself, the transmission should be minimized. Silencers, when feasible, should be used to reduce the fluid-borne noise transmission. Hydraulic silencers have been developed, especially for the oil systems. For small installations elastic hoses and air-vessels can be used.

Reduction of Structure-Borne Noise

Noise of mechanical origin should be kept at a minimum by adequate design and maintenance. All rotating parts should be dynamically balanced. Alignments and eccentricities have to be checked and stiff rotors should be favoured. Certain types of labyrinth seals that are prone to vibration should be avoided. Changes in stiffness, to mismatch the excitation frequency and the natural frequency of the structure, are sometimes necessary to avoid mechanical resonances. If this is not possible, increase in damping is very effective.

Bearings are important noise sources especially if they have some damage or lubrication problems, see Figure 7.7. Transmission of mechanical noise to the piping system and to the floor, should be minimized. Anti-vibration mountings are necessary to reduce the transmission of strong vibrations or to isolate control rooms where people have to stay for many hours.

Reduction of Air-Borne Noise

Not all the structure-borne noise is radiated, but only a part of it. Radiation efficiency depends on several factors. For a rigid panel, it depends on the characteristic length of the body L and on the wavelength λ. When $L < \lambda$, then the behaviour is like a spherical source; the radiation efficiency K_r is low and the sound is omni-directional. When $L > \lambda$, then K_r is large and the sound has a directional character. So partitions of the noise generating surface can reduce noise radiation. Panel resonance f_{cr} considerably affects radiation emission and should not coincide with excitation frequencies.

Significant noise reduction is accomplished with fibre liners and radiation can be reduced. The outer metal cover should not be permitted to be in contact with the machine so as to avoid acoustic bridges. For small machines, enclosures are most effective. Its attenuation can be estimated by the *mass law*. By doubling the weight, a decrease of 5 dB can be achieved. For low and medium frequencies, the performance of the enclosure can be improved with double walls.

If ventilation has to be permitted, as in the case of electric motors, the enclosure must be partly open, thus resulting in the reduction of its attenuation performance. In that case, the inlet and outlet ducts can be fitted with a fan and silencers. Solid barriers placed between the source and receiver can provide some attenuation. They are mainly used when the sound source has small dimensions.

The propagation of noise in rooms can be calculated by the equations given in section 7.4. Acting on the reverberation time by an appropriate distribution of absorbent acoustic materials is the other way to reduce noise. If the noise level is too high, ear muffs and ear plugs should be provided.

References

7.1 Blevins, R. (1977). *Flow induced vibrations*, Van Nostrand Reinhold, London.

7.2 Naudascher, E. and Locher, F.A. (1974). 'Flow induced forces on protruding walls', *ASCE Journal Hydraulics Division*. Vol.100, February.

7.3 Corcos, G.M. (1964). 'The structure of turbulent pressure field in boundary layer flows', *J. Fluid Mechanics*, Vol. 18, pp. 353-378.

7.4 Heller, H.H. and Bliss, D. (1975). 'The physical mechanisms of flow-induced pressure fluctuations in cavities', AIAA paper 75-491.

7.5 Jaeger, C. (1977). *Fluid transients*, Blackie, London.

7.6 Fanelli, M. (1975). 'Les phenomnes de resonance hydraulique', *La Houille Blanche*, 4.

7.7 Naudascher, E. (1982). 'Flow induced vibrations', BHRA, Delft, November.

7.8 Dowell, A. (1989). *A modern course in aeroelasticity*, Kluwer Academic Pub.

7.9 McCroskey, W.J. (1977). 'Some current research in unsteady flow dynamics', *ASME, J. Fluids Eng.*, Vol 99.

7.10 Reynolds, D.D. (1981). *Engineering principles of acoustics*, Allyn and Bacon.

7.11 ISO (International Standards Organization) 3742.

7.12 DIN 45635.

7.13 BS (British Standards) 4196.

7.14 VDI (Verein Deutscher Ingenieur) 3743.

7.15 Ross, D. (1987). *Mechanics of underwater noise*, Peninsula Publishing, California.

7.16 McNulty, P.J. (1986). 'Monitoring the presence of cavitation in pumps by a single fluid-borne noise measurement', Proc., Int. Conf. on Condition Monitoring, BHRA, Brighton, May, pp. 167-182.

7.17 McNulty, P.J. (1981). 'Measurement techniques and analysis of fluid-borne noise in pumps', NEL Report 674.

7.18 Simpson, H.C. et al. (1967). 'A theoretical investigation of hydraulic noise in pumps', *J. Sound and Vibration*, Vol. 5, pp. 456-488.

7.19 *Pumping Manual* (1983). Trade and Technical Press, U.K.

7.20 Pollac, H. (1975). 'A guide to the noise levels in pumps depending on design, duty quality of installation and cost of pump', 6th Conf., British Pumps Manufacturer Association, BPMA.

7.21 France, D. (1977). 'The need for the classification of pump noise and vibration', Proc. Inst Mech. Eng., 247/77.

7.22 Zoog, H. et al. (1983). 'Noise levels from big centrifugal pumps', *Sulzer Technical Review*, Vol.65, No. 3, pp. 17-20.

7.23 Zoog, H. et al. (1987). 'Noise test on single-stage volute pumps by measurement of sound integrity', *World Pumps*, May, pp. 138-140.

7.24 Schmidt, A. and Stoffel, B. (1985). 'Spiralgehäuse Pumpen', *Cavitation*, February.

7.25 Erskine, J.B. (1977). 'A users view on the control of noise from pumps, fans and compressors', Conf. on Limiting Noise, Inst. Mech. Eng., C249/77.

7.26 Palgrave, (1989). 'Diagnosis pump problems from their noise emission signature', 11th Conf. of BPMA, Cambridge, April.

7.27 Houser, D.R. (1988). 'Gear noise state of the art', Internoise Proceedings, Vol. 2, pp. 601-606.

7.28 Egusquiza, E. (1989). 'Condition monitoring of hydraulic machinery by noise analysis techniques' in *The State of the Art in Hydraulic Machinery*, edited by Duan, C.G. and Boldy, A.P.

7.29 Egusquiza, E. and Santolaria, C. (1989). 'On-line control of hydraulic machinery', Proc., Intern. Symposium on Large Hydraulic Machinery, Beijing.

7.30 *Handbook of Noise and Vibration Control*, (1983). Trade and Technical Press, Surrey, U.K.

7.31 Annand, R. (1975). *Gas flow in internal combustion engines*, Foulis.

7.32 Epstein, A.H. (1984). Active suppression of compressor instabilities, AIAA 86-1994.

7.33 Archibald, F.S. (1975). 'Self-excitation of acoustic resonance by vortex-shedding', *J. Sound and Vibration*, Vol. 38, pp. 81-103.

7.34 Dupont, P. et al. (1987). 'Flow analysis for a hydrofoil with and without lock-in', Int. Conf. on Flow Induced Vibrations, BHRA, May.

Chapter 8

Diagnosis

S.K. Bhave

8.1 Introduction

The vibrations experienced by the hydraulic machinery are caused by perturbation forces which could be mechanical, electrical or hydraulic in nature. Many times, these perturbation forces exist together thereby making their identification quite a difficult task. In order to identify the nature of the perturbation forces, it is necessary to use high quality equipment capable of recording the relevant variable quantities with least possible error. These quantities usually are: (1) bearing and shaft vibrations along with the phase information, (2) vibrations of individual elements of the unit, deformations and displacements of components, (3) pressure variations at different sections of the flow passage, (4) noise.

The data obtained through the (appropriate) sensors provide the information about the vibrations, pressure pulsations, noise etc. in the time domain. This time domain information, in general, does not provide sufficient data regarding the likely reasons for the vibration problem, unless the signals are properly processed and converted to the frequency domain. Conversion of the time domain signal to the frequency domain, facilitates identification of distress in the machine since each of the frequencies in the frequency domain is identifiable to a characteristic malfunction in the system. In order to understand the principles of above diagnostic procedures, various instrumentation and signal processing techniques employed, it is worthwhile to review in brief, some of the fundamental concepts of the analysis of periodic signals, such as vibration.

8.1.1 Characteristics of Periodic Vibration

The periodic vibration is an oscillating motion of a particle or body about a reference position such that the motion repeats exactly after a certain time.

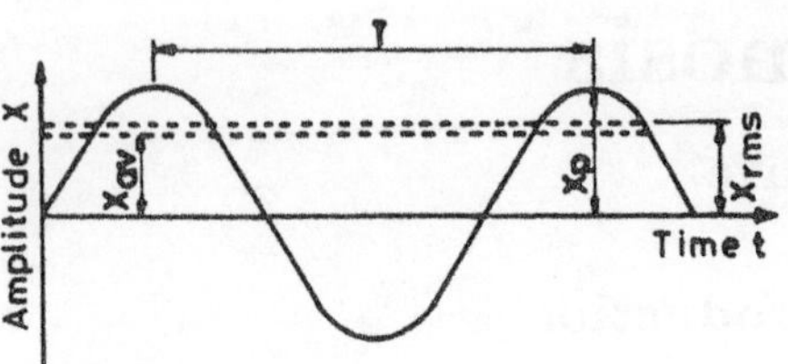

$$\text{Frequency} \quad f = 1/T$$

$$\text{Circular frequency} \quad \omega = 2\pi f$$

$$X = X_p \sin \omega t$$

$$X_{av} = \frac{1}{T} \int_0^T |x| dt$$

$$X_{rms} = \sqrt{\frac{1}{T} \int_0^T X^2 dt}$$

$$\frac{X_{rms}}{X_{av}} = \text{Form factor}, \; f_f = 1.11 \;\; (\simeq 1dB)$$

$$\frac{X_p}{X_{rms}} = \text{Crest factor}, \; f_c = 1.414 \;\; (\simeq 3dB)$$

Figure 8.1 Sinusoidal vibration signal

The simplest of this form of vibration is harmonic motion as shown in Figure 8.1 and can be represented by:

$$X = X_p \sin \omega t \,, \quad \dot{X} = \omega X_p \cos \omega t \;\; \text{or} \;\; \ddot{X} = -\omega^2 X_p \sin \omega t$$

where X corresponds to displacement, $\dot{X}$ velocity and $\ddot{X}$ acceleration of vibration respectively. It is apparent that the form and period of vibration remain the same regardless of whether the vibration is specified in displacement, velocity or acceleration mode. Of course, the phases of these quantities differ i.e., velocity leads displacement by 90 degree. It is therefore enough to characterize the vibration by the peak values of displacement X_p, velocity V_p $(= \omega X_p)$ or acceleration a_p $(= \omega^2 X_p)$ and the frequency ω. The various other quantities such as average absolute value, rms (root mean square) value are shown in Figure 8.1. The figure also shows the parameters such as form factor and crest factor which are used for the indication of wave shape of vibrations.

Most of the vibrations experienced in daily life are not pure harmonic motion as shown in Figure 8.1, even though, many of them may be periodic. Typical non-harmonic periodic motions are shown in Figure 8.2. For the periodic motions shown in Figure 8.2, one can obtain information regarding peak amplitude, average absolute, rms, form factor, crest factor, but these hardly throw any light upon the possible causes for such motion and the forces causing the motion. It was shown over one hundred years ago by Baron Jean Baptiste Fourier that any wave form that one encounters in real life can be generated by adding up sine waves. For example, the wave form shown in Figure 8.2 (a) can be shown to be primarily consisting of harmonically related sine waves as shown in Figure 8.3. Conversely we can break down the real world signal into the same sine waves.

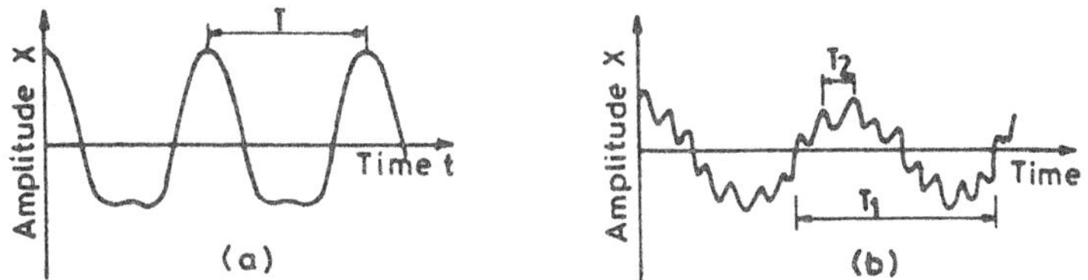

Figure 8.2 Non-harmonic periodic motion

It can be shown that this combination of sine waves is unique and any real world signal can be represented by only one combination of specific sine waves. The mathematical theorem formulated by Fourier states that any periodic wave $f(t)$, no matter how complex, may be looked upon as a combination of a number of pure sinusoidal curves with harmonically related frequencies.

$$f(t) = X_0 + X_1 \sin(\omega t + \phi_1) + X_2 \sin(2\omega t + \phi_2) + \cdots + X_n \sin(n\omega t + \phi_n) \quad (8.1)$$

The harmonic $X_1 \sin(\omega t + \phi_1)$ is known as the first or fundamental harmonic whereas subsequent harmonics having frequency $n\omega$ are known as harmonics of nth order i.e. the harmonics will have frequency equal to $2\omega, 3\omega$, etc.

In practical cases, the vibrations are caused by the simultaneous action of several periodic disturbing forces which need not be harmonically related. Consequently the vibration can be considered as being made up of harmonics with frequencies equal to $\omega_1, \omega_2, \omega_3, \omega_4, \ldots \omega_n$, etc. which may not be multiples of the overall frequency of vibrations.

It is clear that the breaking down of the vibration signal into the constituent sine waves provides yet another dimension of looking at the vibration process. This is illustrated in Figure 8.3.

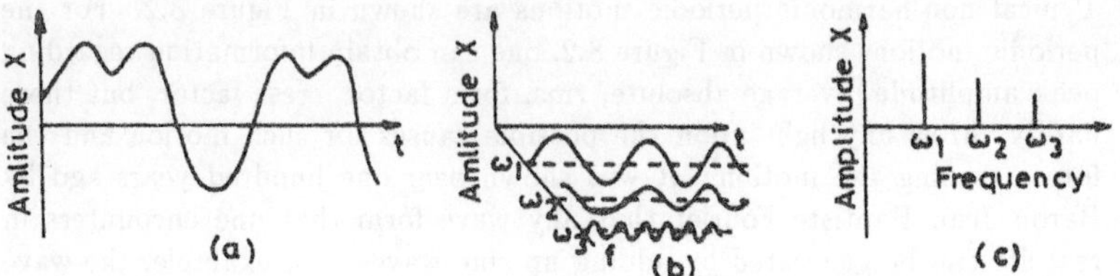

Figure 8.3 Spectrum of periodic vibrations (a) time domain depiction, (b) three-dimensional co-ordinates showing time, frequency and amplitude, (c) frequency domain depiction

The frequency domain representation in Figure 8.3 (c) is called the spectrum of the signal. It is very important to understand that no information of the signal has been lost when viewed in frequency domain. As a mater of fact, it is possible to identify the components (which could ultimately be traced to some physical forces) which constitute the overall signal. This is the reason why spectrum analysis is regarded as a powerful diagnostic tool for analyzing the vibration problems.

The following paragraphs will deal with various signal sensors, signal processors, data reduction/processing and the procedure of diagnosis of vibration problems, through properly collated vibration data and the spectrum analysis. A few case studies are also presented to illustrate application of the diagnostic procedure for the solution of the vibration problem. Finally, the other aspects of vibration problem such as preventive maintenance and safety measures are discussed.

8.2 Sensing and Measurement

8.2.1 General Considerations

As discussed previously, there are three quantities which are of interest in vibration studies, namely the vibratory displacement, velocity and acceleration (peak or rms values). These quantities are related to each other i.e. displacement (amplitude) = acceleration / $(2\pi f)^2$ and velocity= acceleration / $2\pi f$.

The earlier versions of vibration pick-ups producing an electrical output were velocity sensitive and were rather bulky. During the recent years, there has been a marked move towards the use of acceleration sensitive transducers, called accelerometers, on account of their smaller size and also having a wider range of frequency. There has also been a growing interest in high frequency vibration, as a carrier of information on the running condition of the machinery and the corresponding wide range of vibration levels to be detected. The accelerometer transducer caters to this requirement very effectively. Furthermore, the signals from accelerometers could be easily integrated electronically to obtain velocity and displacement, whereas with the displacement/velocity transducer, the electronic differentiation is a more complex and dubious affair.

In theory, it is irrelevant as to which of the three parameters, acceleration, velocity or displacement, should be chosen to measure vibration. However, if the vibration signal is dominated by high frequency components, the displacement or velocity measurement may not be satisfactory. On the other hand, if the vibratory process is predominantly in the low frequency range, as in the case of hydraulic turbines, it is appropriate to select displacement or velocity as the parameter. Displacement is often used as indicator of unbalance of rotating machines because relatively large displacements usually occur at the shaft rotation frequency, while in some situations, they occur at sub-synchronous frequencies. Rms velocity is widely used to determine the vibration severity as it is indicative of vibrational energy and thus a destructive potential of the vibration phenomena.

The frequency range of interest in vibration measurement has been increased steadily during the past one or two decades, with increasing awareness and understanding of dynamic processes occurring at high frequencies in turbomachinery. For example in hydraulic pumps, one of the severest perturbation forces experienced by the component of the pump occurs at vane passing frequency and its harmonic multiples. The detection of the presence of these forces by displacement transducers is very difficult since the associated displacements may be too small. For example a force corresponding to 10 m/s^2 peak to peak acceleration (of the vibrating component) at a vane passing frequency of 525 Hz of a 7 vaned pump running at 75 Hz would produce vibrational displacement of less than 1 μm (peak to peak) which would possibly be missed or overlooked by the vibration analyst in presence of relatively high levels of normal running speed components. The perturbation forces in the high frequency domain may thus escape unnoticed, and at times may cause premature failure of the corresponding components. The failure, once it occurs, immediately manifests into unbalance, bearing failure etc., which produce the perturbation forces in the low frequency domain. It is therefore preferable to carry out vibration measurements in at least two modes, one of which may be acceleration to ensure that no high frequency components are missed.

The vibration associated with fluid flow is, quite often, random in nature and must frequently be measured alone or together with periodic vibration measurements. The measurement technique to be employed in such cases is quite complex.

As mentioned earlier, the diagnosis of vibration problems requires, apart from measurement of overall vibrations, frequency analysis to reveal individual frequency components making up the wide band signal. For this purpose a filter is included in or attached to the vibration measuring instrument, thus making a frequency analyzer. The filter allows only frequency components to be measured which are contained in a specific frequency band. The pass band of the filter is moved sequentially over the whole frequency range of interest so that separate vibration level readings can be obtained at each frequency. The filter section may contain a number of individual fixed frequency filters which are scanned sequentially, or a tunable filter which can be tuned continuously over the frequency range. With the advances in digital signal processing technology, it is now possible to compute and display the instantaneous spectrum on the screens of real time analyzers, which also permit continuous updating of the complete frequency spectrum.

8.2.2 Accelerometers

An accelerometer is an electromechanical transducer which produces at its output terminals, a voltage or charge that is proportional to the acceleration to which it is subjected. Piezoelectric accelerometers are more or less universally preferred for measurements covering a wide frequency range.

The piezoelectric elements, forming the heart of the accelerometer, are made from an artificially polarized ferroelectric ceramic. These elements produce an electrical charge directly proportional to the strain and thus the applied force when loaded either in tension, compression or shear. In the design of the accelerometers, the piezoelectric elements are arranged so that they are loaded by a mass or masses and a preloaded spring or ring. When subjected to vibration, the masses exert a varying force on the piezoelectric element proportional to the vibration acceleration. For frequencies lying well below resonant frequency of the assembly, the acceleration of the masses will be the same as the acceleration of the base and the output signal will be proportional to the acceleration to which the accelerometer is subjected.

Figure 8.4 (a) shows two commonly used configurations of the accelerometer, namely compression and shear type. Figure 8.4 (b) shows the frequency characteristic of a typical piezoelectric accelerometer. Measurements are normally confined to using the linear portion of response curve which at the high frequency end is limited by the accelerometer's natural resonance. As a rule of thumb, the upper frequency limit for measurement can be set to 1/3 of the accelerometer resonance frequency to minimize errors. General purpose accelerometers have resonant frequencies of the order of 30 kHz so that one can obtain upper frequency limit of the order of 10 kHz. The lower limit of measuring frequency is determined by two factors. The first is the low frequency cut off of the associated preamplifier, but this normally is not a problem as the limit is usually well below 1 Hz. The second is the effect of ambient temperature fluctuations to which the accelerometer is sensitive. For normal environments, measurements well below 1 Hz are possible. For details, the reader may refer to the catalogues of accelerometer manufacturers.

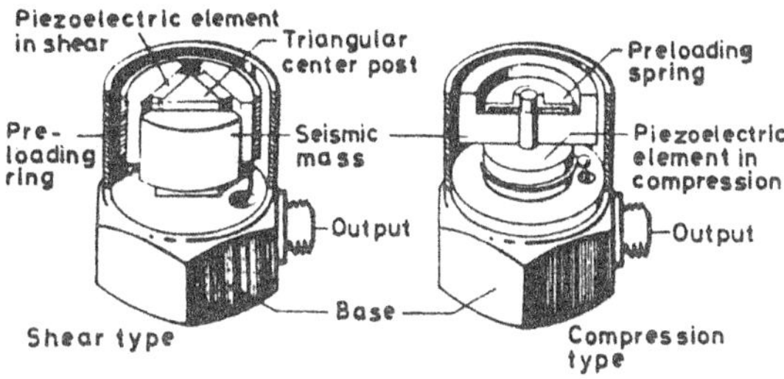

(a) configurations

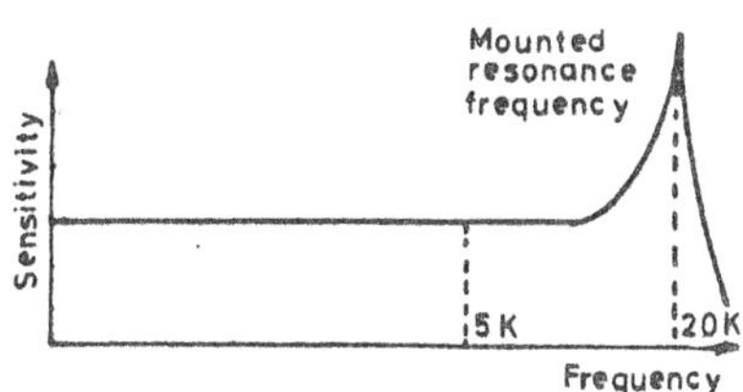

(b) typical frequency response

Figure 8.4 Accelerometer (courtesy of Brüel & Kjaer, Denmark)

The method of mounting the transducer also greatly influences the upper frequency limit. For example, when the accelerometer is mounted using a permanent magnet base, the mounted resonance frequency reduces. The effective frequency range decreases substantially when the accelerometer is hand held; one may have mounted resonance at much lower frequency, such as 1500-2000 Hz so that upper frequency limit drops to 500-600 Hz. The best mounting arrangements are (i) stud mounting with mica washer; (ii) thin film of silicone grease or (iii) fixed using bees wax or epoxy cements.

8.2.3 Shaft Vibration Transducers

The traditional method of assessing the vibration behaviour by monitoring of the bearing cap/casing vibrations, sometimes proves inadequate on account of (variable) transfer impedance between the shaft motion and the casing motion. A vast majority of machine malfunctions manifest themselves (at lower frequencies, usually less than four times the running speed) in the form of high and abnormal shaft vibration behaviour. In the event of bearings providing a large damping ratio, these malfunctions may not appear damaging when measurements are made on the bearing cap/casing only. It is therefore prudent to carry out shaft vibration measurements whenever possible.

Many of the shaft probes operate on the inductive principle and therefore do not touch the object to be observed. These probes are completely unaffected by any material in the probe gap that is not electrically conductive; hence oil, air, gas, etc., between the probe and the observed surface has no effect on the probe output. A shaft probe (proximity probe) provides two signal outputs. They are: (1) voltage proportional to the dynamic motion of the shaft relative to the probe mounting (A.C. signal output), (2) voltage proportional to the average gap between the shaft and probe mounting (the average gap or D.C. signal output). It is customary to mount two mutually perpendicular shaft probes in the bearing housing as shown in Figure 8.5. The D.C. gap voltages from these probes locate the journal position at all speeds, starting from zero speed and also at various load conditions. Using information of position of shaft centre relative to the bearing, one can estimate the attitude angle, thickness of oil film in the bearing and hence the nature of vibration problem encountered. A.C. component of the signals from two shaft probes could be fed into a suitable oscilloscope to obtain the *orbit* of the vibrating shaft.

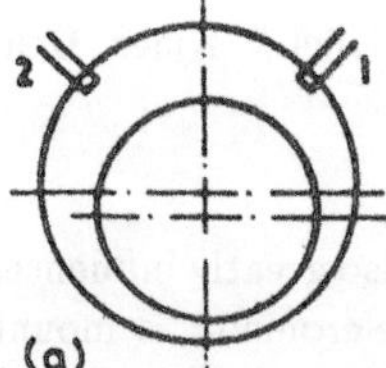

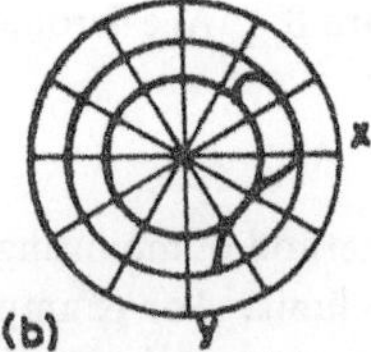

Figure 8.5 Shaft probes (a) two probes to locate journal center, (b) journal center at various speeds on polar plot

A proximity probe by its very nature, when mounted rigidly to the bearing cap or casing of the machine, provides vibration measurement of the relative motion between the shaft and the mounting of the proximity probe. To obtain information about the absolute shaft vibration, vector summation of the bearing absolute and shaft relative vibration must be carried out. The bearing absolute vibration can be measured using accelerometer or other seismic probes.

8.2.4 Phase Measurement

In the analysis, diagnosis and correction of vibration problem, phase analysis plays a very important role. Phase in the simplest terms can be defined as that part of a cycle (0 - 360 degrees) through which one point of machine has travelled relative to another part or a fixed reference. Phase measurements can also be considered as a means of determining the relative motion of various parts of a machine.

In practice, measurement of phase is done using (1) stroboscopic light, or (2) phase meter, or (3) using oscilloscope.

Many of the portable vibration analyzers are furnished with a stroboscopic light as shown in Figure 8.6, which when triggered by the measured vibration, provides a quick and convenient means of obtaining phase readings for analysis and balancing. To obtain phase readings, a reference mark is placed on the rotor at the end of shaft or some other location which can be observed. For convenience, an angular reference scaled in degrees (0 to 360) can be superimposed on shaft to note the phase angle.

Figure 8.6 Use of stroboscopic light for phase measurement (courtesy of IRD Mechanalysis, U.K.)

Using the analyzer, the strobe light can be tuned so that the reference mark appears stationary. Comparative phase readings are obtained by simply moving the vibration pickup to other measurement location but keeping the same reference mark and angular reference used for the initial measurement. In some cases where the end of the shaft is not available for viewing, an angular reference can be placed on the rotating shaft (by placing a graduated tape or marking 0 to 360 in steps of 5 or 10 degrees). The stationary reference used might be the split line between the upper and lower halves of the bearing or any other convenient stationary component.

The measurement of phase as described above, pertains to the phase of rotational component (n component, n = shaft rotational speed (Hz)) of the machine vibration. However, if the vibration of interest is occurring at $2 \times n$, the strobe light will flash twice for each revolution of the shaft and the reference mark will appear at two positions when observed with the light. It would thus be virtually impossible to obtain comparative phase readings with two or more identical reference marks visible under the light. When it is desirable to obtain comparative phase measurements for submultiple, multiple or non-harmonically related vibration frequencies, some other techniques such as using a phase meter or an oscilloscope must be used.

A remote read out of phase requires a voltage reference pulse at the desired vibration frequency. Where it is desired to observe the phase of the vibration occurring at $1 \times n$, a reference pick-up such as a photocell, an electromagnetic pick-up or a noncontact pick-up is mounted close to the shaft, which has been properly prepared to trigger the reference pick-up. While using a photocell, it is customary to wrap black nonreflective tape around the shaft, paint a white line across the tape or attach a small piece of reflective material to ensure that the photocell responds to the changes in reflectivity. When a non contact or electromagnetic pick-up is used, the target area of the shaft is prepared with an abrupt depression or protrusion. An existing key or keyway can be used with excellent results.

The $1 \times n$ voltage pulse from the reference pick-up applied to the analyzer serves two purposes. First, the reference pulse automatically tunes the filter of the analyzer to the rotating speed frequency, so that any speed change during the measurement does not cause errors in the measurements. Second, the reference pulse is electronically compared to the filtered vibration signal to provide a measure of relative phase.

It is also possible to measure the phase of vibration using a reference pick-up mounted at some other point of the machine. If the vibration is predominant at the desired frequency, a signal of the reference pick-up can be fed to a vibration meter shown in Figure 8.7, whose output can be used as reference signal to the analyzer.

This method however cannot be used for balancing, since placement of

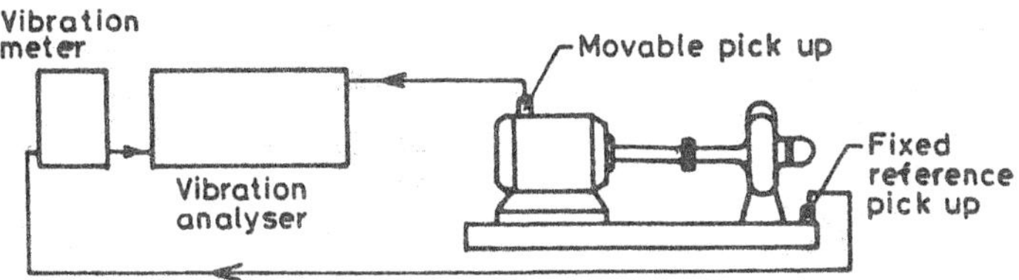

Figure 8.7 Use of vibration meter to provide reference signal to vibration analyzer for phase measurements

trial weights for balancing would also alter the reference signal. The signal from the reference pick-up can also be fed to a vibration analyzer (instead of simple meter) which can be considered as a reference analyzer. The output of the reference analyzer is then connected to the input of the phase analyzer. This method, then, facilitates comparative phase measurements for virtually any frequency of interest: synchronous multiples, sub-multiples, non- harmonic frequencies, etc.

8.2.5 General Purpose Vibration Analyzer

The vibration meter consists of one or more instruments which amplify and condition the signal received from the transducer. The meter is provided with suitable filters to limit the frequency range at the upper and lower end so as to avoid measurement of unwanted signals, noise, etc., an integrator to enable velocity and displacement measurement, and facilities for indicating rms or peak values of the signals. Facilities are provided for connecting to a tracking or tunable filter to enable frequency analysis to be performed, and to a level recorder to plot spectra. A typical instrument may consist of a single band pass filter which may be switched to either 3% or 23% (1/3 octave) band width and which may be tuned over the frequency range 0.2 Hz to 20 kHz in five sub-ranges. Tuning could be done either manually or swept automatically when used in conjunction with a graphic level recorder. The analyzer's versatility can be further increased by adding a photoelectric pick-up and trigger unit/phase meter so that the instrument can be used as balancing analyzer. Figure 8.8 (a) shows a vibration balance analyzer consisting of a meter, tracking filter, phasemeter and photo-optic probe.

The addition of a tracking filter to the general purpose vibration meter, makes a versatile tracking analyzer as shown in Figure 8.8 (b). The tracking filter, in addition to being manually tunable, can be tuned by virtually any

(a) photograph (courtesy of Brüel & Kjaer, Denmark)

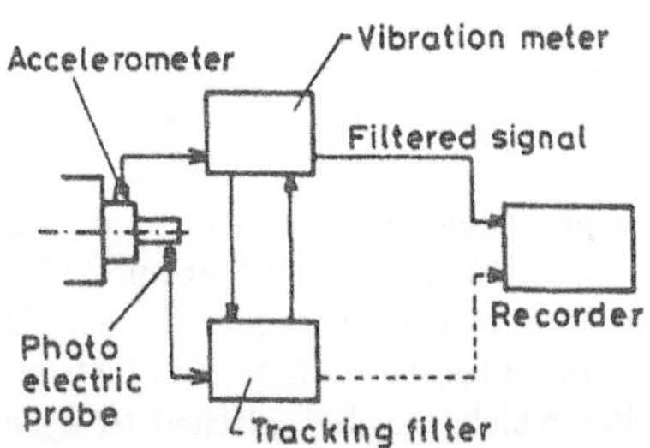

(b) use of tracking filter for engine order analysis

Figure 8.8 Vibration analyzer

periodic signal from, for example, a tachometer probe on the rotating shaft. This tuning facility enables the vibration signal to be analyzed during machine run up/run down. An additional feature of this filter is that it can be tuned to any ratio combination of the tuning (tachometer) signal frequency between 1/99 and 99/1. This enables engine order analysis to be performed, i.e. vibration levels attributable to various harmonics and sub-harmonics of a machine's fundamental rotation frequency are measured as a function of rotation speed.

8.2.6 Tape Recorders

It is often convenient to record the vibration, noise, pressure pulsation signals on magnetic tape for analysis of the signal later and at a greater convenience in the laboratory. This especially is the case when one deals with transient signals. Two recording principles are in common use, direct recording (DR) or frequency modulation (FM). FM recording is normally employed in order to obtain the linearity and low frequency response necessary for many vibration measurement purposes. A typical DR tape recorder would have dynamic range (narrow band) of 70 dB and frequency limits from 2.5 Hz to 50 kHz, while a typical FM tape recorder would have dynamic range of 60 dB and frequency range from DC to 10 kHz.

The tape recorder used in vibration measurements are multichannel so that several data could be simultaneously recorded. Out of these, four channel recorders as shown in Figure 8.9 are quite popular on account of their being portable. Since the tape recorder is likely to be most limiting factor in determining the dynamic range of the system, it is wise to choose the parameter for recording (acceleration, velocity or displacement) which has the flattest spectrum. Conversion between the parameters is quite straightforward once a narrow band spectrum analysis is carried out.

It is convenient, as well as necessary, to provide preamplification and conditioning of the signal input to the tape recorder. The preamplifier enables amplification of the signal so as to achieve full scale level of the tape recorder, even with small signal. The preamplifier usually includes integrator so that acceleration, velocity and displacement measurement can be performed. The preamplifier also provides a choice of high and low pass filters so that unwanted signals can be prevented from influencing the measurements.

8.2.7 Real Time Analyzers

The outstanding advantage of real time frequency analyzers is that they provide analysis, in all frequency bands over their entire analysis range, simultaneously. They give a virtually instantaneous graphical display of analyzed spectrum on a large built-in screen which is continuously updated.

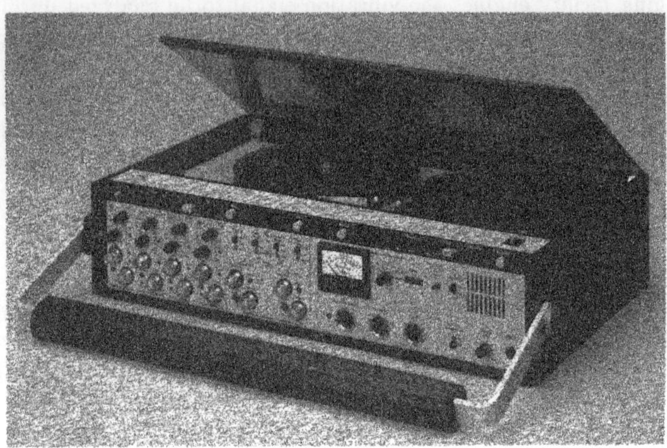

Figure 8.9 Four channel tape recorder with built-in charge amplifiers (courtesy of Brüel & Kjaer, Denmark)

Real time analyzers (RTAs) are also particularly well suited for the analysis of short duration signals such as transients, shock, etc. Read out and display of the analyzed transient takes place practically at the very instance of capture, which with serial frequency analyzing instrumentation is impossible. Most of the RTAs are based upon FFT procedure while some RTAs are based on recursive digital filtering. The former produce constant band width spectrum on linear frequency scale while the latter generate a constant percentage, 1/3 or 1/1 octave band width spectrum. For vibration analysis work FFT analyzers are better suited while analyzers using digital filters are more suited for condition monitoring and quality control work.

The FFT (Fast Fourier Transform) is an algorithm for transferring data from time domain to frequency domain. Because of the computational work involved, the transformation must be implemented on a digital computer. With this, the algorithm transforms digitized samples from time domain to samples in the frequency domain. The governing Discrete Fourier Transforms (DFT) are:

$$\text{Forward:} \quad G(k) = \frac{1}{N}\sum_{n=0}^{N-1} g(n)\, e^{-2\pi jkn/N} \tag{8.2}$$

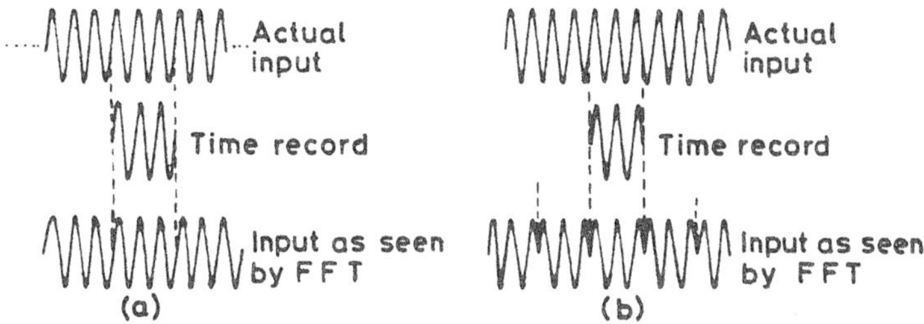

Figure 8.10 Input signals in time record (a) periodic, (b) non-periodic

$$\text{Inverse:} \quad g(n) = \frac{1}{N} \sum_{n=0}^{N-1} G(k)\, e^{2\pi jkn/N} \tag{8.3}$$

where $G(k)$ represents the spectrum values at the N discrete frequencies $k\Delta f$, and $g(n)$ represents samples of time function at the N discrete time points $n\Delta t$. The above equations are finite sums as compared to Fourier transform equations which are infinite integrals of continuous functions. The discrete nature of DFT gives rise to following problems:

Aliasing caused by sampling of the time signal, may result into appearance of lower frequencies which actually do not exist in the signal. This is eliminated by using appropriate anti-aliasing filters and ensuring that sampling rate is greater than twice the highest frequency of the input.

Time window effect resulting from the finite length of time record. For example, if the time record contains an integral number of cycles of the input sine wave, the sampled input to FFT may match actual input. On the other hand, if the input is not periodic in the time record, FFT algorithm is computed on the basis of highly distorted wave form as shown in Figure 8.10. Careful examination of Figure 8.10 (b) shows that most of the distortion is at the edges of the time record, the centre is, however, a good sine wave. If the time record is multiplied by a function that is zero at the ends of the time record and large in the middle, the concentration of FFT would be on the middle of the time record. Such a function is called *window functions* since it forces the data to be seen through a narrow window. There are various types of window functions available such as Hanning window, Rectangular window and Flat top window. In general, windowing brings vast improvement in the FFT result of input which is not periodic in time record as shown in Figure 8.10 (b).

A typical FFT analyzer has a transform size N, in equation (8.3) of 1024 data samples and theoretically gives 1024 frequency values. Moreover, since the data values are real, only 512 positive frequency values are calculated. Not all of the 512 values are used since low pass filters are needed for anti-aliasing purposes. For a commercial analyzer, it is typical to place the filter cut-off so that first 400 lines are valid and are displayed. Thus the frequency resolution Δf is 1/400 of the selected full-scale frequency and the automatically selected sampling frequency is 2.56 times the full-scale frequency. Figure 8.11 shows typical FFT based real time analyzers.

8.2.8 Remote Sensing

In certain situations it is impossible to have direct access to the output of the sensor, for example, a sensor (accelerometer, strain gauge etc.) mounted on a revolving component. The most widely accepted method of collecting signals from the sensors mounted on rotating component, is the use of slip rings. These are mounted at a convenient location on the rotating shaft and using special brushes for collecting the signals brought from the sensors to the slip rings through appropriate wiring. Recently the trend has shifted towards use of FM telemetry system on account of the high signal to noise ratio, longer life and greater simplicity in experimental setup caused by elimination of slip rings.

A typical telemetry system shown in Figure 8.12 comprises a FM transmitter, FM receiver, and a lithium battery (2.9 V) power source. The transmitter collects a suitably conditioned signal from the sensor, performs modulation and transmits the data by radiation through the antenna. The receiver accepts the signal from the receiving antenna and demodulates the signal to retrieve the information in it. Signal conditioners are used to adjust the gain, D.C. off-set, etc. In situations where the diameter of the shaft is small and the transmission distance is small, wire antenna are adequate to radiate and receive the radio frequency power. For larger shaft diameters and transmission distances, dipole antenna, matched to the transmitter output impedance, are used.

8.3 Data Reduction And Processing

There are many ways of obtaining and processing the vibration data for detecting and identifying specific problems in hydraulic machinery. Important amongst them are: 1) amplitude versus frequency, 2) amplitude versus frequency versus time, 3) amplitude and phase versus rpm., 4) time wave form, 5) Lissajous patterns (orbits), 6) mode shape analysis. These techniques are discussed in the following sections with emphasis on practical applications of diagnosing specific machinery malfunctions.

(a) Dual channel analyzer, Brüel & Kjaer, Denmark

(b) Dual channel analyzer, Onno Sokki, Japan

Figure 8.11 FFT analyzer

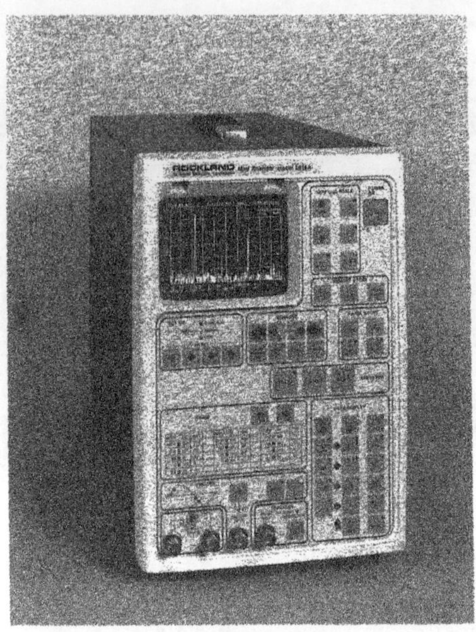

(c) Single channel analyzer, Rockland scientific Corp, USA

Figure 8.11 FFT analyzer

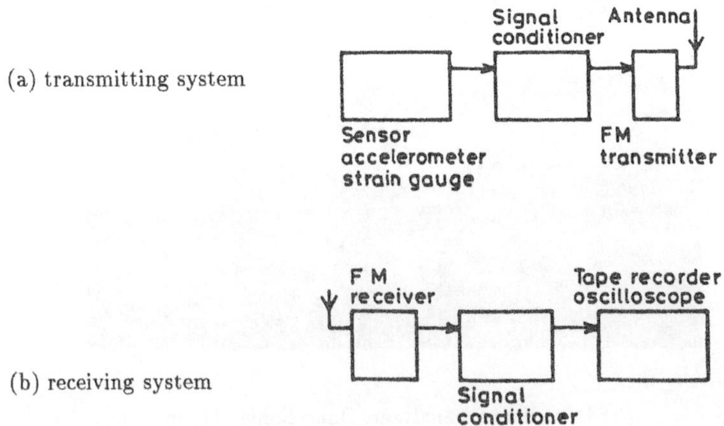

Figure 8.12 Telemetry system

8.3.1 Vibration Amplitudes versus Frequency Analysis

The procedure of obtaining and displaying the amplitude of vibration for all frequencies present, is one of the most useful of all analysis techniques since the majority of the practical problems can be identified through this procedure. However, successful identification of the problems in a machine utilizing this or any other analysis technique requires that complete data be obtained in a systematic way that will simplify interpretation.

It is common practice to record the amplitude versus frequency data measured in the horizontal, vertical and axial pick-up directions at each of the bearings of the machine being analyzed. Wherever it is possible, shaft vibration must also be measured and spectrum analyzed. For vertical machines, the measurements are undertaken in radial, tangential and axial direction. Obtaining measurements in all three directions is extremely important for distinguishing between various mechanical problems. For example, imbalance, misalignment and bent shafts will generally cause vibration at a frequency $1 \times n$ (higher harmonics will also be present for misaligned and bent shafts). However unbalance will almost always produce high amplitudes in the radial (vertical/horizontal) direction and relatively lower vibration in axial direction while misalignment of coupling and bearings or a bent shaft will generally show a relatively high amplitude of vibration in the axial direction along with radial amplitudes.

In addition to comparing the radial and axial readings, it is also important to compare the radial horizontal and radial vertical readings for horizontal machines, and radial and tangential readings for the vertical machines. For vertical hydromachines, the same should also be compared along two directions, namely upstream/downstream side in relation to the penstock/tailrace and across. Much can be learnt about the machine from such comparisons. Horizontal to vertical amplitude ratios of 2:1, 3:1, 4:1 and sometimes even 5:1 in the horizontal machines can usually be considered normal. An abnormally high ratio indicates possibility of resonance in the horizontal direction, while abnormally high vertical vibration may be caused by wiped bearings, extensive bearing clearances or other sources of looseness.

The data obtained through measurements must be displayed in such a way that it facilitates the interpretation. For illustration, Figure 8.13 shows vibration data obtained on a boiler feed pump operating on recirculation mode. The data shows that the vibrations both on Drive end (DE) and Non-Drive end (NDE) bearings are dominated by the vane passing frequency in all the three directions. These vibrations at vane passing frequency may be due to unsymmetrical setting of the pump rotor in the casing, too small tip clearances or even due to resonance condition. The subsequent analysis did reveal that the rotor setting was incorrect and also the impellers were not staggered. The data obtained must also include pertinent information regarding the machine, such as number of blades/vanes, guide vanes, stay vanes, machine sketch, the operating parameters and the vibration pattern at various operating conditions.

Vibration amplitude versus frequency data can be obtained and recorded in several ways. They can be obtained using a basic analyzer with a manually tuned filter wherein the filter is manually tuned over the analyzer frequency ranges, and the subsequent vibration amplitudes and corresponding frequencies identified by carefully observing amplitude and frequency meters. Some analyzers can be connected to a standard $X - Y$ recorder and the filter can be automatically swept over the various frequency ranges to obtain the graphic plots or signatures of the machine as shown in Figure 8.14.

Figure 8.13 shows overall vibration data in displacement, velocity, acceleration mode while the spectrum is shown in the acceleration mode. The analyst may decide which parameter such as displacement/velocity/acceleration to use for the spectrum analysis depending upon the situation. Acceleration mode has been chosen for the example cited above, in view of large acceleration levels in the presence of relatively low levels of vibrational displacement.

Spectrum Averaging

Standard frequency analysis procedures are carried out with the assumption that the vibration being analyzed is reasonably steady state to permit the frequency range to be scanned with the tunable filter. Unfortunately not all analysis situations meet these ideal requirements. Some machines may operate under continuously varying speed, load and temperatures, while some machines have random vibrations resulting from combustion and flow turbulences, for example. In other cases, the vibrations may be present for such a short period of time that the frequency analysis by standard procedures is impossible. In such situations one may use FFT based spectrum analyzer, which provides an instantaneous and continually updated display of the vibration amplitude versus frequency signature, so that the analysis is essentially displayed as it occurs. (Most of the FFT analyzers incorporate built-in strip chart recorders to provide hard copy signatures.) If the vibration spectrum does not fluctuate or change significantly when observed on the analyzer scope display, then a single instantaneous spectrum is adequate to describe the vibration characteristics of the machine. However, when conducting an analysis, it is not unusual to find that while some frequencies have relatively constant amplitudes, others have amplitudes that vary considerably around an average value. In some cases the spectrum may vary so much that it is impossible to accomplish the analysis with any confidence in the results. In such cases, spectrum averaging proves extremely beneficial.

Spectrum averaging in its simplest form is the process of taking a number of samples, adding them together and then dividing the sum by the number

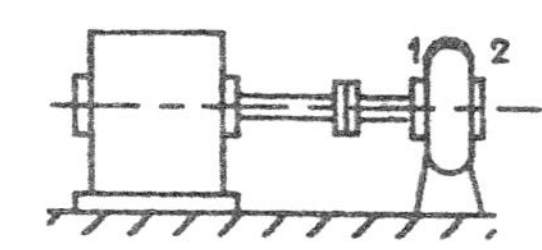

Condition	Probe location		Displacement micron PK-PK	Velocity mm/sec P-P	Acceleration m/sec^2 P-P	Frequency analysis m/sec^2 rms													
						1X	2X	3X	4X	5X	6X	7X	8X	9X	10X	11X	12X	13X	14X
Pump on recirculation discharge pressure 100Kg/cm2 speed 44Hz	Drive end (1)	H	45–50	50–55	130–150	-	-	-	-	-	-	43	-	-	-	-	-	-	3.31
		V	10	15	65	-	-	-	-	-	-	22.7	-	-	-	-	-	-	6.18
		A	12	15	55	-	-	-	-	-	-	5.78	-	-	-	-	-	-	7.24
	Non drive end (2)	H	20–24	28–30	50–55	-	-	-	-	-	-	9.2	-	-	-	-	-	-	19
		V	13–15	10–12	20–22	-	-	-	-	-	-	-	-	-	-	-	-	-	-
		A	12–13	15–17	36–40	-	-	-	-	-	-	12	-	-	-	-	-	-	-

Figure 8.13 Vibration amplitude versus frequency analysis of a 7-vaned pump

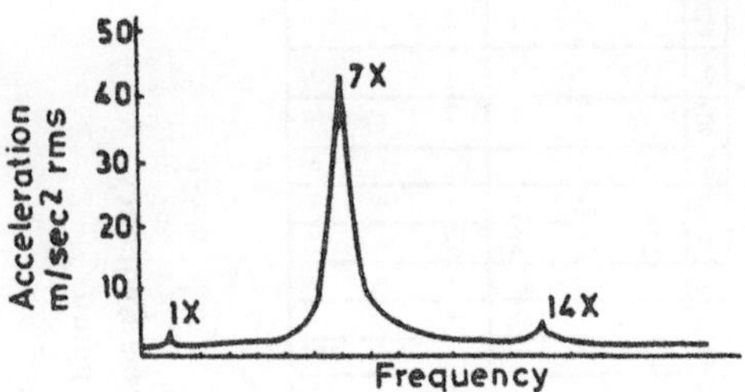

Figure 8.14 Spectrum of vibration (acceleration) on drive end bearing of pump in horizontal direction (Figure 8.13)

of samples. The result will approach the value to be expected if an infinitely large number of samples are taken. The modern analyzers provide sampling ranging from 1 to 512. A good way to determine the number of samples is to begin with a low number, for example, four and then increasing them gradually, each time observing the averaged spectrum on the screen till no significant difference between succeeding averages is found, see Figure 8.15.

8.3.2 Amplitude versus Frequency versus Time Analysis

A single signature can only reveal the characteristics of vibration at a single instant in time with the machine operating at a particular speed and under a specific load condition. However, sometimes it may be important to note the amplitude versus frequency data at various speeds, for example during start up, to obtain information about the resonant conditions or critical speeds being excited by the various forcing frequencies generated by the machine components. Also it may be necessary to evaluate the vibration amplitude and frequency characteristic during a transition in load, temperature or other operating variables. The high speed analysis capability of the real time analyzer is ideally suited for these requirements for the amplitude versus frequency versus time data. For this purpose the real time analyzer is used in conjunction with a high speed recorder to provide chronological signatures at a rate, perhaps of one every second.

Figure 8.16 illustrates one such data obtained during start up of a machine. Plots such as this are called *water fall* diagrams. One can easily see occurrence of an oil whirl during start up. The resonant/critical speeds excited by the rotor imbalance can also be seen. Water fall diagrams are also extremely useful for evaluating the effects of change in the load or other operating parameters of the machine, since the effect of change on each frequency can be detected. The occurrence of new or additional vibration frequencies

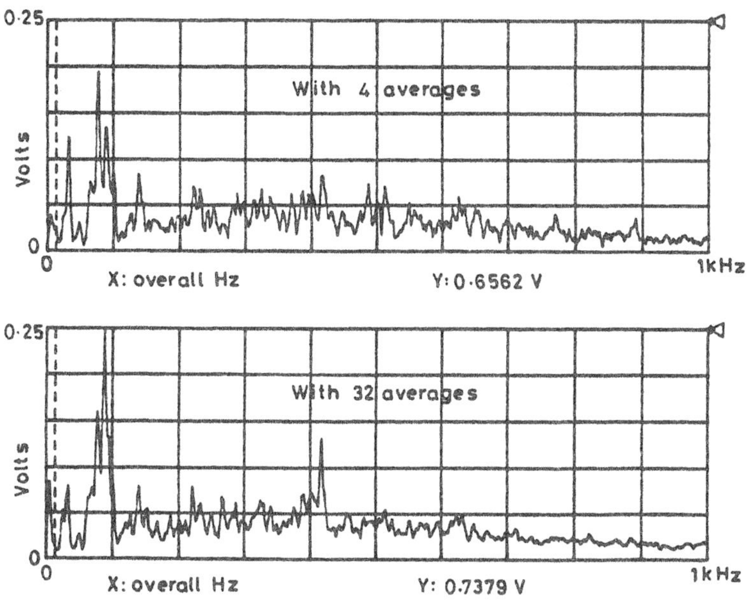

Figure 8.15 Noise spectrum at turbine pit in 50 MW Kaplan turbine

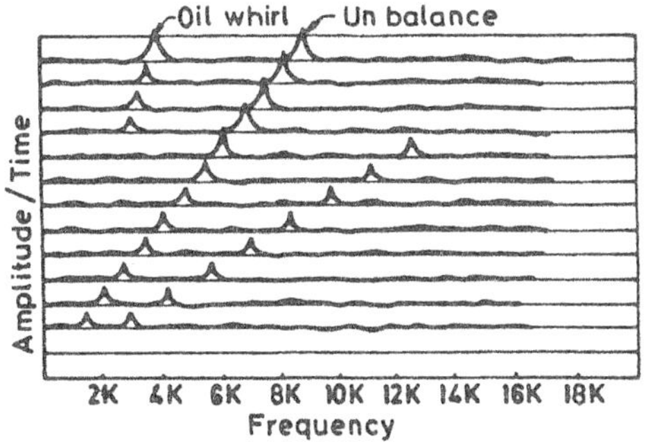

Figure 8.16 Water fall diagram

or the disappearance of existing frequencies can also be detected.

8.3.3 Amplitude/Phase versus rpm Analysis

The machines and their supporting structures have resonant frequencies at which very high amplitudes of vibration can result even from a relatively small exciting force. Since the machines and their structures are generally complex systems consisting of many springs and masses, a large number of resonant frequencies exist and thus resonance is a relatively common problem.

A great deal of information can be learnt about the response of a machine to the forces that cause vibration, by obtaining plots of vibration amplitude and phase as a function of rpm. A typical plot for a utility turbogenerator set is shown in Figure 8.17. Such a plot, sometimes called *Bode plot*, clearly identifies the resonant frequencies by the characteristic peak amplitude and (more importantly) by a corresponding 180 degrees phase shift. From the sample plot in Figure 8.17, it is apparent that the machine in operation has two resonant frequencies - one approximately at 1300 rpm and other at 2800 rpm. Should the machine in its normal running, experience excitation force which corresponds to either of these frequencies (the excitation could be due to unbalance, misalignment, hydraulic perturbation, torque pulse,etc.,), severe vibrations would be experienced. The normal exciting frequencies of vibration inherent to the machine, can be discovered by the amplitude versus frequency analysis of the vibration, as described in the previous section.

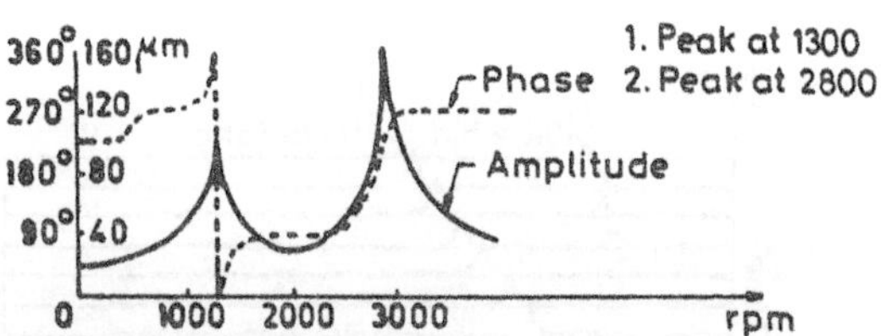

Figure 8.17 Amplitude/phase versus rpm analysis (Bode plot) of turbo generator

The resonant frequencies can also be found out by observing amplitudes of vibration at various points of time during the run-down of the machine. This, of course, has to be supplemented by the data of speed versus time during the run-down, so that the amplitude versus speed data can be generated. This method is quite liable to many errors. For a truly accurate and complete amplitude/phase versus rpm data, such as shown in Figure. 8.17, a vibration analyzer with a tracking filter is used. The instrument utilizes a reference pick-up at the shaft of the machine to provide a voltage pulse in each revolution. The $1 \times n$ signal from reference pick-up, automatically tunes the analyzer filter to the rpm. The $1 \times n$ signal from the reference pick up also controls the reference signal which is actually a D.C. voltage proportional to shaft rpm. This voltage is utilized to drive the X axis of the XY or $XY_1 - Y_2$ recorder for obtaining graphic plots of amplitude and/or phase versus rpm. The reference signal provides fixed reference for comparison with the signal from a vibration pick up, resulting in a D.C. voltage proportional to relative phase between the signals. This D.C. voltage is available to drive Y axis of an $X - Y$ or $XY_1 - Y_2$ recorder for plotting phase versus rpm.(phase versus amplitude).

The passage of a machine through a resonant frequency during run-down/run-up, is accompanied with a peak in the amplitude of vibrations and a characteristic phase shift of approximately 180 degrees. However, at times the signals could be quite confusing in so far as amplitudes are concerned. This is illustrated in Figure 8.18.

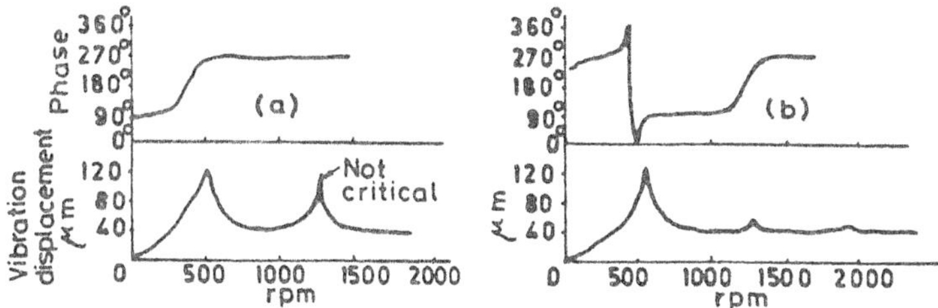

Figure 8.18 Passage through resonant frequencies (a) peak at 500 rpm with +180 degree phase shift showing true resonance but no resonance at 1200 rpm, (b) resonance at 600 and also at 1300 rpm

Figure 8.18 (a) shows two amplitude peaks at 500 rpm as well as at 1200 rpm. However at 1200 rpm, a phase shift of 180 degrees (approx.) does not exist, so that 1200 rpm is not a resonant speed. Figure 8.18 (b) shows a clear resonant peak at 600 rpm and also 180 phase shift at 1300 rpm, but no amplitude peak. The 1300 rpm is also a resonant frequency though the levels are not peaking up at this speed because the pick-up has been perhaps placed at a nodal point. It may therefore be necessary to alter the pick-up position. The absence of a peak at 1300 rpm may also be due to very low level of excitation force or also due to heavy damping.

At times the amplitude and phase versus rpm data may show a amplitude peak accompanied with phase shift of 360 instead of 180 degrees. Such a situation arises in the case of two systems in resonance at or near the same frequency, for example a tuned absorber system would show such behaviour.

The interpretation of amplitude of vibration and phase versus rpm plots obtained using non-contact (shaft) pick ups can be extremely confusing at times, since the non- contact or proximity probe cannot distinguish between actual shaft vibration, eccentricity of the shaft journal and apparent vibration caused by magnetic unevenness of the shaft surface (called runout or glitch). As a result, the signal from the non-contact pick-up would consist of vector sum of runout and actual shaft vibration. It is therefore advisable to carefully measure runout amplitude before response measurements are taken or employ instrumented runout subtraction.

In each of the examples discussed in the previous paragraphs, the recorded vibration amplitude occurred at the rotating speed frequency ($1 \times n$) of the machine. However, these plots alone may not reveal the system response to other vibration frequencies which can affect the overall machine performance. To illustrate this, consider the amplitude versus frequency signature of a typical turbogenerator (TG) set operating at 3000 rpm as shown in Figure 8.19 (a). In this instance, the TG has two significant exciting frequencies, i.e., $1 \times n$ due perhaps to unbalance and another at $2 \times n$ (6000 rpm) probably due to misalignment. It is further assumed that the TG rotor has critical speeds (resonant frequency) at 1300 and 2850 rpm.

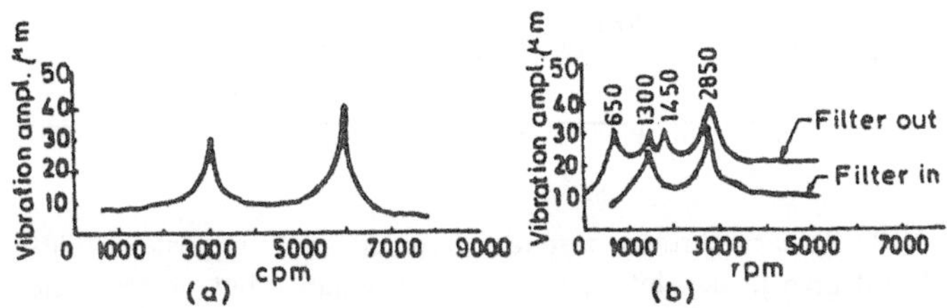

Figure 8.19 Vibration of a turbo generator set (3000 rpm) (a) amplitude versus frequency signature (b) filter- in/filter-out amplitude versus rpm

Obviously when the TG coasts down in speed, the unbalance at $1 \times n$ will excite the resonance when the speed of the TG approaches 2850 rpm. In addition, as the TG continuous to run-down, resonance will once again get excited when the speed reaches 1425 rpm, on account of $2 \times n$ excitation caused by misalignment. This excited resonance will not appear in the Bode plot at 1425 rpm because the analyzer filter is tuned to $1 \times n$ and rejects all other frequencies. Because the filter-in amplitude versus rpm plot does not always give a complete picture of the total system response, it is customary to obtain two plots of amplitude versus rpm; one, filter-in, synchronized to $1 \times n$ and second (filter- out) overall amplitude versus rpm plot as shown in Figure 8.19 (b). These plots are of great value in ascertaining the nature of the distress in the machine.

8.3.4 Time Wave Form Analysis

Although vibration spectrum analysis is generally adequate for identifying most of the machinery problems, sometimes additional information is needed to diagnose a particular defect or to study the dynamic behaviour of a machine under specific operating conditions. One of the additional techniques often used, is the observation of the time wave form of the vibration signal on the oscilloscope. The oscilloscope observations facilitate identification of the spiky nature (if any) of the signal and also enables confirmation as to whether or not the signal is constituted by harmonically related frequencies.

8.3.5 Lissajous Pattern (Orbit) Analysis

It has been mentioned in the section 8.2.3 relating to shaft vibrations, that the signal output from two mutually perpendicularly installed proximity probes around the shaft, can be fed to the horizontal and vertical inputs of the oscilloscope to depict the dynamic motion of the journal centre. The plots or displays, thus obtained, are called Lissajous patterns and are also called orbits. When observing Lissajous patterns on an oscilloscope, it is difficult to obtain frequency information from the display unless some kind of frequency reference such as a synchronous $1 \times n$ pulse is superimposed on the display. The synchronous reference pulse is obtained by installing an electromagnetic pick-up or a non-contact pick-up observing a protrusion or depression on the rotating shaft and the synchronous pulse obtained from electromagnetic pick-up can be applied to the Z axis (intensity) input of the scope. There are many mechanical problems which are readily identified by the characteristic patterns of the shaft orbits. The use of Lissajous patterns for detecting the nature of distress in the system shall be discussed in the following section.

8.3.6 Mode Shape Analysis

The mode shape analysis technique is extremely useful for confirming resonance conditions, identifying the nodal and antinodal points and revealing sources of structural weakness. In this, one determines the vibratory mode shape of the vibrating component, by making measurements of vibration amplitude and phase at various points on the component, plotting them to obtain the mode shape. For example, Figure 8.20 shows mode shape of vibrations of a vertical pump structure. The mode shape clearly indicates that the structure is vibrating at the second flexural resonance. Not only has the resonance condition been confirmed, but the nodal points have been identified as well. This is very important, especially when a decision to stiffen the structure is reached since the mode shape analysis would enable avoiding fixation of stiffening members at the nodal points. If the stiffening members were added to the structure at the nodal points, little or practically no improvement might be seen.

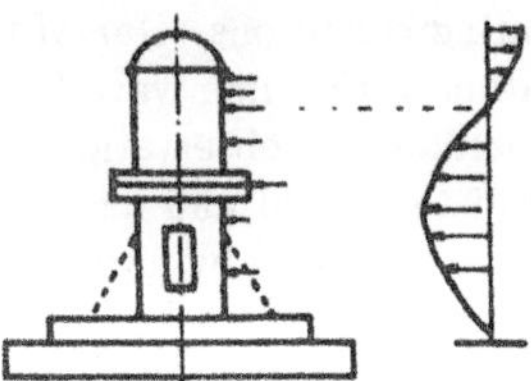

Figure 8.20 Mode shape of vibrations of a vertical pump

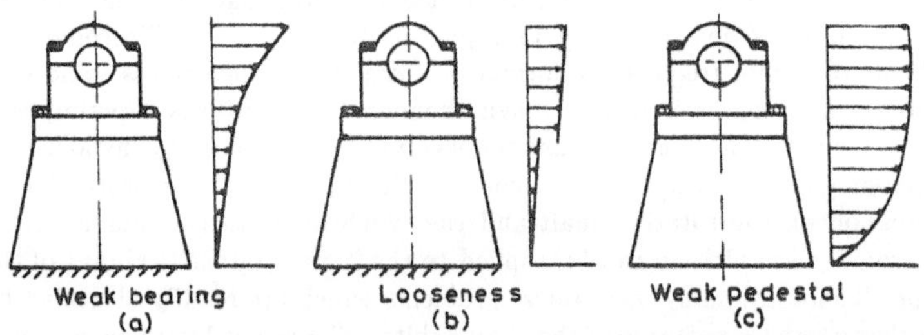

Figure 8.21 Mode shapes for (a) weak bearing, (b) looseness and (c) weak pedestal

The mode shapes in Figure 8.21 illustrate identification of sources of weakness of a bearing pedestal. Figure 8.21 (a) shows low levels of pedestal vibrations and high levels of vibration in the bearing block. Any attempt to strengthen the support would obviously be fruitless. The solution lies in strengthening the bearing. Figure 8.21 (b) illustrates that vibrations on the bearing block are high but do not increase significantly with elevation and also that the bearing block is moving back and forth on top of the pedestal, which is indicative of slackness in the holding down bolts. Figure 8.21 (c) clearly indicates that the vibrations are due to weaknesses of the pedestal. Stiffening of the pedestal in such a case would be highly beneficial.

In short, mode shape analysis can be extremely valuable for identifying sources of structural weakness, resonance, the nodal and antinodal points, before structural modifications are made, in an attempt to solve a problem of excessive vibration. This will ensure avoiding costly and often embarrassing trial and error approaches for solving the problems.

8.4 Diagnosis and Corrective Actions

The perturbation forces responsible for vibration of hydraulic machinery can be classified into two main categories, i.e. (1) periodic in nature under steady state regimes, (2) aperiodic during transition (unsteady) regimes of operation. The perturbation forces for each of the regimes, i.e. steady and unsteady, can further be sub- divided into mechanical, hydraulic and electrical. The basic perturbation forces experienced by hydro units under each of these heads are listed below.

8.4.1 Steady State Operating Regime

(i) Mechanical Perturbation Forces are: (1) centrifugal forces due to imbalance of the rotating parts, (2) elastic forces of the shaft which appear when the centreline is disturbed or distorted, (3) frictional forces, (4) forces caused by incorrect setting and unfavourable operating conditions of guide bearings.

(ii) Hydraulic Perturbation Forces appear due to following causes: (1) Presence of a vortex in the spiral casing, wicket gates, runner and draft tube, (2) non-uniform velocity distribution in flow passages of the turbine, (3) pressure pulsations in the penstock of the hydro unit, (4) flutter of blades of the adjustable blade turbine runner, (5) hydraulic imbalance, (6) operation under cavitation regime.

(iii) Electrical Perturbation Forces are: (1) Unbalanced magnetic pull experienced by generator rotor, (2) forces due to short circuiting of poles, (3) forces created by asymmetrical operating conditions of the generator, (4) forces created by asynchronous operation of the generator, (5) forces and moments caused by magnetic field due to non-uniform air gaps between stator and rotor. The geometrical form of the stator and rotor play an important role in this regard. The perturbation forces in the unsteady state operating regime usually appear during starting/stopping of the unit, load changes, synchronization, etc.

8.4.2 Detection of Perturbation Forces and Corrective Action

Mechanical Imbalance

The components of the hydro units which are vulnerable to imbalance are the turbine runner, generator rotor and the exciter rotor. The imbalance in these components may result from: (a) defects/errors in construction-insufficient rigidity of the hydro unit shaft resulting into bending and the appearance of centrifugal forces in the rotor masses, improper wedging of generator poles leading to non-uniform displacement in operation, (b) manufacturing defects, (c) errors in assembly-inaccurate assembly of components, misalignment of shaft couplings, skewed setting of the labyrinth seal rings, (d) operation of unit-settlement of insulation, non-uniform wear of the runner due to cavitation etc. The imbalance may also develop after the repair and overhaul of the machine (non-uniformity of weld material on the blades, unsatisfactory centring and balancing of the units after maintenance).

The presence of imbalance of the rotating system can primarily be judged by high levels of vibrations at rotating speed frequency in the vibration signature, obtained from pick-ups mounted on the bearing housing and more preferably obtained from shaft pick-ups. As mentioned previously, measurement on bearing housing should be done for vertical machines in three directions, namely radial, transverse and axial and for horizontal machines in radial horizontal, radial vertical and axial directions. Normally the largest amplitude (both in overall and $1 \times n$ component) would occur in the radial direction. However in case of overhung rotors, considerably high levels would be experienced in axial direction also. Similarly a rotor mounted between bearings and having a substantial couple imbalance would also show considerable vibration in axial direction.

Judgement about the nature of imbalance, i.e. static, couple, quasi-static or dynamic imbalance, can be done by the usual signature analysis coupled with phase measurements. Static imbalance would indicate almost identical vibration levels at both bearings and also the phase; couple imbalance would show 180 degrees phase difference; quasi-static imbalance would show unequal levels of vibration with phases either equal or 180 degrees opposite, dynamic imbalance would show unequal levels and also phase difference other than zero or 180 degrees. Predominance of $1 \times n$ frequency in the vibration signature can be, in addition to the imbalance, due to several other reasons. Hence in order to establish that the $1 \times n$ is due to imbalance, additional analysis is required.

In case of vibration due to imbalance, the phase difference between horizontal radial (or radial) and vertical radial (transverse for vertical machines) vibration is 90. In case of problems arising out of resonance, this phase difference would be either zero or 180. Another way to establish the imbalance condition is to watch increase in $1 \times n$ component as a function of the square of the rotating speed and confirm whether or not it is linear function. The shaft orbit analysis also proves to be very effective tool for establishing the condition of imbalance. Figure 8.22 (a) shows a typical shaft orbit when imbalance is a prominent perturbation force.

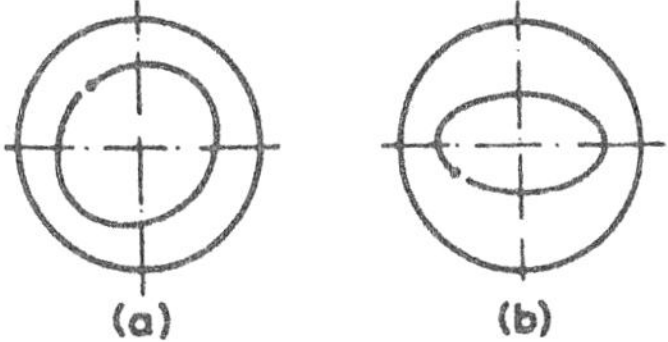

Figure 8.22 Shaft orbit (a) slightly elliptical orbit due to imbalance, (b) highly elliptical pattern due to misalignment, worn-out bearing or resonance

The single reference pulse on the pattern verifies that the shaft motion is occurring at a frequency of $1 \times n$. If the pattern takes on a highly elliptical shape such as the one shown in Figure 8.22 (b) (ratio of major axis to minor axis of the order of 8:1, 10:1 or more), the machine may perhaps be operating at resonance or near resonance. This could be easily confirmed by noting the effect of change of speed. The existence of imbalance (especially dynamic imbalance) in the presence of hydraulic perturbation forces may lead to situations wherein it may be almost impossible to balance the machine and it takes a large number of balancing trials before a satisfactory result is achieved. Barring such situations, the imbalance can be corrected by resorting to single plane or two (or more) plane balancing depending upon the nature of imbalance.

Elastic Forces due to Loss of Centring or Distortion of Shaft Centre Line

The distortion of the shaft centreline and loss of centring of the hydro unit could result from: (1) loss of perpendicularity between shaft axis and the plane of flange coupling giving rise to angular misalignment, (2) plane of the supporting boss of the vertical journal and shaft axis not perpendicular to each other, (3) parallel misalignment between generator and turbine shafts, (4) guide bearings not coaxial with shaft, (5) thrust collar not perpendicular to shaft axis, (6) thrust disc surface not true.

The angularity between axis of coupled shaft (angular misalignment) caused by non-perpendicularity of shaft axis with reference to the plane of flange coupling, gives rise to forces and hence the vibration both in radial and axial directions.

Experience shows that whenever axial vibration is greater than 50% of the highest radial vibration, misalignment is likely to be the one of the major source of perturbation force. As indicated in Figure 8.23 (a), angular misalignment primarily subjects the shaft to axial vibration at the same frequency as shaft rpm.

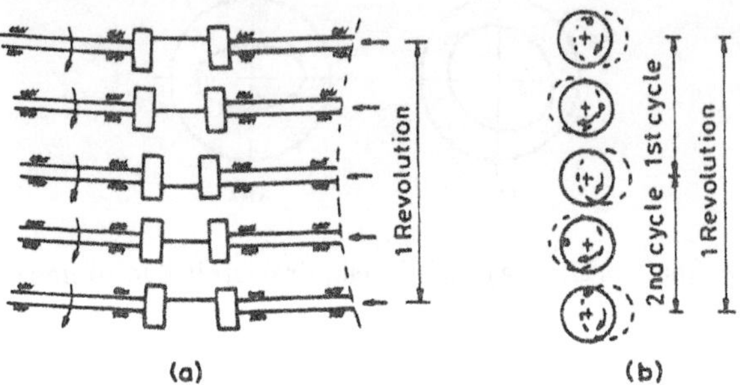

Figure 8.23 Simplified models of (a) angular misalignment (b) off-set alignment

Parallel or off-set misalignment as illustrated in Figure 8.23 (b) produces primarily a radial vibration at twice the rotational frequency of the shaft. It may be appreciated that Figure 8.23 is a highly idealized model using a single pin connection across the coupling. In actual practice multiple pin (for multitooth coupling) connection could produce a highly complex vibration pattern in radial and axial directions and consequently higher frequencies would be generated in the vibration signatures. It is quite usual to see higher harmonics such as $3 \times n$, $4 \times n$ etc., apart from $2 \times n$ component in the vibration signature associated with misalignment.

Figure 8.23 shows that the radial vibration resulting from misalignment will occur predominantly in the direction of misalignment. Thus if the coupling halves are offset vertically (in horizontal machines) the predominant radial vibration is likely to occur in the vertical direction. As a result, the radial vibration of misaligned couplings will often be somewhat directional in nature. The shaft orbit, therefore, may not reveal circular or slightly elliptical patterns characteristic of imbalance, but instead may reveal a pattern shaped like an elongated ellipse or a banana. Sometimes a pattern of a figure eight or still more complicated patterns may develop if higher order frequencies are involved. This is illustrated in Figure 8.24. In some situations, misalignment does not involve couplings but bearings. In such cases, no vibration problem would be encountered if the state of balance is excellent. With imbalance in the system, the vibrations would be experienced both in radial as well as axial direction.

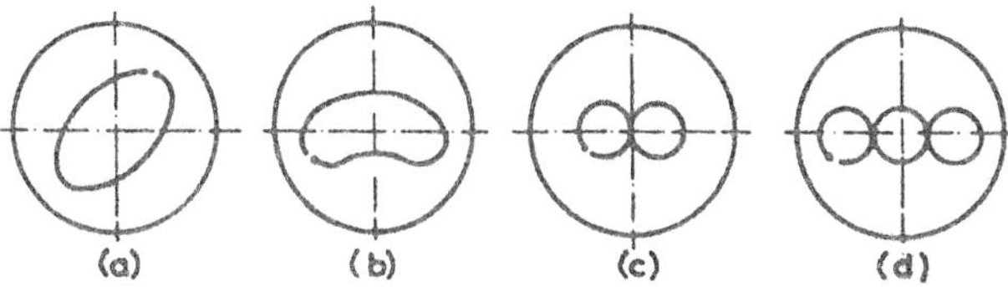

Figure 8.24 Orbit patterns (a) highly elliptical pattern may indicate misalignment, (b), (c) and (d) with heavy misalignment

The misalignment of bearings could be due to errors in the erection of the machine or sometimes due to structural distortions. Misalignment between the shaft and the guide bearing also gives rise to inclination of the shaft from the vertical during rotation, resulting in perturbation forces and thrust on the pad support and in some cases overloading of the bearings causing structural damage. This is also one of the principal causes of the opening of guide bearing clearances and consequent increased vibration problems during operation. The fundamental frequency of vibration in such cases is $1 \times n$ and is also associated with $2 \times n$ component.

The unevenness of the thrust disc surface, especially that caused by elastic and thermal deformation during operation, gives rise to vibrations at frequencies corresponding to $1 \times n$ and $k \times n$ (k = number of thrust pads). In cases, where the thrust disc attains an inclined position on account of unequal deformations of the springs of the spring mattress supporting thrust pads, or disturbed levelling of the thrust pads on account of errors in the erection/assembly, the machine exhibits $1 \times n$ dominated vibration problems. The centring and levelling of the machines are, therefore, the absolute requirements for vibration-free operation of the machine.

The misalignment at coupling flanges can be corrected by scraping the flange faces. In some cases, use of spacer/shims is also made to correct the misalignment. This, however, should never be considered as permanent a solution since spacers quickly breakdown during operation, resulting in greater play in the shaft.

One of the factors often overlooked in the assembly and erection of hydromachines is the centring of the exciter and pilot exciter. The inclination of the axis of the exciter with respect to the generator axis, causes displacement of the centre of gravity of the exciter from the centre line of the generator. This results in perturbation forces which are transmitted to the thrust bearing.

Frictional Forces

The vibration problem caused by the frictional forces has been discussed in section 3.2.2 and hence will not be discussed here in details. However, the following important points may still be mentioned.
(a) The self-excited vibration due to internal friction within shaft material can appear only when the rotational speed of the turbine exceeds critical. (b) The greater the internal friction provided by the shaft, the earlier the appearance of self-excited vibrations. (c) The vibrations occur at the natural frequency of the system. Such vibrations can occur, usually at runaway speed of the turbines.

The self-excited vibrations of the shaft can also result from a film of lubrication in the guide and thrust bearing of the machine.

It has been seen from many tests on operating hydro machines that the hydrodynamic bearings are lubricated not just with oil but a mixture of oil and air, which is present in the form of stable suspension. The air bubbles are usually quite clearly visible and are comparable in diameter to the thickness of the oil film. At times, the diameter may exceed the thickness of the oil film. The degree of saturation of the oil with air in thrust bearings depends upon: (a) discharge of oil from the bearing, (b) turbulence at the surface of the tank, (c) foam at the surface due to agitation of the oil, caused by obstruction in its path (oil coolers,baffles etc.). The foaming sharply increases in the presence of even a smaller amount of water (which can find its way in the system through several means).

The presence of air bubbles in the oil delivered to the thrust bearing of the hydro units, may be one of the causes of direct contact between bearing and shaft surfaces giving rise to vibration problems which are friction induced. Figure 8.25 shows Lissajous patterns when oil whirl and rub phenomena occur.

Hydraulic Perturbation Forces

The hydraulic perturbation forces have already been discussed at length and therefore no detailed discussion on them is presented in this section. It must, however, be borne in mind that the detection of these perturbation forces requires not only the bearing/shaft vibration analysis but also the measurement and analysis of the pressure pulsations in the flow path (starting from penstocks to the draft tube), analysis of noise in the turbine pit, draft tube pit and in the area surrounding the main inlet valve and the study of operating parameters. The measurements, as mentioned, must be accompanied

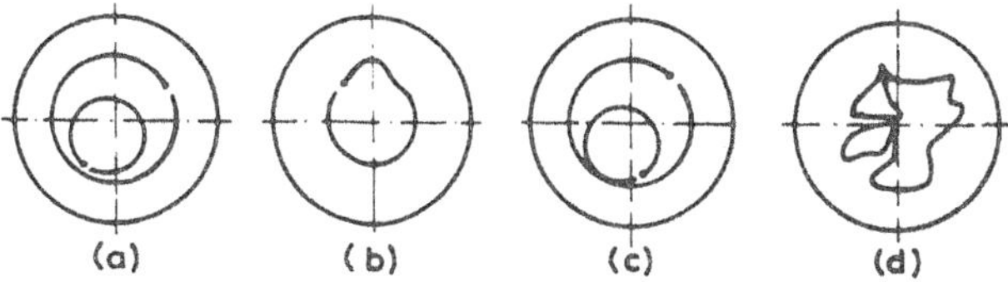

Figure 8.25 Orbit patterns (a) oil whirl, (b) once/revolution mild rub, (c) light rub and (d) heavy rub

with calculations of the vortex-shedding frequencies, generation of interference diagrams for the runner/guide vanes, etc.

An example of such vibration analyses is demonstrated in the following case study.

Case Study: High vibration/noise and blade failure problems on 60 MW, 167.7 rpm hydromachine, specific speed 263 (metric). A large number of cracks were detected in the runner blades of a medium head Francis runner in a power house where six turbines were installed. The design head of the machines was 66 meters, however due to changes in the hydrological conditions of the region, the machines had operated at a head of higher than 90 meters for a considerable amount of time. During the operation of the machines, the phenomena of high vibrations as well as noise was noticed. The failure of the blades was observed between 13000-15000 hours of operation of the machines. The cracks in the blades had propagated from the trailing edges, travelled along the crown for some distance, then dropped down at 45 degrees and finally progressed towards the leading edge. Scanning Electron Microscopic examination of the failed blades confirmed the failure to be taking place due to fatigue.

Considering the fact that the machines had been exhibiting high vibrations and noise from the time they were commissioned, it was necessary to study the vibration/noise behaviour of the machine at two head conditions namely 90 meters and 62 meters at various loads. The analysis of vibrations, draft tube and turbine pit noise showed that at a 90 meter head, the behaviour of the machines is quite rough, whereas at a 62 meter head, a marked improvement was exhibited. It was also seen that the noise levels decreased from 112 dB to 105.2 dB (at 90m) and from 106 dB to 103.7 dB (at 62m head) when air is injected below the runner. The vibration spectra at a 90 meter operating head indicate prominent perturbation forces at 42 Hz (which corresponds to runner vane passing frequency for 15 runner vanes) and 245 Hz, while the noise spectra showed 42Hz as the prominent frequency of perturbation force. The vibration/noise spectra at design head however showed extremely low levels of excitation at 42 Hz and 245 Hz.

One of the uncracked runners was subjected to a modal test to determine the various natural frequencies and the mode shapes. These tests revealed that the runner has two natural frequencies 46 Hz and 245 Hz, in the domain of our interest. Out of these, 46 Hz and 245 Hz frequencies correspond to bending and torsional modes. It may be noted that these frequencies were detected on a runner in air and thus, they may come down slightly, in water due to added mass. The calculations showed the vortex-shedding frequency to be lying in the range of 235 Hz (90 meter head) to 248 Hz (65 meter head).

Based upon these findings it was concluded that the fatigue failure of the blades had occurred on account of resonance at 42 Hz and 245 Hz (blade passage frequency and vortex-shedding frequencies in the close vicinities of natural frequencies) under prolonged operation at off- design head conditions. To solve the problem, the trailing edges were thinned down and chamfered to 45. The problem of the cracking of the runner ceased after this modification.

Electrical Perturbation Forces

The electrical perturbation forces normally encountered are: (a) periodic component of magnetic pull, (b) forces due to non-uniform air gap between the stator and rotor, (c) forces due to short-circuiting of rotor windings, (d) forces due to asymmetrical operation.

The frequency of electrical perturbation forces is fairly high (50 Hz and above) under steady state operating conditions. Their impact is largely confined to generator stator and to a lesser extent to the turbine. The errors in the construction of the stator and rotor, and unequal impedances of poles are the usual source of vibration.

There have been cases where the state of balance of the machine becomes significantly changed when the machine is excited and loaded. This sometimes leads to the necessity of carrying out compromise balancing, so that the machine exhibits somewhat higher vibrations in the mechanical run, but smooth vibration behaviour in the excited and loaded condition.

8.4.3 Preventive Maintenance

The usual maladjustments experienced on hydro units are:(a) slackening of assembled components and bolted joints – slackening of joints occur at the supporting lugs of upper spider of the generator, housing of generator guide bearings, lower spider, hub of the turbine runner and labyrinth seal ring, generator rotor poles, runner blades of adjustable blade turbines, holding down bolts of the buckets and at various bolted joints; (b) increase in the clearance in guide bearings of turbines and generators; (c) silt erosion of turbine blades, guide blades, etc.; (d) cavitation erosion of the blades; (e) wear out of labyrinth.

Monitoring of the behaviour of the machine from noise and vibration points of view, always helps in obtaining an idea about the deteriorating condition of the unit. Gradual increase in the levels of shaft vibration may indicate loss of balance, on account of wear of the runner or opening of the clearances at the bearings. One of the powerful indicators is the shift of the journal centre with respect to pads; this can be known from the data of D.C. voltage of the proximity probes as indicated earlier. The gradual increase of noise in the generator pit is usually associated with electromagnetic phenomena and state of assembly of the stator core. Any abnormal increase in vibration/noise of the machine should be regarded as a warning of deteriorating condition and an alarm for maintenance.

It is advisable to continuously monitor the temperature of the oil and thrust pads, the depth of foam and the presence of air bubbles. An increase in the thrust pad temperatures may indicate additional hydraulic thrusts caused by rotor lift. The changes in the axial thrust are caused by changes in the seal gaps and unloading effectiveness of the upper rim. This, of course, will be associated with appearance of vibration problems.

One of the problems often encountered on the hydrogenerators, is the loosening of the stator wedges during operation of the machine. The loose bars tend to vibrate causing wear of the insulation layer and ultimately leading to failure of insulation. It is therefore prudent to check the tightness of wedges whenever opportunities are available during short period shut downs.

8.4.4 Expert Systems

Vibration measurement forms an integral part of turbomachinery diagnosis. As seen in previous sections, the proper interpretation of vibration data and its utilization for diagnosing problems in the machine, is a complex task requiring analytic skills and a wealth of experience. The traditional diagnosis process therefore requires the presence of an experienced engineer, even for relatively simple problems. Thus there is an increasing need for computer-based consultation systems which can help maintenance personnel diagnose the problems in absence of the experts. Of late, such systems have appeared for medical diagnosis, electronic trouble-shooting, diesel locomotives trouble-shooting and for turbomachinery.

Most of the computer-based systems are also called knowledge based systems because of the way they are designed. These systems separate knowledge about the problem domain from the control strategy needed to utilize the knowledge. Knowledge can be represented in a number of ways, one of the best known methods employ simple *if* premise *then* conclusion kind of rules. The other method is to generate and store common situations of distress, behaviour during distress and identify the situation at hand. Another form of knowledge representation is decision trees.

The rule-based system can easily incorporate special case reasoning of the expert. Also each of the rules represents a separate chunk of knowledge, thus either existing knowledge can be modified or added to easily. Furthermore a rule-based system can very conveniently explain the reasoning by displaying rules which were used to arrive at the conclusion. Rule-based systems as mentioned above, consist of simple *if condition* then *action* kind of rules. These rules are utilized by the interpreter which links together in chains of inference. The interpreter is basically of two types (a) forward chaining (event driven) or (b) backward chaining (goal driven). A forward chaining interpreter starts from a list of facts and tries to find all the rules whose premises are satisfied. If it succeeds in finding such rules, it asserts the conclusions of those rules as new facts and repeats the whole process. This method continues until no more new rules can be found to be true.

These systems are most appropriate for tasks that respond to a stream of data input such as the one during condition monitoring. A backwards chaining interpreter employs a different approach. It starts from a goal and tries to find out all the rules whose conclusions are the same as the goal. For each of these rules it makes the premise of the rule as the new goal and repeats the process, until either all the premises are proved to be true when the conclusion is asserted, or one of the premises is found to be false/known when the rule fails. Medical diagnosis systems utilize this approach.

The diagnosis requires the following steps. (a) Knowledge of basic mechanical characteristics: physical parameter of machinery and details of components like seals, bearings, coupling, etc., are important. Other important data are natural frequencies, mode shapes, etc. (b) Basic knowledge of the problems likely to occur and their corresponding symptoms. This knowledge can very conveniently be represented by heuristic rules of thumb which relate the presence of symptoms to incipient problems in the machine. The problems occurring in the machine could be broken into a small number of major categories; these categories could be further broken down into subcategories. Thus a hierarchy could be imposed on the diagnostic task. Table 8.1 summarizes this approach. The first process of expert diagnosis consists of forming preliminary diagnosis about general problem categories. The second step consists of refining the initial diagnosis with the help of additional information.

The key parameters that act as symptoms for diagnosing the problems are: (a) machine characteristics - rotor/stator vibration, frequency, direction, phase, orbit and bearing temperature, (b) process characteristics - load, rpm, head, tailrace level, guide apparatus opening and draft tube pressure.

The expert systems utilize computer languages such as Prolog or LISP. The rules are represented as clauses in the following manner:

Check clause (A, B, C)
B = X, Y, Z

A is the problem name/category, B is collection of statements/premises which need to be true for A to be true. X, Y, Z represent individual premises, C denotes certainty associated with a rule. Thus the following rule in English, 'if the predominant frequency is $1 \times n$ and the amplitude is highest in radial, then the initial problem is imbalance with certainty 0.5', would be expressed as check clause (iprob (imbalance), B, 0.5):-

B = Pred-Frequency (1)
Pred-ampl.(dir.(radial))

The facts are mainly represented as attribute value pairs as shown above. The Pred-frequency (1) is a fact in which pred-frequency is the attribute and 1 is the value of that attribute.

For the malfunctions depicted in Table 8.1 rules can be made. For example, if predominant frequency is $1 \times n$ and vibration radial and on the rotor, then the problem is imbalance with belief 0.7, or if predominant frequency is $2 \times n$ and vibration radial, problem is mechanical looseness with belief 0.4. It is evident that in practical situations one may come across more than one likely cause of a problem. Consider for example that there are four rules which add to belief in imbalance and misalignment as follows:- Rule 1 belief in misalignment 0.4, Rule 2 belief in misalignment 0.2, Rule 3 belief in imbalance 0.6, rule 4 belief in imbalance 0.5. These beliefs could be combined by Dempster's rule of combination resulting in total belief in misalignment as 0.48 and in imbalance as 0.2.

The expert systems do provide opportunities to quickly self-assess the nature of problems encountered but, at the present level of development, cannot replace the human expert.

Table 8.1 Problems in hydro units

Resonances	Support structure Rotor bearing system critical Static part
Instability	Oil whirl Hysteritic whirl Oil whip Seals
Hydraulic Problem	Misaligned runner Acoustic resonance Vane passing frequency Pressure pulsation Vortex shedding Draft tube vortex Stator-runner interference Hydraulic imbalance Flutter of vanes Rotor lift
Mechanical Problem	Imbalance Coupling misalignment Bearing misalignment Incorrect levelling and centring Looseness Wear of runner Rubs Excessive bearing clearance Foaming of oil Excessive thrust bearing temp.
Electrical	Imbalanced Magnetic pull Unequal air gap Stator/rotor not found Shorted pole

References

8.1 Bently, D.E. (1974). 'Forced subrotative speed dynamic action of rotating machinery', ASME Publications, 74-Pet-16, Petroleum Mechanical Engineering Conference, Dallas.

8.2 Bently, Nevada (1989). *Rotating machinery systems and service*, Bently Nevada, Minden.

8.3 Bhave, S.K, Murthy, Ch.B. and Goyal, S.K. (1986). 'Investigation into blade failures of Francis Turbines', *Water Power & Dam construction*, January.

8.4 Brüel & Kjaer (1989). *Master catalogue electronic instrument*, Brüel & Kjaer, Denmark.

8.5 Clapis, A., Lapini, G. and Rossini, T. (1977). 'Diagnosis in operation of bearing misalignment in Turbogenerator', Proc. ASME Design Engineering Technical Conference, Chicago, September.

8.6 Den Hartog J.P. (1956). *Mechanical Vibrations*, McGraw-Hill, New York.

8.7 Franklin, D.E., and Moysa, N. (1978). 'Investigations into Francis turbine runner blade cracking', B.C. Hydro Power Authority, Canada, Research report. Also in Proc., IAHR Symposium on Hydraulic Machinery, Equipment and Cavitation, Fort Collins, 1978.

8.8 Hewlet Packard (1985). 'The Fundamentals of signal analysis application', Note 243, Hewlett Packard Co.

8.9 Jackson, C. (1983). *The Practical vibration primer*, Gulf Publishing Co., Houston.

8.10 Kito, F. (1959). 'The vibrations of penstocks', *Water Power*, October.

8.11 Muszynska, A. (1984). 'Synchronous and self excited rotor vibration caused by a full annular rub', Eighth Machinery Dynamics Seminar, Halifax, Natural Research Council, Canada.

8.12 Netsch, H. and Giacometti, A. (1982). 'Axial flow induced vibrations in large high head machines', *Water Power and Dam Construction*, August.

8.13 Ofterbro, I. and Lonning, A. (1966-'67). 'Pressure oscillations in Francis turbines', Proceedings of Institution of Mechanical Engineers, Vol 181, Part 3A.

8.14 'Product Catalogue', Endevco Corporation. San Juan, Capistrand, USA.

8.15 'Product Catalogue', IRD Mechanalysis. Wigan WN22DX, UK.

8.16 'Product Catalogue', Rockland Scientific Corporation. Rockleigh, NJ07647, USA.

8.17 Radhakrishna, H.C., Bhave, S.K. and Murthy, Ch.B. (1989). 'Vibration and noise measurements in Francis turbine power plants', International Symposium on Large Hydraulic Machinery and Associated Equipments, Beijing, May.

8.18 'Rotating Machinery Information Systems and Services', Bently Nevada (1988-'89). Minden, USA.

8.19 Timoshenko, S. and Young, D.H. (1964). *Vibration problems in engineering*, East West student edition, Affiliated East West Press.

8.20 Vladislavlev, L.A. (1979). *Vibration of hydro units in hydro electric power plant*, 2nd Edition, Amerind Publishing Co., New Delhi, New York.

Appendix I

Nomenclature of Symbols

1. Introduction

This document is a unified nomenclature, accepted by all authors, to ensure coherence and credibility for the hydraulic machinery book series.

We have tried as far as possible to conform to IEC and ISO, but have accepted some differences which seemed necessary: the main differences are pointed out in the nomenclature.

It is of course evident that any attempt to unify a system of symbols or nomenclature in any field of study will meet with strong reactions due to the habits already established in particular domains. We are conscious of this, but feel that the system based on the principles exposed here has one major advantage: it comes from a coherent internal logic, which proves satisfactory after 20 years of use within the Hydraulic Machines and Fluid Dynamics Institute of the Swiss Federal Institute of Technology in Lausanne, and also by several Swiss manufacturers.

2. Symbols

2.1 Guidelines

Symbols are Latin and Greek letters, both capital and small, and Arabic and Roman numerals.

The system's internal logic lies in the distinct meaning given to each of these forms of symbols.

Exceptions are accepted only when supported by widespread use.

The lists of symbols given below are not exhaustive. Additional symbols may be defined for the description of particular topics, but this must be done in agreement with the system's internal logic.

Different variables may be represented by the same letter. This situation is acceptable when other solutions would be too far-fetched, and the variables are not likely to be found within the same equation.

We suggest "high pressure side" and "low pressure side" to replace the confusing upstream/downstream reference, which is always a cause of misunderstanding, specially in dealing with reversible pump-turbine units.

2.2 Latin and Greek capital letters

Latin and Greek capital letters stand for dimensional variables in absolute value.
Capital Latin letters are also used as subscripts for elements of a plant (see 2.5.1.).

2.3 Latin and Greek small letters

Latin small letters stand for non-dimensional variables, called "factors". They are more often the ratio of a dimensional variable to a reference value for this variable. The symbol for this factor corresponds, in small letters, to the capital letter assigned to the variable.

Examples

hydraulic specific energy transformed by runner blades to mechanical specific energy	E_t	→	energy efficiency factor	$e_t = E_t/E$
flow velocity	C	→	velocity factor	$c = C/\sqrt{2E}$
radius of the runner blade at hub side of high pressure side edge	R_{1i}	→	radius factor	$r_{1i} = R_{1i}/R_{\bar{1}e}$

Small Latin letters are also used as subscripts for the elements of a machine (see 2.5.2.), for the characteristic points for machine power (see 2.5.3.), for the directions (Velocity components,see 2.5.4) and for the operating conditions of the machine (see 2.5.5). Small Greek letters stand for non-dimensional variables, called "coefficients", characteristic of the working conditions of a machine: specific speed, flow and hydraulic energy coefficients, efficiency, Thoma number.

2.4. Roman and Arabic numerals

Roman numerals are used as subscripts for particular points of a plant (see 2.5.1.).

Arabic numerals are used as subscripts for particular points of a machine (see 2.5.2.).

2.5. Subscripts

2.5.1. Elements and particular points of the plant

Elements of a hydraulic plant have capital Latin letters. For elements located on the low pressure side of the machine, letters are topped with the sign (¯). (Figure 1)

A	headwater basin
B	intake trash rack
T	power (headrace) tunnel
S	high pressure side surge tank
V	high pressure side valve
P	high pressure penstock
R	manifold with high pressure side valve
M	machine

$\overline{R}$	manifold with low pressure side valve
$\overline{P}$	low pressure penstock
$\overline{V}$	low pressure side valve
$\overline{S}$	low pressure side surge tank
$\overline{T}$	low pressure (tailrace) tunnel
$\overline{B}$	low pressure side trash rack
$\overline{A}$	tailwater basin

Particular points within a hydraulic plant are denoted by Roman numerals. For points located on the low pressure side of the machine, the numerals are topped with the sign ($\overline{\ }$). (Figure 1)

VII	headwater side of headwater trash rack
VI	machine side of headwater trash rack
	headwater section of head race tunnel
V	machine section of headrace tunnel
	headwater section of high pressure side surge tank
IV	machine section of high pressure side surge tank
	headwater section of high pressure side valve
III	machine section of high pressure side valve
	headwater section of high pressure penstock
II	machine section of high pressure penstock
	headwater section of high pressure manifold
I	machine section of high pressure manifold
	high pressure reference section of machine
$\overline{I}$	low pressure reference section of machine
	machine section of low pressure manifold
$\overline{II}$	tailwater section of low pressure manifold
	machine section of low pressure penstock
$\overline{III}$	tailwater section of low pressure penstock
	machine section of low pressure side valve
$\overline{IV}$	tailwater section of low pressure side valve
	machine section of low pressure side surge tank
$\overline{V}$	tailwater section of low pressure side surge tank
	machine section of tailrace tunnel
$\overline{VI}$	tailwater section of tailrace tunnel
	machine side of tailwater trash rack
$\overline{VII}$	tailwater side of tailwater trash rack

Note: the performance guarantees of the machine are referred to the reference sections I and $\overline{I}$ (1 and 2 for IEC Codes, respectively)

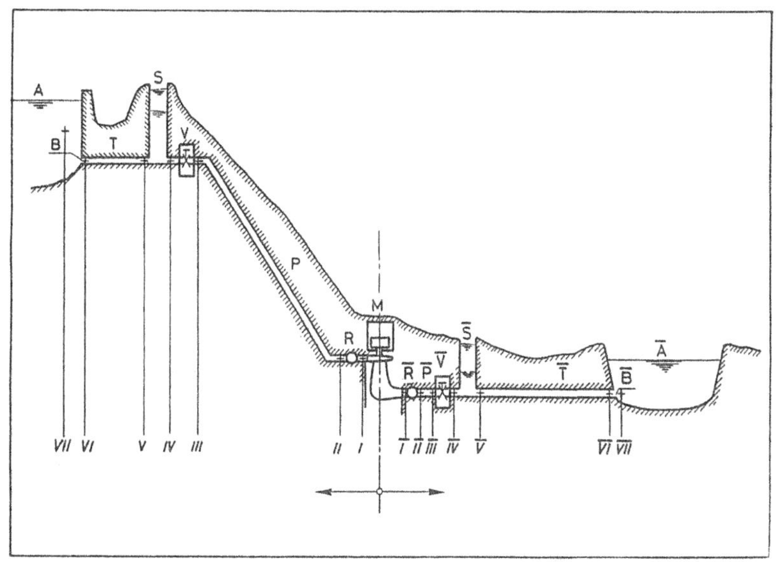

Figure 1 Particular points within a hydraulic plant

2.5.2. Elements and particular points of the machine

Elements of a machine are denoted by small Latin letters.

machine		turbine	pump
	Subscript	Terms	
	c	spiral case nozzle	casing
high pressure side components	v	stay vanes	diffusor
	g	wicket gate	wicket gate
runner	b	runner blades	impeller blades
	i	internal, at the hub	
	e	external, at the band	
low pressure side components	d	draft tube	suction pipe

Arabic numerals denote particular points within the machine. For points located on the low pressure side of the runner blades, the numerals are topped with the sign (¯). (Figure 2)

	turbine		pump
6	spiral case high pressure (inlet) section	6	casing high pressure (outlet) section
5	leading edge of stay vanes	5	trailing edge of diffusor vanes
4	trailing edge of stay vanes	4	leading edge of diffusor vanes
3	leading edge of wicket gate	3	trailing edge of wicket gate
2	trailing edge of wicket gate	2	leading edge of wicket gate
1	high pressure (inlet or leading) edge of runner blades	1	high pressure (outlet or trailing) edge of impeller blades
$\bar{1}$	low pressure (outlet or trailing) edge of runner blades	$\bar{1}$	low pressure (inlet or leading) edge of impeller blades
$\bar{2}$	inlet edge of draft tube fins or stabilizing device	$\bar{2}$	outlet edge of suction pipe fins or stabilizing device
$\bar{3}$	outlet edge of draft tube fins	$\bar{3}$	inlet edge of suction pipe fins
$\bar{4}$	inlet section of draft tube elbow	$\bar{4}$	outlet section of suction pipe elbow
$\bar{5}$	outlet section of draft tube elbow	$\bar{5}$	inlet section of suction pipe elbow
$\bar{6}$	suction (outlet) section of draft tube	$\bar{6}$	suction (inlet) section of suction pipe

2.5.3. Characteristic points for machine power

Small Latin letters denote characteristic points for machine power (Fig.2).

Subscript	Definition and Term
h	hydraulic power turbine : P_h hydraulic power available for producing mechanical power pump : P_h hydraulic power imparted to the water
t	transformed by runner / impeller blades turbine: P_t hydraulic power converted to mechanical power pump: P_t mechanical power converted to hydraulic power
m	at runner (impeller) coupling flange P_m mechanical power of runner
no letter	at machine coupling flange P mechanical power of the machine

Note : IEC Codes do not take into consideration P_t,for sake of simplicity

2.5.4. Directions (see Figure 3)

Subscript	Term
a	axial
m	meridional
r	radial
u	peripheral

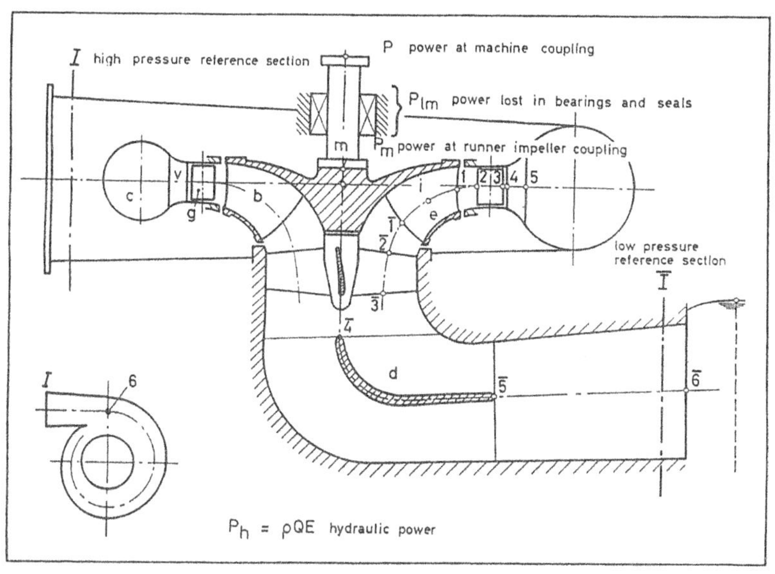

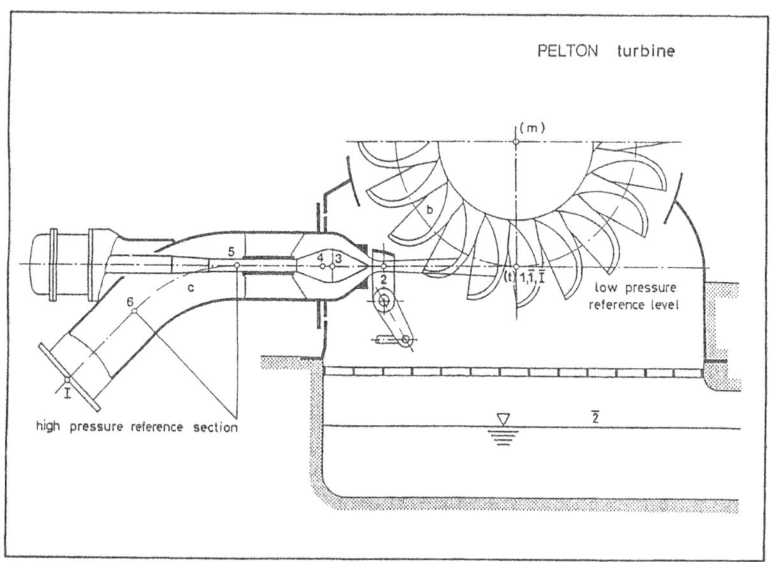

Figure 2 Characteristic points for machine power

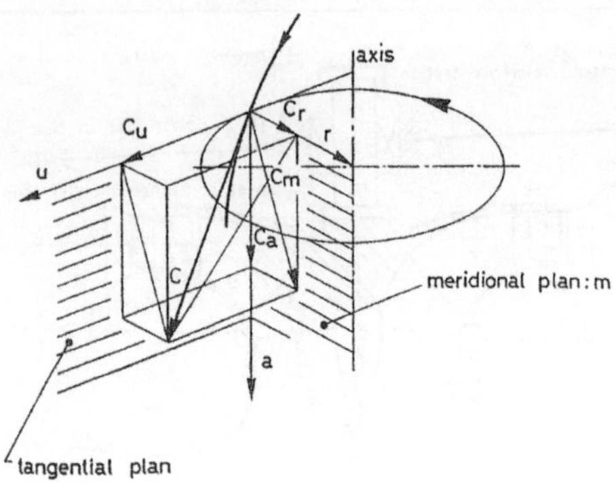

Figure 3 Flow direction

2.5.5. Operating conditions of a machine

Subscript	Term
v	turbine's zero load discharge
∧	machine's best efficiency
sp	machine's specified discharge, specific hydraulic energy, power...
max	maximum of specified variable
min	minimum of specified variable
r	turbine's runaway speed

2.5.6. Other subscripts

Subscript	Term	Example	
a	atmospheric	p_a	atmospheric pressure
c	cavitation	ψ_c	cavitation energy coefficient
d	drag	c_d	drag coefficient
f	friction	F_f	friction force, drag
k	kinetic	E_k	kinetic hydraulic specific energy
ℓ	discharge loss	Q_ℓ	discharge loss
p	potential	E_p	potential hydraulic specific energy (position and pressure)
	pressure	F_p	pressure force
r	losses	h_r	specific energy loss of turbine
		q_r	specific volumic loss of turbine
ref	reference	E_{ref}	reference hydraulic energy
s	shear	F_s	shear force
va	vapour	p_{va}	vapour separation pressure
z	elevation	E_z	position hydraulic specific energy

3. Symbol lists

3.1 Alphabetic list of capital Latin letters

Symbol	Term	Definition	Units
A	cross-section		m^2
	opening (of guide vane ,valve or needle)		m
B	guide vane height,bucket width (Pelton)		m
C	absolute velocity		m/s
$\tilde{C}$	compliance	$\tilde{C} = \partial V/\partial gH$	m.s
C_a	axial component of absolute velocity		m/s
C_m	meridional component of absolute velocity		m/s
C_r	radial component of absolute velocity		m/s
C_u	peripheral component of absolute velocity		m/s
D	diameter		m
E	specific (massic) hydraulic energy of machine :		
	delivered to turbine	$E = g_I H_I - g_{\bar{I}} H_{\bar{I}}$	J/kg
	delivered by pump	$E = g_I H_I - g_{\bar{I}} H_{\bar{I}}$	J/kg
E	elasticity (YOUNG) modulus		N/m^2
E_b	specific hydraulic energy at runner / impeller blades		
	turbine: specific hydraulic energy delivered to runner blades	$E_b = E_t + E_{rb}$	J/kg
	pump: specific hydraulic energy delivered by impeller blades	$E_b = E_t - E_{rb}$	J/kg
E_{kd}	specific hydraulic energy		
	delivered to turbine draft tube	$E_{kd} = \left(C_{\bar{1}}^2 - C_{\bar{I}}^2\right)/2$	J/kg

Symbol	Term	Definition	Units
E_r	lost specific hydraulic energy		
	turbine: from I to Ī	$E_r = E - E_t$	J/kg
	pump: from Ī to I	$E_r = E_t - E$	J/kg
E_{rb}	specific hydraulic energy lost within runner / impeller blades		
	turbine	$E_{rb} = E_b - E_t$	J/kg
	pump	$E_{rb} = E_t - E_b$	J/kg
E_{rd}	specific hydraulic energy lost within turbine draft tube	$E_{rd} = gH_{\bar{1}} - gH_{\bar{I}}$	J/kg
E_t	transformed specific energy		
	turbine: specific hydraulic energy converted to specific mechanical energy	$E_t = E - E_r$	J/kg
	pump: specific mechanical energy converted to specific hydraulic energy	$E_t = E + E_r$	J/kg
E_{td}	specific hydraulic energy transformed (kinetic → potential) by turbine draft tube	$E_{td} = E_{kd} - E_{rd}$	J/kg
E_s	suction specific hydraulic energy		J/kg
F	force		N
F_d	drag		N
F_l	lift		N
F_p	pressure force		N
F_f	friction force, resistance force		N
Fr	FROUDE number	$Fr = C / \sqrt{gL}$	-
gH	specific hydraulic energy		
	in a section	$gH = g.Z + \frac{p}{\rho} + \frac{C^2}{2}$	J/Kg
gH_k	kinetic specific hydraulic energy in a section	$gH_k = C^2/2$	J/kg

Symbol	Term	Definition	Units
gH_p	potential specific hydraulic energy in a section	$gH_p = g.Z+p/\rho$	J/kg
H	head (hydraulic energy per unit weight)		m
	height		m
H_a	representative height of atmospheric pressure		m
H_r	head loss		m
H_s	suction head		m
H_{va}	representative height of vapour pressure		m
J	clearance		m
K	sand roughness		m
L	length		m
	airfoil/blade length		m
	wave length	$L = S/f$	m
$\tilde{L}$	pipe inertance	$\tilde{L} = \int dL/A$	m^{-1}
M	mass		kg
Ma	MACH number	$Ma = C/S$	-
M_f	bending moment		N.m
M_t	twisting torque		N.m
NPSE	net positive suction energy	$NPSE = \frac{\left(\frac{p_a - p_{va}}{\rho}\right) - g.H_s}{E}$	J/Kg
NPSH	net positive suction head	$NPSH = \frac{NPSE}{g_{\bar{I}}}$	m

Symbol	Term	Definition	Units
P	power		
	mechanical power of the machine (at the machine coupling flange)		
	turbine	$P = P_h.\eta$	W
	pump	$P = P_h/\eta$	W
P_h	hydraulic power		
	turbine hydraulic power available for producing mechanical power		
	pump hydraulic power imparted to the water	$P_h = (\rho.Q_1).E$	W
$P_{\ell m}$	mechanical losses within the turbine (seals, bearings,.)	$P_{\ell m} = P_m\text{-}P$	W
	mechanical losses within the pump (seals, bearings.)	$P_{\ell m} = P\text{-}P_m$	W
P_m	mechanical power of the runner (at the runner / impeller coupling flange)		
	turbine	$P_m = P_h.\eta_m$	W
	pump	$P_m = P_h/\eta_m$	W
P_t	Power transformed by runner / impeller blades		
	turbine: hydraulic power converted to mechanical power	$P_t = \rho.Q_t.E_t = P_h.\eta_h$	W
	pump: mechanical power converted to hydraulic power	$P_t = \rho.Q_t.E_t = P_h/\eta_h$	W
$P_{h\text{-}t}$ (1)	total hydraulic losses within the turbine (volumic and energetic)	$P_{h\text{-}t} = \rho.Q_\ell.E_r$	W
	total hydraulic losses within the pump (volumic and energetic)	$P_{h\text{-}t} = \rho.Q_\ell.E_r$	W
P_{rm}	mechanical and disc friction losses		
	within the turbine	$P_{rm} = P_t\text{ -}P$	W
	within the pump	$P_{rm} = P\text{-}P_t$	W

(1) The IEC Codes do not take into consideration P_t(see note at page 8). Therefore $P_{h\text{-}t}$ becomes $P_{h\text{-}m}$, total hydraulic losses within the machine, including $P_{t\text{-}m}$ assumed as internal "hydraulic" and not mechanical losses within the machine, for simplicity.

Symbol	Term	Definition	Units
P_{t-m} (1)	disc friction and labyrinth losses		
	within the turbine	$P_{t-m} = P_t - P_m$	W
	within the pump	$P_{t-m} = P_m - P_t$	W
Q_I	discharge, volume-flow rate	$Q_I = dV/dt$	m^3/s
	delivered to turbine	$Q_I = Q_t + Q_\ell$	m^3/s
	delivered by pump	$Q_I = Q_t - Q_\ell$	m^3/s
Q_ℓ	discharge loss		
	turbine	$Q_\ell = Q_I - Q_t$	m^3/s
	pump	$Q_\ell = Q_t - Q_I$	m^3/s
Q_t	discharge concerned with energy transformation		
	turbine	$Q_t = Q_I - Q_\ell$	m^3/s
	pump	$Q_t = Q_I + Q_\ell$	m^3/s
R	radius		m
	degree of reaction		-
$\tilde{R}$	resistance	$\tilde{R} = \partial E/\partial Q$	$m^{-1}s^{-1}$
Re	REYNOLDS number	$Re = C.L/\upsilon$	-
R_h	hydraulic radius	$R_h = 2.A/U$	m
S	sound (wave) propagation speed		m/s
St	STROUHAL number	$St = C.t/L$	-
T	torque		N.m
	external torque (at machine coupling flange)		N.m
	spacing		m
	period	$T = 1/f$	s
T_h	hydraulic torque		N.m
T_m	internal torque (at runner / impeller coupling flange)		N.m

Symbol	Term	Definition	Units
U	wetted perimeter		m
	peripheral velocity	$U = R.\omega$	m/s
V	volume		m^3
W	relative velocity	$\vec{W} = \vec{C} - \vec{U}$	m/s
W_a	relative axial velocity		m/s
W_m	relative meridional velocity		m/s
W_r	relative radial velocity		m/s
W_u	relative peripheral velocity		m/s
X	abscissa		m
Y	ordinate		m
	admittance	$Y = 1/Z$	m.s
Z	elevation		m
	altitude		m
Z	impedance	$Z = dgH/dQ$	$m^{-1}s^{-1}$
Z_{ref}	reference impedance	$Z_{ref} = S/A$	$m^{-1}s^{-1}$

3.2. Alphabetic list of capital Greek letters

Symbol	Term	Definition	Units
Γ	circulation	$\Gamma_{1-2} = \int_1^2 C.dL = \Phi_2 - \Phi_1$	m^2/s
Δ	amplitude		-
Φ	velocity potential	$\vec{W} = \overrightarrow{grad}\,\Phi$	m^2/s
θ	temperature		K or °C
Ψ	stream function		m^2/s

3.3. Alphabetic list of small Latin letters (see 2.3.)

Symbol	Term	Definition	Units
a	cross-section factor	$a = A/R_{\bar{1}e}^2$	-
b	guide vane height factor	$b = B/R_{\bar{1}e}$	-
c	absolute velocity factor	$c = C/\sqrt{2E}$	-
c_m	meridional component of absolute velocity factor	$c_m = C_m/\sqrt{2E}$	
c_p	pressure coefficient	$c_p = \dfrac{\Delta p}{\frac{1}{2}\rho C^2}$	-
	pressure energy factor	$c_p = (p/\rho)/E$	-
c_{pa}	atmospheric pressure energy factor	$c_{pa} = (p_a/\rho)/E$	-
c_{pva}	vapour pressure energy factor	$c_{pva} = (p_{va}/\rho)/E$	-
c_u	peripheral velocity factor	$c_u = C_u/\sqrt{2E}$	-
c_d	drag coefficient		-
c_l	lift coefficient		-
c_r	head loss coefficient, resistance		-
e	specific hydraulic energy factor of machine, delivered to turbine or delivered by pump	$e = E/E = 1$	-
e_b	specific hydraulic energy factor at runner (impeller) blades - turbine: specific hydraulic energy delivered to blades (1)	$e_b = \dfrac{E_b}{E} = e_t + e_{rb}$	-

(1) Reference specific energy: for turbine: E
for pump: E_t

Symbol	Term	Definition	Units
	- pump: specific hydraulic energy delivered by blades [(1)]	$e_b = \frac{E_b}{E_t} = 1 - e_{rb}$	-
e_{kd}	kinetic specific hydraulic energy factor, delivered to turbine draft tube	$e_{kd} = c_{\bar{1}}^2 - c_{\bar{I}}^2$	-
e_p	specific potential hydraulic energy factor	$e_p = z + p$	-
e_r	specific hydraulic energy loss factor of turbine	$e_r = 1 - e_t$	-
e_{rt}	specific hydraulic energy loss factor in pump	$e_{rt} = E_r / E_t$	-
e_{rb}	specific hydraulic energy loss factor within runner / impeller blades turbine pump	 $e_{rb} = e_b - e_t$ $e_{rb} = 1 - e_b$	 - -
e_{rd}	specific energy loss factor within turbine draft tube	$e_{rd} = e_{\bar{1}} - e_{\bar{I}} = e_{\bar{1}\text{-}\bar{I}}$	-
e_{rr}	specific residual velocity energy factor at turbine runner outlet	$e_{rr} = c_{\bar{1}}^2$	-
e_t	specific energy transformation factor: turbine : specific hydraulic energy to specific mechanical energy conversion factor pump : specific mechanical energy to specific hydraulic energy conversion factor	 $e_t = E_t/E = 1 - e_r$ $e_t = E/E_t$	 - -
e_{td}	specific hydraulic energy transformation factor (kinetic → potential) by turbine draft tube	$e_{td} = e_d - e_{rd}$	-
e_{tdd}	turbine draft tube energy efficiency factor	$e_{tdd} = e_{td} / e_d = 1 - e_{rd}/e_d$	-
k	relative sand roughness	$k = K/L$	-

Symbol	Term	Definition	Units
$q^{(1)}$	discharge factor		
	delivered to turbine	$q = Q_l/Q_I = 1$	-
	delivered by pump	$q = Q_l/Q_t$	
q_ℓ	specific volumic loss factor		
	turbine	$q_\ell = Q_\ell/Q_I = 1 - q_t$	-
	pump	$q_\ell = Q_\ell/Q_t = 1 - q_t$	-
q_t	volumic efficiency		
	of turbine	$q_t = Q_t/Q_I = 1 - q_\ell$	-
	of pump	$q_t = Q_I/Q_t = 1 - q_\ell$	-
r	radius factor	$r = R/R_{\bar{1}e}$	-
u	peripheral velocity factor	$u = U/\sqrt{2E}$	-
w	relative velocity factor	$w = W/\sqrt{2E}$	-
z	position energy factor	$z = g.Z/E$	-
z_b	number of runner (impeller) blades or buckets		-
z_g	number of adjustable guide vanes		-
z_{gf}	number of fixed guide vanes		-
z_v	number of fixed stay vanes in spiral case		-
z_n	number of nozzles (impulse turbine)		-

3.4 Exceptions

Exceptions due to widespread use are listed hereafter.

Symbol	Term	Definition	Units
a	acceleration		m/s^2
f	frequency	$f = 1/T$	$s^{-1} = Hz$
g	acceleration due to gravity		m/s^2

(1) Reference discharge: for turbine: Q_I
for pump: Q_t

Symbol	Term	Definition	Units
g_n	standard gravity acceleration	$g_n = 9.80665$	m/s^2
i	incidence angle		°
n	rotational speed	$n = \omega/2\pi$	s^{-1}
p	pressure		N/m^2
p_a	atmospheric pressure		N/m^2
p_{va}	vapour pressure		N/m^2
s	needle opening (impulse turbine)		m
$\tilde{s}$	complex pulsation		s^{-1}
t	time		s

3.5 Alphabetic list of small Greek letters

Symbol	Term	Definition	Units
ζ	transient energy ratio	$\zeta = \sqrt{\frac{E}{E_{ref}}}$	-
η	efficiency		-
	of turbine	$\eta = P/P_h = \eta_m \cdot \eta_{em} = \eta_h \cdot \eta_t$	-
	of pump	$\eta = P_h/P = \eta_m \cdot \eta_{me} = \eta_h \cdot \eta_t$	-
η_{em}	mechanical efficiency of turbine (seals and bearings losses)	$\eta_{em} = P/P_m$	-
η_{me}	mechanical efficiency of pump (seals and bearings losses)	$\eta_{me} = P_m/P$	-
η_t	total mechanical efficiency		
	of turbine	$\eta_t = P/P_t$	-
	of pump	$\eta_t = P_t/P$	-
η_m	hydraulic efficiency		
	of turbine	$\eta_m = P_m/P_h = \eta_h \cdot \eta_{mt}$	-
	of pump	$\eta_m = P_h/P_m = \eta_{tm} \cdot \eta_h$	-

Symbol	Term	Definition	Units
η_{mt}	internal mechanical (disc friction losses) efficiency of turbine	$\eta_{mt} = P_m/P_t$	-
η_{tm}	internal mechanical (disc friction losses) efficiency of pump	$\eta_{tm} = P_t/P_m$	-
η_h	internal efficiency (1)		
	of turbine	$\eta_h = P_t/P_h = e_t \cdot q_t$	-
	of pump	$\eta_h = P_h/P_t = e_t \cdot q_t$	-
κ	valve opening	$\kappa = (Q/A_{ref})/\sqrt{2E}$	-
λ	power coefficient	$\lambda = \varphi.\psi = c_m/u^3$	-
		$\lambda_{\bar{1}e} = \dfrac{2P_h}{\rho\pi R^5_{\bar{1}e}\omega^3}$	-
	head loss coefficient	$gH_r = \lambda L/(2DA^2).Q^2$	-
μ	complex form of STROUHAL number	$\mu = \sqrt{Cs(Ls+R)}$	-
	friction coefficient	$\mu = F_r/F_p$	-
υ	specific speed	$\upsilon = \varphi^{1/2}_{\bar{1}e} / \psi^{3/4}_{\bar{1}e} = \dfrac{\omega.(Q_I/\pi)^{1/2}}{(2E)^{3/4}}$	-

Note: The values $n_q = \dfrac{60\, n(Q_I)^{1/2}}{H^{3/4}}$ or $n_s = \dfrac{60\, n\, P^{1/2}}{H^{5/4}}$ can be used as exceptions but there are so many other definitions of the specific speed that we recommend strongly the use of υ.

Symbol	Term	Definition	Units
ρ	ALLIEVI number	$\rho = Z_{ref}Q/2E_{ref}$	-
σ	THOMA number	$\sigma = \dfrac{NPSE}{E}$	-

(1) IEC Codes define $\eta_h = P_m/P_h$ for turbine and $\eta = P_h/P_m$ for pump, considering the disk friction losses and leakage losses as hydraulic losses (see footnote at page 14)

Symbol	Term	Definition	Units
φ	flow coefficient	$\varphi = c_m/u;\ \varphi_{\bar{1}e} = \dfrac{Q_I}{\pi R^3_{\bar{1}e}\omega}$	-
	isopotential lines		-
ψ	hydraulic energy coefficient	$\psi = 1/u^2;\ \psi_{\bar{1}e} = \dfrac{2E}{\left(R_{\bar{1}e}\omega\right)^2}$	-
	stream lines		-

3.6 Exceptions

Exceptions due to widespread use are listed hereafter.

Symbol	Term	Definition	Units
α	absolute velocity angle		rad
β	relative velocity angle		rad
β_b	runner / impeller blade setting angle	$\beta_b = \beta + i$	rad
γ	guide vane opening (rotating stroke)		rad
ε	lift/drag ratio	$\varepsilon = c_l / c_d$	rad
ζ	dynamic viscosity	$\zeta = \upsilon . \rho$	kg/m/s
υ	kinematic viscosity	$\upsilon = \zeta / \rho$	m^2/s
ρ	density (volumic mass)	$\rho = M/V$	kg/m^3
φ	polar angle		rad
	phase angle		rad
ω	angular speed		rad/s
	pulsation		rad/s

Appendix II

Reference List of Organisations

AMERICAN HYDRO CORPORATION:
130 Derry, Court, York, PA 17402, USA **Product Range:** Turbines: Francis, Kaplan, pump turbine. Impellers. Propeller. **Service:** Engineering, repairs, rehabilitation: analysis, upgrade. **Contact Name:** William Colwill, V P Adv. Tec.
Tel. No. (717) 764-3587
Fax. No. 717-764-0848

BALAJU YANTRA SHALA(P) LIMITED:
P O Box 209, BID Balaju, Katmandu, Nepal **Product Range:** Turbines up to 400 KW. Penstock and accessories. Control and instrumentation equipment. **Service:** Installation, commissioning, survey and design for small hydro plants. **Contact Name:** S L Vaidya, Head of Hydro Power Dept.
Tel. No. 412379
Fax. No. 2429 BYS NP

BERGERON SA:
155 Boulevard Haussmann 75008 Paris, France **Product Range:** Hydraulic engineering. Water pumping systems. Centrifugal, mixed flow and axial flow pumps for large capacities. **Service:** Transient flow studies. Turnkey pumping stations. **Contact Name:** Jean Louis Bloch, Commercial Director
Tel. No. (33) 145619555
Telex. No. F643557 Bergron

BOETTICHER Y NAVARRO, SA:
Ctra. De Andalucia, KM 9 28021 Madrid, Spain **Product Range:** Gates, valves, manual and self-cleaning trashracks, stoplogs, penstocks, bridge cranes, sliding gates. **Service:** Consulting, engineers and design. **Contact Name:** Javier Castellanos Ybarra, Director
Tel. No. 7978200/7979000
Telex. No. 47964 Bynsae

BOMBAS ELECTRICAS SA:
Ctra.
Mieras, s/n P O Box 47, 17820 Banyoles, Girona, Spain **Product Range:** Electric pumps for fluids, centrifugal pumps, submersible pumps, multi-stage pumps, swimming pool pumps. **Service:** Swimming pools, whirl-pool baths, irrigation, household and gardens. **Contact Name:** Josep Planas, Export Manager
Tel. No. 34-72 570662
Telex. No. 57218

BOVING FOURESS PVT. LTD:
Plot. No. 2, Phase II Peenya Industrial Area Bangalore 560 058, India **Product Range:** Small hydro turbines (up to 5000 KW of both reaction and impulse type in either horizontal or veritical shaft configuration). Governors and associated auxiliaries. **Service:** Design, manufacture, supply erection & commissioning of small hydro turbine systems. Refurbishing of small hydro installations. **Contact Name:** D R Bhutani, Plant Director
Tel. No. 385734 Ex. 30
Fax. No. 0812-385176
Telex No. 0845-5086

CHENGDU HYDROELECTRIC INVESTIGATION AND DESIGN INSTITUTE OF MWREP:
Chengdu, Sichuan Province, People's Republic of China 610072 **Service:** The institute comprises five specialised divisions, computer centre and a scientific research department. The hydraulic machinery section engages in designing hydraulic mechanical parts of various hydro-

electric power stations, pressure vessels and scientific researches as well as operating personnel training. **Contact Name:** Hu Dun-Yu, Chief of Electro Mechanical Department
Tel. No. 24023
Telex. No. 60158 CSDI CN

D M W CORPORATION:
28-4 Kamata 5-chome, Ota-Ku, Tokyo 144, Japan **Product Range:**Pumps. Fans, blowers and compressors. Valves. Water treatment system. Underwater dredging robot. Jet cutter. **Contact Name:** T Fujii, Manager of Sales Department
Tel. No. 03 739-9312
Telex. No. 02466391 DGSTOK J

EBARA CORPORATION:
11-1 Haneda-Asahi-Cho Ohta-Ku, Tokyo 144, Japan **Product Range:** Pumps: centrifugal, axial mixed, submersible, motor, self-priming. Turbines: Francis, propeller, Kaplan, Pelton. Pumps. Valves. Filters. **Service:** Engineering and construction for various pumping stations. **Contact Name:** International Sales Div. **Address:** Asahi Bldg, 6-7 Ginza 6-Chome, Chuo-Ku, Tokyo 104, Japan
Tel. No. 03 572 5611 Telex. No. TOEBARA J26976

ELC-ELECTROCONSULT SPA:
20151 Milan - Via Chiabrera, N.8, Italy **Product Range:** Power plants: hydroelectric, thermoelectric, geothermal, nuclear. Transmission and distribution lines. **Service:** Master plans, feasibility studies, contract and tender documents. Construction supervision, project management training, environmental studies. **Contact Name:** G E Casartelli, Business Development Manager
Tel. No. 02-30071
Telex. No. 331103 MILELC I

ELECTROWATT ENGINEERING SERVICES LTD:
Bellerivestrasse 36, PO Box CH-8022 Zurich, Switzerland **Service:** Consulting engineering services for all types of hydraulic machinery, hydro mechanical and power plant equipment, and related fields: planning, design and specification plant and equipment rehabilitation operation and maintenance planning quality control and factory inspection execution supervision and acceptance tests technical assistance **Contact Name:** Helmut Muller, Vice President
Tel. No. 01/385 22 61
Telex. No. 815 115

ENERGOPROJEKT: CONSULTING AND ENGINEERING CO. WATER RESOURCES DEVELOPMENT DEPT.
11070 N Beograd, Lenjinov Bulevar 12, PO Box 20, Yugoslavia **Service:** Investigation, design, consulting and engineering services in waterpower, water economics and infra- structure facilities and systems. **Contact Name:** R Zivojinovic, Deputy Director
Tel. No. 011 144 491
Telex. No. 11181 ENERGO

ENERSA:
Poligono Industrial de Malpica, Calle A, Parcela 20 50016 Zaragoza, Spain **Product Range:** Turbines: Kaplan, Francis, Pelton. Gates and valves. Speed increasers. **Service:** Construction. Installation and starting. **Contact Name:** Xarier Segui Puntas, Director General

Tel. No. 76 57 07 84
Telex. No. 58 163

ERHARD-ARMATUREN:
Postfach 1280, D-7920 Heidenheim, West Germany **Product Range**: Valves: butterfly, gate, flap, tapping, diaphragm, float, check, needle, control, hydrants, non-return, air, cone outlet, flow guards, knife gate. Penstocks. **Service:** Water hammer and cavitation calculations. Valve calibration. **Contact Name:** H Hahnel, Advertising Manager
Tel. No. 7321 320196
Telex. No. 714872

FU CHUN JIANG HYDRAULIC MACHINERY WORKS:
Tonglu, Zhejiang, People's Republic of China 311504 **Product Range:**Various types of hydraulic turbines. Electric generator set. Electro-hydraulic governor. Static thyristor excitation equipment. Gates, cranes and hoists. **Contact Name:** Xu Xiao Hua
Tel. No. 181 Cable. 1381.

GE CANADA INC:
795 First Avenue, Lachine, Quebec H8S 2S8, Canada **Product Range**: Hydraulic turbines. Hydro generators. Exciters. Bus ducts. **Service:** Design, manufacturing, testing laboratory, installation, diagnostic analysis. **Contact Name:** G E Drew, Manager - Marketing
Tel. No. 514 634 3411
Telex No. 05 821673

HANG ZHOU PUMP WORKS:
Qing Tai Men Wai, Hang Zhou City, People's Republic of China **Product Range:** Pumps: submersible, centrifugal. Mechanical seals. **Service:** Various advanced equipment and facilities for manufacture and test. Research institute with more than 70 engineers and technicians. **Contact Name:** Jiang Wenhai
Tel. No. 26192

HARBIN ELECTRICAL MACHINERY WORKS (HEMW):
35 Daqing Rd, Harbin, People's Republic of China 150040 **Product Range**: Hydro generator, governor and oil pressure equipment, turbogenerator, synchronous condenser, large capacity synchronous machine, large and medium capacity AC and DC machine set, automation system and excitation system. **Contact Name:** Wu Xin Run, Deputy Chief Engineer
Tel. No. 52871
Telex. No. 87015 HEMW CN

HIDROELECTRICA ESPANOLA SA:
Heimosilla 3, 28001 Madrid, Spain **Product Range:** Production & distribution of electric energy **Contact Name:** El Secretario General
Tel. No. 34 1 4024020
Telex. No. 23786

HIDROWATT SA:
C/Aragon, No 295, 7a Planta, 08009, Barcelona, Spain
Product Range: Screening equipment. Gates, valves. Control and communication equipment. Electrical equipment. Flow meters. **Service:** Consulting services: small hydro specialists, feasibility studies, turnkey services, project & construction management. **Contact Name:** Joan Fajas, Manager
Tel. No. 93/215 02 09
Telex. No. 50439-E

HYDROART SPA:
Via Stendhal 34, 20144, Milano, Italy **Product Range:** Governors. Pump turbines. Storage pumps. Valves, Water turbines. **Contact Name:** Ing. S. Moroni, Export Sales Manager
Tel. No. 02 479 104
Telex. No. 332281 HY ART I

IMPSA INTERNATIONAL INC:
Manor Oak II - Suite 536, 1910 Cochran Rd, Pittsburgh PA 15220, USA **Product Range:** Gates. Valves. Power house cranes. Turbines: Kaplan, Francis, pit, bulb, tubular. **Contact Name:** Raul Chaluleu, Director of Marketing
Tel. No. 412 344-7003
Telex. No. 710 664 2025

DEPARTMENT OF HYDRAULIC MACHINERY. INSTITUTE OF WATER CONSERVANCY & HYDRO ELECTRIC POWER RESEARCH (IWHR):
Al FuXing Rd, Beijing, 100038, People's Republic of China **Service:** Co-operative research & development on model hydraulic turbine, pump and pump turbines. Providing optimised model of turbine for hydro power project. International acceptance test on model of hydraulic turbine. Consultancy service on hydraulic turbine R & D and laboratory technology. **Contact Name:** Wang Haian, Deputy Head
Tel. No. 86.7078
Telex. No. 22786 ITCES-CN

INSTRUMENTATION LTD A GOVERNMENT OF INDIA ENTERPRISE:
Kanjikode West, Palghat, 678 623, Kerala, India **Product Range:** Butterfly valves, safety relief valves, orifice plates, flow nozzles, process control valves, electric and pneumatic actuators, positioners, accessories and electro magnetic flow meters. **Service:** Complete design, engineering, manufacture supply and consultancy for selection, application and training. **Contact Name:** Shri R.G. Kini, General Manager
Tel. No. 24452
Telex. No. 0852 205 ILP IN

IRRIGATION AND DRAINAGE MACHINERY RESEARCH INSTITUTE OF CAAGM:
No. 1 Beishatan, Deshengmen Wai, Beijing, People's Republic of China 100083 **Product Range:** Pumps: large axial, mixed, centrifugal, deep well, submersible, hand. Equipment for sprinkler and drip irrigation. **Service:** Designs of engineering for drainage or irrigation water, sprinkler irrigation. Consultant and training management, operation and maintenance of pump station. **Contact Name:** Nie Jinhuang, Director
Tel. No. 441331 2366

ISHIKAWAJIMA-HARIMA HEAVY INDUSTRIES CO. LTD:
2-1 Ohtemachi 2-Chome, Chiyoda-Ku, Tokyo, Japan **Product Range:** Pumps: centrifugal, mixed flow, volute, volute type mixed flow and axial flow. **Service:** Pump engineering (including modification), design, manufacture, installation and after sales service. **Contact Name:** M Watanabe, Manager
Tel. No. 03-244-5483
Telex. No. J22232 IHICO

KIRLOSKAR BROTHERS LTD:
Udyog Bhavan, Tilak Road, Pune 411 002, India **Product Range:** Pumps: rotody-

namic, end suction, double suction, vertical turbine, vacuum, axial flow. Valves: sluice, check, gate, butterfly. **Service:**Irrigation, water works,fire protection, sewage handling, process industries, chemical industries, thermal power stations, mining, refineries. **Contact Name:** Mr S C Kirloskar, Managing Director
Tel. No. 58133
Telex. No. 0145-247 KBPN IN

KSB, KLEIN SCHANZI & BECKER, AG:

Joh-Klein-Str. 9, PO Box 225 D-6710 Frankenthal, West Germany **Product Range:** Centrifugal pumps for: power stations, thermal, nuclear water supply, irrigation, drainage, process industry, environmental. engineering, domestic and general industrial engineering Valves: cast iron, steel. **Contact Name:** Peter Hergt, Manager hydraulic research development
Tel. No. 06233-86-2442
Telex. No. 465211-0 KS

KUBOTA LTD:

1-3 Nihonbashi-Muromachi 3-Chome. Chuo-Ku, Tokyo 103, Japan **Product Range:** Pumps: mixed flow, double suction volute, volute type mixed flow, single suction volute, multi-stage single suction, submersible, vaneless. **Service:** Pumping system from design to turnkey completion for city water, industrial water, desalination, chemical, sewage treatment, irrigation. **Contact Name:** Mr Masaru Tsuboi, Manager Pump Export Dept.
Tel. No. 03-245-3456
Telex. No. 222-3922

KVAERNER BRUG A/S:

Kvaerner V.10, PO Box 3610 GB 0135 Oslo 1, Norway **Product Range:** Hydro turbines, valves, governors, gates. **Service:** Repair. Refurbishment and upgrading. Training. **Contact Name:** Knut Pettersen, Sales Manager
Tel. No. 472 666020
Telex. No. 71650 KBN

LABEIM (LABORATORIOS DE EUSAYOS E INVESTIFACING INDES):

C/Westa De Olabeaga 16, 48013 Bilbao, Spain **Service:** Research and development. Model test on turbo- machinery **Contact Name:** Andomi Larreategui, Head of hydraulic machinery section
Tel. No. 34 4 4419300

MECANICA DE LA PENA SA:

Aita Gotzon No. 37, 48610 Urduliz (Vizcaya) Apartado 1.308 - 48080 Bilbao, Spain. **Product Range:** Turbines: Kaplan, bulb, Francis, Pelton, pump turbines. Transfer pumps. Valves: ball, butterfly. Coefferdams, penstock. Small hydraulic plants. **Service:** Design, construction and installation. Maintenance. **Contact Name:** Jesus Urquidi, Director
Tel. No. 4 676.10.11
Fax. No. 4 676.28.81
Telex No. 3301 MELPE-E

MITSUBISHI HEAVY INDUSTRIES LTD:

5-1 Marunouchi 2-Chome, Chiyoda-Ku, Tokyo, Japan **Product Range:**Water turbine. Pump turbine. Compressor. Pump and mechanical drive turbine. **Service:** Power plant, process and industrial plant, water works, sewage irrigation, flood control. **Contact Name:** H Nish-

ioka, Manager
Tel. No. 03-212-3111
Telex. No. J22443 HISHIJU

NEYRPIC:

75 Ave General Mangin BP 75-38041 Grenoble Cedex, France **Product Range:** Large water turbines from 15 to 1000 MW. Small water turbines from 0.1 to 15 MW. Standardised mini- turbines from 0.1 to 5 MW. Spherical and butterfly valves. Gates. Automatic systems and speed governors. **Service:** Rehabilitation and modernisation of power plants. **Contact Name:** M F de Vitry, Chairman M Y Couchet, Deput General Manager
Tel. No. 76 39 30 00
Telex. No. 320750 F

NORTH CHINA INSTITUTE OF WATER CONSERVANCY AND POWER:

Handan, Hebei, People's Republic of China 056021 **Service:** Internal flow analysis of hydraulic machinery. Cavitation mechanism and detection of hydraulic machinery. Hydraulic transient computer simulation of hydraulic machinery. Computer monitoring and control of hydraulic machinery. CAD of hydro-power station. **Contact Name:** Z Y Liu, Professor
Tel. No. 25951

O'HAIR GROUP:

7 Victoria Terrace, Bowen Hills, Brisbane, Queensland, Australia **Product Range:** Pumps: centrifugal, axial, mixed-flow, gear, screw, piston/plunger. Water turbines. Power recovery turbines. Blowers. Turbo-compressors. Sewage and fish hatchery aerators. Mechanical seals. **Service:** Consultants for specifications, design, installation and testing all hydraulic and fluid machinery and associated civil facilities. Can arrange supply of machinery and installation and testing, agencies, joint ventures, import/export finance. Spare and repair & re-designs submersible motor units. CAD software. **Contact Name:** J Brian O'Hair, Director
Tel. No. 61-7-2528001
Fax. No. 61-7-1257
Telex No. AA140472 Attention O'Hair Group

QIAN JIANG PUMP WORKS:

Xiao Shan City, Zhejiang Province, People's Republic of China **Product Range:** ISO 2858 Standard: IB IS type single-stage centrifugal pump. High quality. High efficiency. **Contact Name:**
Tel. No. 22328

RADE KONCAR:

Fallerovo Setaliste 22 41000 Zagreb, Yugoslavia **Product Range:** Generators and transformers. Electrical equipment for power plants. Control and instrumentation for power plants. **Service:** Design, production, erection, contracting of complete electrical equipment for power plants. **Contact Name:** Kozina Josip, Sales Manager
Tel. No. 041 316726
Telex. No. 21-159, 21-104

SHANGHAI PUMP WORKS:

Jiang Chuan Road, Min Hang, Shanghai, 200240 People's Republic of China **Product Range:** Pump: auxiliary for nuclear power station, boiler feed, circulating, condensate, drainage for power station, process, hot water circulation (West Germany KSB licence) for petrochemical enterprises, big variable or invariable

mixed and axial flow for irrigation, sewage for city engineering (USA DRESSER licence); 50 or 60 HZ marine (West Germany KSB licence). ISO standard mechanical seal and welded metal bellow seal (USA Sealol licence). **Contact Name:** Gu Xian
Tel. No. 358191
Telex. No. 33546 SMUDC CN

SHI SHOU PUMP WORKS:
Zhong Shan Road, Shi Shou City, Hubei Province, Peoples's Republic of China **Product Range:** Pumps: impurity, screw, mine drainage multi-stage, boiler feed, single-stage centrifugal clean water. **Service:** Technical services offered: development of erosion-resistant materials, design of special pumps for consumer. **Contact Name:** Sun Dong He
Tel. No. 2391

SHIN NIPPON MACHINERY CO. LTD:
Seio Bldg, 1-28 Shiba 2-Chome, Minato-Ku, Tokyo 105, Japan **Product Range:** Pumps: axial flow, mixed flow, centrifugal, ring section, barrel, liquid ring vacuum, slurry and screw. **Service:** Irrigation, water treatment, water supply, boiler feed water, process pump and chemical pump. **Contact Name:** Mr T Ikeshita, General Manager
Tel. No. 03 454-1412
Telex No. 242 4302 SNZOKIJ

SIGMA KONCERN SE:
Kosmonautu 6, 772 23 Olomouc Czechoslovakia **Product Range:** Pumps. Valves. Irrigation, water treatment. Waste water purification. **Contact Name:** J Holada, Director
Tel. No. 02 235 77 48
Fax. No. 02 265616
Telex No. 12 12 05 C

SOCIETE HYDROTECHNIQUE DE FRANCE SHF:
199 rue de Grenelle, 75007 Paris, France **Product Range:** Scientific association concerned with the development of knowledge and techniques for the engineering of fluids and water management. **Contact Name:** P Constans
Tel. No. 1 47051337

STORK POMPEN BV:
Lansinkesweg 30, PO Box 55, 7550 AB Hengelo, The Netherlands **Product Range:** Centrifugal pumps for: process and petrochemical duties, irrigation and drainage, drydocks, industrial applications, drinking water supply, power stations. **Service:** Industrial measurements and consultancy on pumps and pumping systems. Facilities for pump tests at works (model) and at site. **Contact Name:** N Van Vuren, Manager Sales Department
Tel. No. 074 404000
Fax. No. 074 425696
Telex No. 44324+ SPH +

SULZER-ESCHER WYSS LTD:
Escher Wyss Platz, 8023 Zurich, Switzerland **Product Range:** Equipment for hydro-electrical power plants. All types and sizes of water turbines, Storage pumps. Governors. Pump turbines. Shut-off valves. Penstock and manifolds **Service:** Maintenance, overhaul, modernisation, modification **Contact Name:** Ch. Habegger, Asst. Vice-President
Tel. No. 1-278-22-11
Telex No. 822 900 11 SECH

TEXMO INDUSTRIES:
MTP Rd, G N Mills, P O Coimbat-

ore 641 029, Tamil Nadu State, India **Product Range:** Pumps: irrigation and agricultural, shallow and deep well, borehole, household water supply and sewage. Sprinkler supply systems. **Contact Name:** Mr C Balaram, Marketing Manager
Tel. No. 33455

THOMPSONS, KELLY & LEWIS LTD:
26 Faigh Street, Mulgrave, Victoria 3170, Australia **Product Range:** Pumps: axial flow, propellor, boiler feed, centrifugal, mixed flow, multi-stage, vertical turbine, vertical inline, concrete volute. **Contact Name:** A Grage, Sales Director
Tel. No. 03 562 0744
Telex No. AA31365

TORISHIMA PUMP MFG. CO. LTD:
1-1-8 Miyata-cho, Takatsuki City, Osaka, Japan **Product Range:** Pumps. Mechanical seals. Cast products **Service:** Installation of pumps and their relative equipments. Engineering of pumping stations and after sales service. **Contact Name:** Mitsuma Kitajima, Manager General Affairs Dept.
Tel. No. 0726 95 0551
Telex No. 5336568 TORIPU J

TOSHIBA CORPORATION:
1-1 Shibaura 1-Chome Minatoku, Tokyo 105, Japan **Product Range:** Hydraulic turbines. Pump turbines. Inlet valves. Governors. **Service:** Engineering design, supply, finance **Contact Name:** Mr Hiroji Morimoto, Senior Manager
Tel. No. 03 457 4828 31
Telex No. J22587

DIVISION OF HYDRAULIC MACHINERY, DEPARTMENT OF HYDRAULIC ENGINEERING, TSINGHAU UNIVERSITY:
Beijing, People's Republic of China **Service:** Internal flow investigation and analysis on hydraulic machinery. Cavitation and flow-induced vibrations. Two-phase flow. Facilities in the hydraulic machinery laboratory include two closed-circuit test stands for turbine and pump, open-flow test stand and a slurry pump test stand. **Contact Name:** Lin RuChang, Professor Dept. of Hydraulic Engineering
Tel. No. 282451-2251
Telex 22617 QHTSC CN Fax. 86-01-2562768

VEVEY ENGINEERING WORKS LTD:
CH 1800 Vevey, Switzerland **Product Range:** Vevey specialises in the field of energy (hydro-power products) high, medium and low head water turbines, e.g. Pelton, Francis, Kaplan and bulbs, reversible pump turbines, single- and multistage and isogyre-type. **Contact Name:** Michon, Sales Manager
Tel. No. 4121 9257111
Telex No. 451104 VEYCH

WATER CONSERVANCY AND WATER POWER ENGINEERING RESEARCH INSTITUTION:
Dalian Institute of Technology, Dalian, People's Republic of China 116024 **Service:** Vibration problems of turbo- generator units, turbine, Penstock pumps. Hydropower house, pump house, equipment of hydropower station and pumping stations. **Contact Name:** Dong Yu Xin, Professor

Tel. No. 471511-519
Cable No. 7108

XIN CHANG SPRINKLER IRRIGATION WORKS:
Cheng Guan, Xin Chang, Zhejiang Province, People's Republic of China **Product Range:** Sprinkler irrigation systems. Self priming pumps. Centrifugal pumps. Sprinklers. Low and high temperature oil pumps. **Contact Name:**
Tel. No. 23600-22545

ZHEJIANG RESEARCH INSTITUTE OF MECHANICAL SCIENCE:
122 Laodong Rd, Hang Zhan, Zhejiang Province, People's Republic of China 310002 **Service:** Over 30 year's research experience in pumps, hydraulics and automatic control etc. The test centre is equipped with fully computerised, high precision test stands. **Contact Name:** Cheng Ji Zhong
Tel. No. 27778

Appendix III

List of Educational Establishments

AUSTRALIA

UNIVERSITY OF ADELAIDE
Department of Civil Engineering, Adelaide, S.A. Dr. A. Simpson. Water column separation

UNIVERSITY OF MELBOURNE
Department of Civil & Argicultural Engineering, Grattan St., Parkville, Victoria 3052. Tel: (03) 344 6789. Dr. H.R. Graze. Waterhammer, Air chambers, Water column separation

UNIVERSITY OF QUEENSLAND
Department of Civil Engineering, St. Lucia, Queensland. Prof. C. Apelt. Waterhammer

UNIVERSITY OF TASMANIA
Department of Civil and Mechanical Engineering, Hobard, Tasmania. Dr. S. Montes. Surge tanks

The following is a supplementary list of Engineering Faculties in Australia:

UNIVERSITY OF ADELAIDE
Dept. of Electrical and Electronic Engineering, Tel: (08) 228 5277: Fax: (08) 224 0464. Dr. Donald W. Griffin

AUSTRALIAN MARITIME COLLEGE
School of Engineering, Tel: (003) 260 757: Fax: (003) 260 717. Mr. John J. Seaton, Head

SOUTH AUSTRALIAN INSTITUTE OF TECHNOLOGY
Faculty of Engineering, Tel: (08) 343 3219: Fax: (08) 349 6939. Prof. K.J. Atkins, Dean

UNIVERSITY OF WESTERN AUSTRALIA
Faculty of Engineering, Tel: (09) 380 3105/3106: Fax: (09) 382 4649. Prof. Alan R. Billings, Dean

BALLARAT COLLEGE OF ADVANCED EDUCATION
Faculty of Engineering, Tel: (053) 339 100: Fax: (053) 339 545. Mr. Derek Woolley, Dean

BENDIGO COLLEGE OF ADVANCED EDUCATION
School of Engineering, Tel: (054) 403 339: Fax: (054) 403 477. Dr. Tim Dasika, Head

CANBERRA COLLEGE OF ADVANCED EDUCATION
Dept. of Electronics and Applied Physics, Tel: (062) 522 515: Fax: (062) 522 999. Dr. Paul Edwards, Head

CAPRICORNIA INSTITUTE OF ADVANCED EDUCATION
Faculty of Engineering, Tel: (079) 360 543: Fax: (079) 361 361. Mr. Frank Schroder, Dean

CHISHOLM INSTITUTE OF TECHNOLOGY
Faculty of Engineering, Tel: (03) 573 2162: Fax: (03) 572 1298. Dr. Brian Jenney, A/g Dean

CURTIN UNIVERSITY OF TECHNOLOGY
Faculty of Engineering, Tel: (09) 350 7093: Fax: (09) 458 4661. Assoc. Prof. Lachlan Millar, Dean

DARLING DOWNS INSTITUTE OF ADVANCED EDUCATION
School of Engineering, Tel: (076) 312 527:

Fax: (076) 301 182. Dr. Tom Ledwidge, Dean

DEFENCE ACADEMY
Dept. of Mechanical Engineering, Tel: (062) 688 274. Dr. Ray Watson

FOOTSCRAY INSTITUTE OF TECHNOLOGY
Faculty of Engineering, Tel: (03) 688 4244: Fax: (03) 689 4069. Mr. Ivan A. Bellizzer, Dean

GIPPSLAND INSTITUTE OF ADVANCED EDUCATION
School of Engineering, Tel: (051) 220 461: Fax: (051) 222 876. Dr. Ken Spriggs, Head

JAMES COOK UNIVERSITY
Faculty of Engineering, Tel: (02) 697 5001. Prof. Noel L. Svensson, Dean

ROYAL MELBOURNE INSTITUTE OF TECHNOLOGY
Faculty of Engineering, Tel: (03) 660 2523: Fax: (03) 663 2764. Dr. Bill Carroll, Dean

UNIVERSITY OF MELBOURNE
Faculty of Engineering, Tel: (03) 344 6619: Fax: (03) 347 1343. Assoc. Prof. Bill W.S. Charters, Dean

MONASH UNIVERSITY
Faculty of Engineering, Tel: (03) 563 3400: Fax: (03) 565 3409. Prof. Peter Le Darvall, Dean

UNIVERSITY OF NEWCASTLE
Faculty of Engineering, Tel: (049) 685 395: Fax: (049) 674 946. Prof. Alan W. Roberts, Dean

UNIVERSITY OF NSW
School of Chemical Engineering & Industrial Chemistry, Prof. Chris J.D. Fell

QUEENSLAND INSTITUTE OF TECHNOLOGY
Faculty of Engineering, Tel: (07) 223 2415: Fax: (07) 229 1510. Dr. John J.B. Corderoy, Dean

UNIVERSITY OF QUEENSLAND
Faculty of Engineering, Tel: (09) 380 3105/3106: Fax: (09) 382 4649. Prof. Alan R. Billings, Dean

SWINBURNE INSTITUTE OF TECHNOLOGY
Faculty of Engineering, Tel: (03) 819 8282: Fax: (03) 819 5454. Dr. Murray M Gillin, Dean

UNIVERSITY OF TECHNOLOGY, SYDNEY
Faculty of Engineering, Tel: (02) 218 9272: Fax: (02) 282 2498. Dr. Ken Faulkes, Dean

UNIVERSITY OF TECHNOLOGY, SYDNEY
Faculty of Engineering, Tel: (02) 218 9272: Fax: (02) 281 2498. Prof. J. Paul Gostelow, Dean

UNIVERSITY OF SYDNEY
Faculty of Engineering, Tel: (02) 692 2329: Fax: (02) 692 2012. Prof. John R. Glastonbury, Dean

TASMANIAN STATE INSTITUTE OF TECHNOLOGY
School of Engineering, Tel: (003) 260 576. Mr. Peter Crewe, Head

UNIVERSITY OF TASMANIA
Faculty of Engineering, Tel: (002) 202 129: Fax: (002) 202 186. Mr. Eric Middleton, Dean

UNIVERSITY OF WOLLONGONG

Faculty of Engineering, Tel: (042) 270 491: Fax: (042) 270 477. Prof. Brian H. Smith, Dean

CANADA

BC HYDRO HEAD OFFICE AND GENERAL ADMINISTRATION

970 Burrard, Vancouver, BC V6Z 1Y3. Tel: (604) 663 2212: Fax: (613) 663 3597. Mr H Lang, P. Eng.

CANADIAN DEPARTMENT OF ENERGY, MINES AND RESOURCES

7th Floor, 580 Booth Street, Ottawa, Ont. K1A 0E4. Tel: (613) 996 6119: Fax: (613) 996 6424. Mr T Tung, P. Eng. Operation Engineer Hydro

CENTRE DE RECHERCHES DE L'HYDRO-QUÉBEC

Varennes, PQué, Boîte Postale 1000. Tel: (514) 652 8090: Fax: 05-2677486. Dr B Dubé, Ing.

2687 Dautrive, Ste Foy, Qué, G1W 2C8. Tel: (418) 651 2189: Fax: (418) 656 6425. Dr H Netsch, Ing

ECOLE POLYTECHNIQUE

Dept. de Mathématiques Appliqués, Université de Montréal, Box 6079, Montréal, P.Q. H3C 3A7. Tel: (514) 340 4639: Fax: 340 4440. Prof. Dr. R Camerero

HYDRO-QUÉBEC

5655 Rue Marseille, Montréal, Qué, H1N 1J4. Tel: (514) 251 7043: Fax: (514) 251 7117. M J M Lévesque, Ing., Ingénieur Specialiste, Expertises Techniques

LABORATOIRE HYDRAULIQUE

Montréal Rd, Ottawa, Ont, K1A 0R6. Tel: (613) 993 6650: Fax: (613) 952 7679. M Thierry D Faure, Ing

2368 Chemin des Foulons, Sillery, Qué, G1T 1X4. Tel: (418) 656 6394: Fax: (418) 656 6425. M Y Jean, Ing M.Sc., Spécialiste

UNIVERSITÈ LAVAL

Dept de Génie mécanique Québec, Qué G1K 7P4. Tel: (418) 656 5359: Fax: (418) 656 5902. Professeure Dr C Deschaînes, Ing

CHINA

BEIJING INSTITUTUE OF ECONOMIC MANAGEMENT OF WATER POWER ENGINEERING

Postgraduate Division, Xizhimen Wei, Ziahuyuan, Beijing, 100044 Tel: 8414894. Prof. Duan Chang Guo, Prof. Shen Zhenju. Silt erosion; Hydraulic transients of hydraulic machinery; Measurement technology of hydraulic system

BEIJING UNIVERSITY OF ARGRICULTURAL ENGINEERING

Department of Hydraulic Engineering, East Qinghua Road, Beijing, 100083 Tel: 2017622. Prof. Liu Shankun - Director of Teaching and Research Division. Turbodynamics and design of pumps; Transients of turbine; Sprinkler irrigation facility and systems

GEZHOBA INSTITUTUE OF HYDRAULIC ENGINEERING

Water Power Engineering Department, Yichang City, Hubei Province. Tel: 22011-3615. Assoc. Prof. Zeng Dean. Hydraulic machinery. Selection, design and operation of hydraulic turbines (including auxiliary equipment).

HUAZHONG UNIVERSITY OF SCIENCE AND TECHNOLOGY

Department of Electrical Engineering, Wuhan, Hubei, 430074. Tel: Wuhan 870154: Telex: 40131 HBU. Prof. Tan Yuechan. Turbodynamics: Optimisation design theory; Cavitation, vibration, structure and strength design of hydraulic machinery.

JIANGSU INSTITUTE OF TECHNOLOGY

Department of Power Machinery, Zhenjiang City, Jiangsu Province, 212013. Tel: 34071: Telex: 2894. Prof. Cha Sen. Hydraulic performance of hydraulic machinery; Two-phase flow pump; Erosion resistant material, Sprinkler and drip irrigation system; Pump station and irrigation and drainage engineering.

NORTH CHINA INSTITUTE OF WATER CONSERVANCE AND POWER

Department of Power Engineering, Handan, Hebei, 056021. Tel: 22775. Assoc. Prof. Liu Z.Y. - Instrumentation. Prof. Li Shengcai - Hydraulic Machinery. Internal flow of turbines: Cavitation mechanism and its detection: CAD of hydropower station: Hydraulic transients of turbines: Computer control of power station.

NORTH-EAST COLLEGE OF HYDRAULIC AND ELECTRICAL ENGINEERING

Department of Electrical Engineering, Changchun City, Julin Province. Tel: 55991-95. Mr. Shi Zhensheng - Lecturer. CAD of selection design of hydraulic machinery; Test rig and data logger system

NORTH-WEST UNIVERSITY OF AGRICULTURE

Hydraulic Engineering Department, Yangling City, Shaanxi Province. Professor Yang Songpu. Operation, cavitation and erosion resistance of hydraulic machinery

TSINGHUA UNIVERSITY

Department of Hydraulic Engineering, Beijing, 400084 Tel: 282451: Telex: 22617 QHTSC CN. Prof. Lin Ruchang. Internal flow of hydraulic machinery; Regulation and cavitation; Erosion and two-phase flow of hydraulic machinery; Cavitation and performance test of turbine and pump; Solid-liquid two-phase flow pumps

SHAANXI INSTITUTE OF MECHANICAL ENGINEERING

College of Hydro-electrical Engineering, Xian City, Shaanxi Province, 710048. Tel: 721236; Cable: Xian 8503. Assoc. Prof. Chen Jiamo. Cavitation and regulation system of turbine automation of hydro power station

SICHUAN INSTITUTE OF TECHNOLOGY

Power Engineering Department, Changdu City, Sichuan Province, 611744. Tel: 68271/68287: Cable: 6922. Prof. Dutong - Prof. of Hydraulic Machinery. Cavitation, erosion and corrosion of hydraulic machinery; Development of the turbine and pump; Feasibility study of small pumping storage power station

WUHAN INSTITUTE OF HYDRAULIC AND ELECTRICAL ENGINEERING

Research Institute of Irrigation and Drainage Engineering, Luojia Hill, Wuchang City, Hubei Province, 430072. Tel: Wuhan 812212-280: Telex:

40170 WCTEL CN

YUNNAN INSTITUTE OF TECHNOLOGY
Department of Electrical Engineering, Kunming City, Yunnan Province. Tel: 29031. Assoc. Prof. Huang Fenjie. Impulse turbine; Operation of high head Francis turbine.

EGYPT

UNIVERSITY OF ALEXANDRIA
Faculty of Engineering, Mechanical Dept. Alexandria, Horria Str. Prof. Esam Salem. Fluid mechanics and lubrication technology.

UNIVERSITY OF ASSIAT
Faculty of Engineering, Mechanical Dept. Assiat. Prof. A. Huzzien. Fluid mechanics and pumping machinery

UNIVERSITY OF CAIRO
Dept. of Mech. Power Eng, Giza. Prof. A. Mobarak. Experimental and theoretical analysis of flow in turbomachines

UNIVERSITY OF EIN-SHAMS
Dept. of Mech. Eng, Abbasia. Cairo. Prof. M. El Sebaie. Fluid mechanics and hydraulic machines

FRANCE

ENSEEIHT HYDRAULIC LABORATORY
2 Rue Camichel, 31071 Toulouse. Tel: (61) 588200. Fax: (61) 620976. Telex: 530171. Prof. J Gruat, Prof. C Thirriot, Dr.P Crausse

ENSHMG DOMAINE UNIVERSITY, GRENOBLE
38402 St. Martin d'Heres, Grenoble. Tel: (76) 825000: Fax: (76) 825001: Telex: 980668 Hymegre. Prof. P Bois, Prof. P Trompette

INDIA

INDIAN INSTITUTE OF SCIENCE
Department of Civil Engineering, Bangalore 560 012, India. Tel: (0812) 344411 Ex. 245. Telex: (0845) 8349. Rama Prasad. Research and teaching - hydraulic turbines and pumps

INDIAN INSTITUTE OF TECHNOLOGY, DELHI
Department of Applied Mechanics, Hauz Khas, New Delhi 110 016, India. Tel: 653458: Telex: 31-3687 IIT IN. Prof. V. Seshadri. Under graduate and post graduate testing; Research in analysis of flow through hydraulic machines. Performance tests on hydraulic machinery; Microhydel devices and their development; Flow instrumentation and calibration.

INDIAN INSTITUTE OF TECHNOLOGY
Hydroturbines Laboratory, Madras 600 036, India. Tel: 415342 Ex 274. Telex: 041 21062. Prof. H.C. Radha Krishna, Prof. M. Ravindran, Prof. P.A. Aswathanarayana, Dr. V. Balabaskaran. Flow research through hydroturbomachines; Cavitation, Vibration and Noise; Fully reversible axial flow hydroturbo- machines, bulb turbines, Allied research for hydraulic machines

MOTILAL NEHRU REGIONAL ENGINEERING COLLEGE
Mech. Eng. Department, Allahabad

211004 (U.P.) India. Tel: 53520: Telex: 0540 269 MNEC IN. Dr. Y.V.N. Rao - Principal, Dr. R. Yadav, Professor, Dr. S.K. Agrawal, Reader, Dr. V.K. Nema, Reader. Theoretical and experimental investigation in the following fields: centrifugal pumps, centrifugal and axial flow compressors

ITALY

UNIVERSITY OF BOLOGNA

Dept. of Mechanical Engineering, Viale Risorgimento, 2 - 40136. Tel: 051 582162/Bologna 582163. Prof. C. Bonacini

UNIVERSITY OF CAGLIARI

Dept. of Mechanical Engineering, Piazza D'Armi - 09100 Cagliari. Tel: 070 2900071/72/73. Prof. R. Masala

UNIVERSITY OF FIRENZE

Dept. of Civil Engineering, Via. S. Marta, 3 - 50139 Firenze. Tel: 055-47961. Prof. G. Federici

UNIVERSITY OF FIRENZE

Dept. of Enegetics, Via S. Marta, 3 - 50139 Firenze. Tel: 055-47961. Prof. E. Carnevale

UNIVERSITY OF GENOVA

Dept. of Hydraulics, Via Montallegro, 1 - 16145 Genova. Tel: 010-303416. Prof. F. Siccardi

UNIVERSITY OF GENOVA

Dept. of Mechanical Engineering, Via Montallegro, 1 - 16145 Genova. Tel: 010-303416. Prof. O. Acton, Prof. A. Satta

POLITECHNIC OF MILANO

Dept. of Energetics, Piazza Leonardo da Vinci, 32 - 20133 Milano. Tel: 23991. Prof. E. Macchi, Prof. F. Bassi

UNIVERSITY OF PADOVA

Dept. of Mechanical Engineering, Via Venezia, 1 - 35100 Padova. Tel: 049-8071988. Prof. G. Ventrone, Prof. V. Quaggiotti

UNIVERSITY OF PAVIA

Dept. of Hydraulics, Piazza Leonardo da Vinci - 27100 Pavia. Tel: 0382-31325/0382-21636. Prof. R. Sala

UNIVERSITY OF POTENZA

Dept. of Hydraulics, Via N. Sauro, 85 - 85100 Potenza. Tel: 0971-334611. Prof. B. De Bernardinis

UNIVERSITY OF ROMA

Dept. of Mechanical and Aeronautical Eng. Via Eudossiana, 18 - 00184 Roma. Tel: 06-4687314. Prof. C. Caputo, Prof. U. Pighini

JAPAN

EHIME UNIVERSITY

Dept. of Mechanical Engineering, 3 Bunkyo-cho, Matsuyama-shi 790. Tel: (0899) 24 7111. Prof. K. Ayukawa. Internal flow of turbomachines

UNIVERSITY OF FUKUI

Dept. of Mechanical Engineering, 3-9-1 Bunkyo, Fukui-shi 910. Tel: (0776) 23 0500. Prof. I. Ashino. Numerical analysis of flow in turbomachines

HIROSHIMA INSTITUTE OF TECHNOLOGY

Dept. of Mechanical Engineering, 725 Miyake, Itsukaishi-cho, Saeki-ku, Hiroshima 731-51. Tel: (0829) 21 3121. Prof. H. Murai. Cavitation, min-hydroturbines

UNIVERSITY OF HOKKAIDO

Dept. of Mechanical Engineering, Kita 13 Jo, Nishi 8 Chome, Kita-ku, Sapporo-shi 060. Tel: (011) 711 2111. Prof. M. Kiya. Numerical analysis and measurement of flow in turbomachines

KANAGAWA INSTITUTE OF TECHNOLOGY

Dept. of Mechanical Engineering, 1030 Hagino, Atsugi-shi 243-02. Tel: (0462) 41 1211. Prof. S. Akaike. Mini-hydroturbines, flow analysis of hydraulic turbines

KANAGAWA UNIVERSITY

Dept. of Mechanical Engineering, 3-22 Rokkakubashi, Kanagawa-ku, Yokohama-shi 221. Tel: (045) 481 566. Prof. T. Ida. Cavitation, similarity law of pumps and hydraulic turbines

KANAZAWA UNIVERSITY

Dept. of Mechanical Engineering, 2-40-20 Kotatsuno, Kanazawa-shi 920. Tel: (0762) 61 2101. Prof. S. Miyae. Performance of pumps

KEIO UNIVERSITY

Dept. of Mechanical Engineering, 3-14-1 Hiyoshi-sho, Kohoku-ku Yokohama-shi 223. Tel: (044) 63 1141. Prof. T. Ando. Analysis and measurement of flow in turbomachines

KOBE UNIVERSITY

Dept. of Mechanical Engineering, 1-1 Rokkadai-cho, Nada-ku, Kobe-shi 657. Tel: (078) 881 1212. Prof. T. Iwatsubo. Two-phase flow, rotodynamic problems of turbomachines

KYOTO UNIVERSITY

Dept. of Mechanical Engineering, Yoshida-Honmachi, Sakyo-ku, Kyoto-shi 606. Tel: (075) 751 2111. Prof. H. Akamatsu. Cavitation and two-phase flow phenomena

KYUSHU INSTITUTE OF TECHNOLOGY

Dept. of Mechanical Engineering II, Sensui-cho, Tobata-ku, Kitakyushu-shi 804. Tel: (093) 871 1931. Prof. M. Nishi. Flow measurement by small 5-hole probes, unsteady performances

UNIVERSITY OF KYUSHU

Dept. of Mechanical Engineering, 6-10-1 Kakazaki, Higashi-ku, Fukuoka-shi 813. Tel: (092) 641 1101. Prof. Y. Takamatsu. Numerical analysis and measurement of flow in turbomachines, cavitation

MIE UNIVERSITY

Dept. of Mechanical Engineering, 1515 Kamihama-cho, Tsu-shi, Mie-ken 514. Tel: (0592) 32 1211. Prof. Y. Shimizu. Flow in draft tube, self-priming pumps, jet pumps

MURORAN INSTITUTE OF TECHNOLOGY

Dept. of Mechanical Engineering II, 27-1 Mizumoto-cho, Muroran-shi 050. Tel: (0143) 44 4181. Prof. T. Watanabe. Cavitation, performance of pumps and hydraulic turbines

NAGOYA INSTITUTE OF TECHNOLOGY

Dept. of Mechanical Engineering, Gokiso-cho, Showa-ku, Nagoya-shi 466. Tel: (052) 7322111. Prof. Y. Yamada. Disk friction loss, friction on rotating cones

NAGOYA UNIVERSITY

Dept. of Mechanical Engineering, Furo-cho, Chijusa-ku, Nagoya-shi 464. Tel: (052) 781 5111. Prof. K. Kikuyama. Two-phase flow in pumps, turbulent boundary layer and secondary flow

UNIVERSITY OF OSAKA

Faculty of Eng., Dept of Mechanical Eng. 2-1 Yamodaoka, Suita-shi, Osaka 565. Tel: (06) 877 5111. Prof. Y. Miyaka. Numercial analysis of flow in turbo-machines, turbulence modelling

UNIVERSITY OF OSAKA

Faculty of Eng. Science, Dept. of Mechanical Eng. 1-1 Machikaneyana-cho, Toyonaka-shi 560. Tel: (06) 844 1151. Prof. Y. Tsujimoto. Rotating stall, unsteady phenomena in centrifugal impellers

SOPHIA UNIVERSITY

Dept. of Mechanical Engineering, 7 Kioi-cho, Chiyoda-ku, Tokyo-to 102. Tel: (03) 238 3305. Prof. K. Takahashi. Slurry pumps, unsteady radial and axial thrust of pumps

TOHUKU UNIVERSITY

Dept. of Mechanical Engineering, Aza-Aoba, Aramaki, Sendai-shi 980. Tel: (0222) 22 1800. Prof. H. Daiguji. Numerical analysis of flow in turnbomachines, cavitation

TOHOKU UNIVERSITY

Institute of High Speed Mechanics, 2-1-1 Katahira, Sendai-shi 980. Tel: (0222) 27 6200. Prof. R. Ohba. Cavitation and supercavitation, performance of hydraulic turbines

TOKAI UNIVERSITY

Dept. of Production Engineering, 1117 Kita-Kaname-cho, Hiratsuka-shi 259-12. Tel: (0463) 58 1211. Prof. Y. Nakayama. Flow visualisation of flow in pumps and hydraulic turbines, cavitation

UNIVERSITY OF TOKUSHIMA

Dept. of Mechanical Engineering, 2-1 Mishima-cho, Janjo, Tokushima-shi 770. Tel: (0886) 23 2311. Prof. T. Nakase. Cross-flow turbines, tidal and wave power turbines

TOKYO DENKI UNIVERSITY

Dept. of Mechanical Engineering, 2-2 Nishiki-cho, Kanda, Tokyo-to 101. Tel: (03) 294 1551. Prof. Y. Hosoi. Draft tube surge, performance of pumps and hydraulic turbines

TOKYO METROPOLITAN UNIVERSITY

Dept. of Mechanical Engineering, 2-1-1 Fukazawa, Setagaya-ku, Tokyo-to 158. Tel: (03) 717 0111. Prof. H. Kato. Flow in diffusers, turbulent boundary layer of turbomachines

TOKYO INSTITUTE OF TECHNOLOGY

Dept. of Mechanical Engineering, 2-12-1 Ohokayama, Meguro-ku, Tokyo-to 152. Tel: (03) 726 1111. Prof. R. Yamane. Cavitation, internal flow in turbomachines

SCIENCE UNIVERSITY OF TOKYO

Dept. of Mechanical Engineering, 2641 Higashi-kameyama, Yamazaki, Noda-shi 278. Tel: (0471) 24 1501. Prof. H. Maki. Flow analysis and measurement of radial diffusers, flowmeters

UNIVERSITY OF TOKYO

Dept. of Mechanical Engineering, Hongo 7-3-1, Bunkyo-ku, Tokyo 113. Tel: (03) 812 2111. Prof. H. Ohashi. Cavitation, transient phenomena in conduit, two-phase cascade flow

UNIVERSITY OF TOKYO

Institute of Industrial Science, 7-22-1 Roppongi, Minato-ku, Tokyo-to 153. Tel: (03) 294 1551. Prof. T. Kobayashi. Numerical analysis and flow visualisation of flow in turbomachines

TOYOTA INSTITUTE OF TECHNOLOGY

Dept. of Mechanical Engineering, 2-12 Hisakata, Tenpaku-ku, Nagoya-shi 468. Tel: (052) 802 1111. Prof. S. Murata. Numerical analysis and measurement of flow in impellers

UNIVERSITY OF TSUKUBA

Institute of Engineering Mechanics, 1-1-1 Tennodai, Sakura-mura, Niihari-gun, Ibaraki 305. Tel: (0298) 53 2111. Prof. H. Tahara. Performance of centrifugal pumps, fluid force on whirling impellers

WASEDA UNIVERSITY

Dept. of Mechanical Engineering, 3-4-1 Ohkubo, Shinjuku-ku, Tokyo-to 160. Tel: (03) 209 3211. Prof. K. Yamamoto. Flow instability of pumping system, similarity law of suction reservoir

YOKOHAMA NATIONAL UNIVERSITY

Department of Mechanical Engineering II, 2-31-1 Oooka, Minami-ku, Yokohama-shi 233. Tel: (045) 335 1451. Prof. T. Toyokura. Study of annular cascades, internal flow and axial thrust of pumps

NORWAY

THE NORWEGIAN INSTITUTE OF TECHNOLOGY

Division of Hydro - and Gas - Dynamics, Water Power Laboratory. Tel: 47 7 593856: Fax: 47 7 593854 Alfred Getz eg 4, N7034 Trondheim NTH. Prof. H. Brekke. Numerical analysis and experimental tests of unsteady flow; Numerical analysis and model test of turbomachinery

SOUTH AMERICA

UNIVERSITY OF LA PLATA

Faculty of Engineering, Calle 2/7, No. 200, 1900 - La Plata. Telex: 31151 Rpca. Argentina BULAP AR. Head of Laboratory: Prof. Eng. Fernando Zarate. Hydraulics Laboratory "Guillermo Cèspedes"

UNIVERSIDAD NICIONAL AUTÒNOMA DE MEXICO

Apdo. 770472, Deleg. Coyoacàn 04510, Mexico D.F., Mexico. Head of the Hydromechanical Section: Prof. Eng. A. Palacios. Instituto de Ingenieria

UNIVERSITY OF SAO PAULO

Cidade Universitaria, CXP 11014, 05508, SP, Brasil. Head of the Laboratory: Prof. Eng. Giorgio Brighetti. Centro Tecnològico de Hidràulica

UNIVERSITY OF URUGUAY

Faculty of Engineering, J. Herrera y Reissig 565, Montevideo, Uruguay. Tel: 71 03 61. Head of the Institute: Prof. Dr. Rafael Guarga. Instituto de Mecànica de los Fluìdos e Ingenierìa Ambiental

SPAIN

UNIV. AUTÒNOMA DE BARCELONA

Fac. de Ciencias, Campus Univ. de Bellaterra, Barcelona 08071. Tel: (93) 2038900.

UNIV. DE CÀDIZ
E U Ing. Tècn. Industrial, C/Sacramento, 82, Càdiz 11071. Tel: (956) 224359.

UNIV. POL. DE CANARIAS
E T S Ing. Industriales, Cuarto Pab. Seminario de Tarifa Baja, P G Canaria 35071. Tel: (928) 350004

UNIV. DE CANTABRIA
E T S I Caminos, Canales y Puertos, Avda. de los Castros, s/n, Santander 39071. Tel: (942) 275600.

UNIV. DE CAST - LA MANCHA
E U Ing. Tècn. Industrial, Avda. de Portugal, s/n, Toledo 45071. Tel: (925) 223400.

UNIV. POL. DE CATALUNYA
E T S Ing. Caminos, Canales y Puertos, C/Jordi Girona, 31, Barcelona 08071. Tel: (93) 2048252

UNIV. PONTIF. DE COMILLAS
E T S Ing. Industriales, Alberto Aguilera, 23, Madrid 28015. Tel: (91) 2483600

UNIV. SAND. DE COMPOSTELA
E T S Ing. Industriales, C/La Paz, s/n, Vigo 36071. Tel: (986) 373012

UNIV. DE CÒRDOBA
E U Ing. Tècn. Industrial, Avda. Menendez Pidal, s/n, Còrdoba 14071. Tel: (957) 291555

UNIV. NAC. EDUC. DISTANCIA
E T S Ing. Industriales, Ciudad Universitaria - UNED, Madrid 28071. Tel: (91) 4493600

UNIV. DE EXTREMADURA
E U Pol. de Mèrida, C/Calvario, 2, Mèrida 06071. Tel: (924) 318712

UNIV. DE GRANADA
E U Ing. Tècn. Industrial, Avda. de Madrid, 35, Jaen 23071. Tel: (953) 250448

UNIV. DE LEÒN
E U Ing. Tècn. Industrial, C/Jesùs Rubio, 2, Leòn 24071. Tel: (987) 204052

UNIV. POL. DE MADRID
E T S Ing. Industriales, C/Josè Gutierrez Abascal, 2, Madrid 28071. Tel: (91) 2626200

UNIV. DE MÀLAGA
E U Ing. Tècn. Industrial, Plaza El Ejido, s/n, Màlaga 29071. Tel: (952) 250100

UNIV. DE MURCIA
E U Politècnica, P. Alfonso XIII, 34, Caragena 30071. Tel: (968) 527378

UNIV. DE NAVARRA
E T S Ing. Industriales, Urdaneta, 7, San Sebastian 20006. Tel: (943) 466411

UNIV. DE OVIEDO
E T S Ing. Industriales, Ctra. de Castiello, s/n, Gijòn 33071. Tel: (985) 338680

UNIV. DE SALAMANCA
E U Ing. Tècn. Industrial, Av. Ferna 'ndo Ballesteros, s/n, Bejar 37071. Tel: (923) 402416

UNIV. DE SEVILLA
E T S Ing. Industriales, Avda. Reina Mercedes, s/n, Sevilla 41071. Tel: (954) 611150

UNIV. POL. DE VALENCIA
E T S Ing. Industriales, Camino de Vera, s/n, Valencia 46071. Tel: (96) 3699862

UNIV. DE VALLADOLID
E T S Ing. Industriales, Avda. Santa Teresa, 30, Valladolid 47071. Tel: (983) 353608

UNIV. DEL PAIS VASCO
E U Ing. Tècn. Ind. (CEI), Avda. Bilbao, 29, Eibar 20071. Tel: (943) 718444

UNIV. DE ZARAGOZA
E T S Ing. Industriales, Ciudad Universitaria. Zaragoza 50071. Tel: (976) 350058

SWITZERLAND

SWISS FEDERAL INSTITUTE OF TECHNOLOGY
Hydraulic Machines and Fluid Mechanics Institute, 1015 Lausanne. Tel: 19 06 1988: Fax: 19 06 1988: Telex: 455 806. Prof. P Henry. Numerical analysis of flow in turbomachines, cavitation, boundary layer, friction losses, unsteady behaviour of hydraulic machines

U. K.

UNIVERSITY OF BIRMINGHAM
Chemical Engineering, PO Box 363, Birmingham B15 2TT. Tel: 021-472-1301 Ext. 2105. Dr N Thomas. Hydrodynamics, multi-phase flows, fluid mechanics

UNIVERSITY OF BRADFORD
Civil Engineering, Bradford, West Yorkshire BD7 1DP. Tel: 0274-733466 Ext. 8391. Prof. R A Falconer. Modelling, hydrodynamics, turbulence, water quality, tides, sediment transport, dispersion, circulation

UNIVERSITY OF BRISTOL
Civil Engineering, University Walk, Bristol, Avon BS8 1TR. Tel: 0272-303280. Dr R H J Sellin. Rivers, drag reduction, hydraulic models, flood plains, sewers

CAMBRIDGE UNIVERSITY
Engineering, Trumpington Street, Cambridge CB2 1PZ. Tel: 0223-332632. Dr J F A Sleath. Beaches, coastal engineering, erosion, offshore structures, pipelines, sediment transport

CITY UNIVERSITY
Dept of mechanical engineering, London EC1. Dr P A Lush. Cavitation, fluid machines

CRANFIELD INSTITUTE OF TECHNOLOGY
School of mechanical engineering, Cranfield, Bedford MK43 0AL. Tel: 0234 750111: Fax: 0234 750728. Prof. R Elder. Computer-aided pump studies

UNIVERSITY OF DUNDEE
Civil Engineering, Dundee, Scotland DD1 4HN. Tel: 0382-23181 Ext. 4340. Prof. A E Vardy. Unsteady flow, pressure transients, surge, numerical analysis

UNIVERSITY OF GLASGOW
Civil Engineering, Oakfield Avenue, Glasgow, Scotland G12 8QQ. Tel: 041-8855 Ext. 7210. Dr D A Ervine. Aeration, air entrainment, channels, waterways dams, floods, spillways

IMPERIAL COLLEGE OF SCIENCE AND TECHNOLOGY
Civil Engineering, Imperial College Road, London SW7 2BU. Tel: 01-589-5111 Ext. 4864. Dr. J D Hardwick. Vibration, hydraulic structures, hydroelastic modelling

KING'S COLLEGE LONDON
Civil Engineering, Strand, London WC2R 2LS. Tel: 01-836-5454 Ext. 2723. Mr J H Loveless. Sediment, drainage conduits,

hydraulic structures cavitation, coastal hydraulics, air-entrainment

UNIVERSITY OF LIVERPOOL

Civil Engineering, Brownlow Street, PO Box 147, Liverpool L69 3BX. Tel: 051-709-6022 Ext. 2461. Mr T S Hedges. Hydrodynamics, waves, wave-current interaction, nearshore processes, mathematical modelling

UNIVERSITY COLLEGE LONDON

Civil and Municipal Engineering, Gower Street, London WC1E 6BT. Tel: 01-387-7050 Ext. 2709. Dr A J Grass. Boundary layer turbulence, sediment transport, fluid loading on structures and pipelines

LOUGHBOROUGH UNIVERSITY OF TECHNOLOGY

Dept. of Mechanical Engineering, Loughborough, Leics LE11 3TU. Tel: 0509 223206: Telex: 34319: Fax: 0509 232029. Mr R K Turton, Senior Lecturer. Gas/liquid pumping: Pump inducer studies

LOUGHBOROUGH UNIVERSITY OF TECHNOLOGY

WEDC Dept of Civil Engineering, Loughborough, Leics LE11 3TU. Tel: 0509 222390: Telex: 34319. Prof. J A Pickford. Water engineering for developing countries

UNIVERSITY OF NEWCASTLE UPON TYNE

Civil Engineering, Newcastle upon Tyne NE1 7RU. Tel: 091-232-8511 Ext. 2399. Dr C Nalluri. Sediment transport, flood channels, sewers, sea outfalls

UNIVERSITY OF NEWCASTLE UPON TYNE

Mechanical Engineering, Stephenson Building, Claremont Road, Newcastle upon Tyne NE1 7RU. Tel: 091-232-8511. Dr A Anderson. Cavitation, hydroelectric power, penstocks, pumped storage, pumps, turbines, valves, water hammer

UNIVERSITY OF NOTTINGHAM

Civil Engineering, University Park, Nottingham, Nottinghamshire Ng7 2RD. Tel: 0602-566101 Ext. 3537. Dr C J Baker. Sediment transport, sandwaves (fluvial and aeolian), Scour around structures

UNIVERSITY OF NOTTINGHAM

Dept of Mechanical Engineering, University Park, Nottingham. Dr A Lichtarowicz. Cavitation, erosion studies

UNIVERSITY OF OXFORD

Engineering Science, Parks Road, Oxford OX1 3PJ. Tel: 0865-273000. Dr. R E Franklin. Cavitation, inception, noise, nuclei, bubble dynamics, bubbly flows, gas content

UNIVERSITY OF READING

Engineering, Whiteknights Park, Reading, Berkshire RG6 2AY. Tel: 0734-875123 Ext. 7315. Dr J D Burton. Turbine, pump, installation, micro-hydro, inertia flow, alternating flow hydraulics

UNIVERSITY OF SALFORD

Civil Engineering, Salford, Lancashire M5 4WT. Tel: 061-736-5843 Ext. 7116. Dr R Baker. Concrete revetment blocks, spillways, models, sediment transport

UNIVERSITY OF SALFORD

Civil Engineering, Salford, Lancashire M5 4WT. Tel: 061-736-5843 Ext. 7122. Prof.

E M Wilson. Hydrology, hydroelectricity, tidal energy

UNIVERSITY OF SHEFFIELD

Civil and Structural Engineering, Mappin Street, Sheffield S1 3JD. Tel: 0742-768555 Ext. 5418 or 5059. Dr F A Johnson. Dams, failures, floods, routing, waves, reservoirs control

UNIVERSITY OF SOUTHAMPTON

Dept. of Mechanical Engineering, Southampton, Hants SO9 5NH. Prof. S P Hutton, Prof. M Thew. Fluid metering, fluid machines

UNIVERSITY OF STIRLING

Environmental Science, Stirling, Scotland FK9 4LA. Tel: 0786-73171. Dr R I Ferguson. Sediment transport, gravel, non-uniform flow, braided rivers, field measurement

UNIVERSITY OF STRATHCLYDE

Civil Engineering, 107 Rottenrow, Glasgow, Scotland G4 0HG. Tel: 041-552-4400 Ext. 3168. Prof. G Fleming. Hydrology, erosion, simulation, management, sedimentation, reservoirs, land-use, dredging

THAMES POLYTECHNIC

Civil Engineering, Oakfield Lane, Dartford, Kent DA1 2SZ. Tel: 0322-21328 Ext. 318. Mr A Grant. Sediment transport, urban drainage

UNIVERSITY OF WARWICK

Dept of Engineering, Hydrotransient Simulation Unit, Coventry CV4 7AL. Tel: 0203-523086. Dr A P Boldy. Simulation, transients, turbines, hydroelectric

UWIST

Civil Engineering and Building Technology, Colum Drive, Cardiff CF1 3EU. Tel: 0222-42588 Ext. 2802. Dr P W France. Finite difference, hydraulic structures, flow measurement, weirs

U S A

CALTECH, DIVISION OF ENGINEERING AND APPLIED SCIENCE

Thomas Laboratory, 104-44 Pasadena, California 91125. Tel: (818) 356-4106 M E Dept.: Fax: (818) 568-2719. A J Acosta, C E Brennen. Experimental work on cavitation, rotor dynamic forces, unsteady flow effects on axial and centrifugal hydraulic machines, primarily pumps. Rotor-stator blade interactions, inlet shear flows, shroud boundary layer flow. The experimental test facility is capable of cavitation and dynamic testing at rotor power levels of about 15 kW with rotational speeds up to 7000 rpm. Unsteady force and flow instrumentation.

GEORGIA INSTITUTE OF TECHNOLOGY

School of Civil Engineering, Atlanta, Georgia 30332, U. S. A. Tel: 404-894-2224: Telex: 542507: Fax: 404-894-2224. Professor C Samuel Martin. Research and consultation regarding pump-turbine characteristics and hydraulic transient analysis

UNIVERSITY OF MINNESOTA

St Anthony Falls Hydraulic Laboratory, Mississippi River at 3rd Avenue S E. Tel: (612) 627-4010: Fax: (612) 627-4609. Dr Roger E A Arndt, Charles C S Song, John Gulliver. An independent turbine test stand for turbine acceptance testing. Physical and mathematical modelling works on intake structure, turbine rotor and stator, draft tube, cavitation and bubble dynamics are being conducted. Mathematical model for economic evaluation of small hydropower development.

U S S R

MOSCOW INSTITUTE OF HYDROTECHNICAL ENGINEERING AND LAND RECLAMATION

Moscow 127550, Prianisnikova St, 19. Tel: 216-11-85. Prof. Vadim Phirsovitch Chebaevski. Cavitation in blade pumps; design of reclamation pumping plants

RESEARCH INSTITUTE OF POWER MACHINE-BUILDING OF THE MOSCOW HIGHER TECHNICAL SCHOOL

Moscow 107005, Second Bauman St, 5. Tel: 261-59-89. Vladimir Ivanovitch Petrov, Senior Researcher. Cavitation in blade pumps

WEST GERMANY

TECHNISCHE UNIVERSITÄT BERLIN,

Dept. of Civil Eng. Institut für Wasserbau und Wasserwirtschaft, Strasse des 17. Juni 142-144, 1000 Berlin 12. Prof. Dr Ing. P Franke. Numerical analysis

TECHNISCHE UNIVERSITÄT BERLIN,

Dept. of Mech. Eng. Lehrstuhl für Maschinenkonstruktionen, Hydraulische Maschinen und Anlagen, Steinplatz 1, 1000 Berlin 1. Prof. Dr Ing. E Siekmann. Numerical analysis and measurement of flow in turbomachines

TECHNISCHE UNIVERSITÄT BOCHUM

Lehrstuhl für Regelsysteme und Steuerungstechnik, Dept. of Mech. Eng. Postfach 10 21 48, 4630 Bochum. Prof. Dr Ing. K H Fasol. Numerical analysis

TECHNISCHE HOCHSCHULE BRAUNSCHWEIG PFLEIDERER

Institut für Strömungsmaschinen, Dept. of Mech. Eng. Langer Kamp 6, 3300 Braunschweig. Prof. Dr Ing. H Petermann, Prof. Dr Ing. G Kosyna. Numerical analysis and measurement of flow in turbomachines

TECHNISCHE HOCHSCHULE DARMSTADT

Institut für Hydraulische Machinen, Dept. of Mech. Eng. Magdalenenstr. 8 - 10, 6100 Darmstadt. Prof. Dr Ing. B Stoffel. Numerical analysis and measurement of flow in turbomachines

TECHNISCHE UNIVERSITÄT HANNOVER

Institut für Strömungsmechanik, Dept. of Civil Eng. Callinstr. 32, 3000 Hannover. Prof. Dr Ing. W Zielke. Numerical analysis

TECHNISCHE UNIVERSITÄT HANNOVER

Institut für Strömungsmaschinen, Dept. of Mech. Eng. Appelstr. 9, 3000 Hannover. Prof. Dr Ing. M Rautenberg. Numerical analysis and measurement of flow in turbomachines

TECHNISCHE UNIVERSITÄT KAISERSLAUTERN,

Dept. of Mech. Eng. Erwin-Schrödinger

strasse, 6700 Kaiserslautern. Prof. Dr Ing. F Eisfeld. Numerical analysis and measurement of flow in turbomachines

TECHNISCHE UNIVERSITÄT KARLSRUHE

Institut für Hydromechanik, Dept. of Civil Eng. Kaiserstr. 12, 7500 Karlsruhe 1. Prof. Dr Ing. E Naudascher, Prof. Dr Ing. H Thielen. Numerical analysis and measurement of flow

TECHNISCHE UNIVERSITÄT KARLSRUHE,

Deptartment of Mechanical Eng. Institut für Strömungslehre und Strömungsmaschinen, Kaiserstr. 12, 7500 Karlsruhe 1. Prof. Dr Ing. K O Felsh, Dr Ing. J Zierep. Numerical analysis and measurement of flow in turbomachines

TECHNISCHE UNIVERSITÄT MÜNCHEN,

Dept. of Civil Eng. Lehrstuhl für Hydraulik und Gewässerkunde. Arcisstr. 21, D8000 München 2. Prof. Dr Ing. F Valentin. Numerical analysis

TECHNISCHE UNIVERSITÄT MÜNCHEN,

Dept. of Mech. Eng. Lehrstuhl und Laboratorium für Hydraulische Maschinen und Anlagen. Arcisstr. 21. D8000 München 2. Tel: (089) 2105 3453. Prof. Dr Ing. Habil Joachim Raabe. Analysis and dynamic measurement of quick response of real flow in diffuser, draft tube. rotor channel and pipe, including waterhammer and two-phase flow

TECHNISCHE UNIVERSITÄT MÜNCHEN,

Dept. of Mech. Eng. Lehrstuhl und Laboratorium für Hydraulische Maschinen und Anlagen, Arcisstr. 21, D8000 München 2. Prof. Dr Ing. R Schilling. Numerical analysis and measurement of flow in turbomachines

UNIVERSITÄT STUTTGART

Institut für Hydraulische Strömungsmaschinen. Dept. of Mech. Eng. Pfaffenwaldring 10, 7000 Stuttgart 80. Prof. Dr Ing. G Lein. Numerical analysis and measurement of flow in turbomachines

UNIVERSITÄT STUTTGART

Institut für Wasserbau. Dept of Civil Eng. Pfaffenwaldring 10, 7000 Stuttgart 80. Prof. Dr Ing. J Giesecke, Dr Ing. H B Horlacher. Numerical analysis and measurement of flow

YUGOSLAVIA

UNIVERSITY OF BELGRADE

Faculty of Mechanical Engineering, 27 Marta 80, 11000 Belgrade. Tel: 329 021. Prof. Stanislav Pejovic. Education, test facilities, turbines, pumps, transients, hydropower and pumping systems, measurements, consultancy.

For Product Safety Concerns and Information please contact our EU representative GPSR@taylorandfrancis.com Taylor & Francis Verlag GmbH, Kaufingerstraße 24, 80331 München, Germany

Printed and bound by CPI Group (UK) Ltd, Croydon, CR0 4YY
08/05/2025
01864511-0001